Lecture Notes in Artificial Intelligence   13170

Subseries of Lecture Notes in Computer Science

Series Editors

Randy Goebel
University of Alberta, Edmonton, Canada

Wolfgang Wahlster
DFKI, Berlin, Germany

Zhi-Hua Zhou
Nanjing University, Nanjing, China

Founding Editor

Jörg Siekmann
DFKI and Saarland University, Saarbrücken, Germany

Jie Chen · Jérôme Lang · Christopher Amato ·
Dengji Zhao (Eds.)

Distributed Artificial Intelligence

Third International Conference, DAI 2021
Shanghai, China, December 17–18, 2021
Proceedings

 Springer

Editors
Jie Chen
Tongji University
Shanghai, China

Christopher Amato
Khoury College of Computer Sciences
Northeastern University
Boston, MA, USA

Jérôme Lang
Lamsade Bureau
Université Paris-Dauphine
Paris Cedex 16, France

Dengji Zhao
ShanghaiTech University
Shanghai, China

ISSN 0302-9743   ISSN 1611-3349 (electronic)
Lecture Notes in Artificial Intelligence
ISBN 978-3-030-94661-6   ISBN 978-3-030-94662-3 (eBook)
https://doi.org/10.1007/978-3-030-94662-3

LNCS Sublibrary: SL7 – Artificial Intelligence

This Springer imprint is published by the registered company Springer Nature Switzerland AG
The registered company address is: Gewerbestrasse 11, 6330 Cham, Switzerland

Preface

Lately, there has been tremendous growth in the field of artificial intelligence (AI) in general and in multi-agent systems research in particular. Problems arise where decisions are no longer made by a center but by autonomous and distributed agents. Such decision problems have been recognized as a central research agenda in AI and a fundamental problem in multi-agent systems. Resolving these problems requires that different scientific communities interact with each other, calling for collaboration and raising further important interdisciplinary questions. Against this background, a new conference, the International Conference on Distributed Artificial Intelligence (DAI), has been organized since 2019. DAI aims at bringing together international researchers and practitioners in related areas including general AI, multi-agent systems, distributed learning, computational game theory, etc., to provide a high-profile, internationally renowned forum for research in the theory and practice of distributed AI.

This year, we received 31 submissions. Despite the relatively small number of submissions, the authors represented various countries including China, the USA, Israel, Italy, Germany, the UK, and New Zealand. Each paper was assigned to three Program Committee (PC) members. We ensured that each paper received at least three reviews. Given the reviews, the final decisions were made based on the discussion and consensus of the Program Committee with the chairs.

After the first round of reviewing, 10 out of 31 papers were accepted. Since there was a good number of borderline papers with reasonably good support from the reviewers, we decided to give conditional acceptance to another five borderline papers which were significantly revised by the authors according to the reviews and were reviewed again by the chairs before being accepted. The topics of the accepted papers include reinforcement learning, multi-agent learning, distributed learning systems, deep learning, applications of game theory, multi-robot systems, human-agent interaction, signaling and information design, etc.

We were delighted to have Craig Boutilier (Google Research, USA), Bart Selman (Cornell University, USA), and Julie Shah (Massachusetts Institute of Technology, USA) to offer us great keynote, and also to have Adam Tauman Kalai (Microsoft Research, USA), Olga Megorskaya (Tolaka, Russia), and Osher Yadgar (Amdocs, Israel) to offer us great talks from industry.

Lastly, we would like to sincerely thank the conference committee and the Program Committee for their great help in making DAI 2021 another successful event.

December 2021

Christopher Amato
Dengji Zhao
Jie Chen
Jérôme Lang

Organization

General Chairs

Jie Chen   Tongji University, China
Jérôme Lang   Université Paris-Dauphine, France

Program Committee Chairs

Christopher Amato   Northeastern University, USA
Dengji Zhao   ShanghaiTech University, China

Program Committee

Bo An   Nanyang Technological University, Singapore
Haris Aziz   University of New South Wales, Australia
Haitham BouAmmar   Huawei Noah's Ark Lab and University College London, UK
Mithun Chakraborty   University of Michigan, USA
Jilles Dibangoye   INSA Lyon, France
Maria Gini   University of Minnesota, USA
Jianye Hao   Tianjin University and Huawei Noah's Ark Lab, China
Guifei Jiang   Nankai University, China
Priel Levy   Bar Ilan University, Israel
Xudong Luo   Guangxi Normal University, China
Reuth Mirsky   University of Texas at Austin, USA
Weiran Shen   Renmin University of China, China
Zheng Tian   University College London, UK
Chongjun Wang   Nanjing University, China
Feng Wu   University of Science and Technology of China, China
Chao Yu   Sun Yat-sen University, China
Chongjie Zhang   Tsinghua University, China
Weinan Zhang   Shanghai Jiao Tong University, China
Zongzhang Zhang   Nanjing University, China
Ming Zhou   Shanghai Jiao Tong University, China

Contents

The Power of Signaling and Its Intrinsic Connection to the Price of Anarchy

Jamie Nachbar[1] and Haifeng Xu[2(✉)]

[1] Yale University, New Haven, CT, USA
jamie.nachbar@yale.edu
[2] University of Virginia, Charlottesville, VA, USA
hx4ad@virginia.edu

Abstract. Strategic behaviors often render the equilibrium outcome inefficient. Recent literature on information design, a.k.a. *signaling*, looks to improve equilibria by selectively revealing information to players in order to influence their actions. Most previous studies have focused on the *prescriptive* question of designing optimal signaling schemes. This work departs from previous research by considering a *descriptive* question, and looks to quantitatively characterize the *power of signaling* (PoS), i.e., how much a signaling designer can improve her objective at equilibrium.

We consider four types of signaling schemes with increasing power: full information, optimal public signaling, optimal private signaling, and optimal *ex-ante* private signaling. Our main result is a clean and tight characterization of the additional power each signaling scheme has over its predecessors above in the general classes of cost-minimization and payoff-maximization games where: (1) all players minimize non-negative cost functions or maximize non-negative payoff functions; (2) the signaling designer (naturally) optimizes the sum of players' utilities. We prove that the additional power of signaling—defined as the worst-case ratio between the equilibrium objectives of *any* two signaling schemes in the above list—is bounded precisely by the well-studied notion of the *price of anarchy* (PoA) of the corresponding games. Moreover, we show that all these bounds are *tight*.

1 Introduction

A basic lesson from game theory is that strategic behaviors often render the equilibrium outcome inefficient. That is, the objective function value of an equilibrium outcome may be far from that of an optimal outcome in the absence of strategic behaviors. To reduce such inefficiency, one can "tune" the game equilibrium towards a more desirable outcome, and there are two primary ways to achieve this goal: through providing *incentives* or providing *information*. The

This work is done while Nachbar is visiting the University of Virginia as a summer research intern. Xu would like to thank Yu Cheng and Shaddin Dughmi for inspiring discussions at the early stage of this work, including a variant of the tight example for PoS(Pri:Pub) in Fig. 2. This work is supported by an NSF grant CCF-2132506.

© Springer Nature Switzerland AG 2022
J. Chen et al. (Eds.): DAI 2021, LNAI 13170, pp. 1–20, 2022.
https://doi.org/10.1007/978-3-030-94662-3_1

former approach has been widely studied in the celebrated field of mechanism design [7,8,26,28]. This paper, however, focuses on the second approach, namely, improving equilibrium by providing carefully designed information to influence players' decisions. This falls into the recent flourishing literature on information design, a.k.a., *signaling* or *persuasion* [10,18]. Researches in this literature so far have mainly focused on computing optimal signaling schemes for either fundamental setups [6,12,13,32] or models motivated by varied applications including auctions [3,14,23], public safety and security [29,33], conservation [34], privacy protection [35], voting [5], congestion games [2,4,9], recommender systems [24], robot design [20], etc.

Departing from the theme of all these previous works, this paper considers a different style of question. We look to characterize the *power of signaling* (PoS)—*how much a signaling designer can improve her objective function of the equilibrium and can we quantitatively characterize this power?* To our knowledge, this *descriptive* question has not been formally examined before in the literature of signaling, except for a few studies which implicitly show certain PoS-type results in the special case of non-atomic routing games [9,10,25]. Nevertheless, the study of the PoS is extremely well-motivated. It not only deepens our understanding about signaling as an important "knob" to influence equilibrium, but also justifies the value of previous prescriptive studies of optimal signaling design—after all, the designed optimal signaling schemes are useful in practice only when the power of signaling is not negligible. Thus PoS is an important measure when determining the adoption of a signaling scheme, especially when its tradeoff with other potential drawbacks such as communication costs [16] and fairness concerns [17] need to be balanced.

We focus on the general classes of cost-minimization games and payoff-maximization games, where each player minimizes a non-negative cost function or maximizes a non-negative payoff function. These classes of games are often studied in the literature of the price of anarchy (PoA) [30], and include many widely studied examples such as routing games, congestion games, most formats of auctions, valid utility games [31], etc. Like all standard models of signaling, players' utilities depend on a common random state of nature θ, which is drawn from a publicly known prior distribution. A signaling designer, referred to as the *sender*, has an informational advantage and can access the realized state θ. The sender is equipped with the natural objective of optimizing the total welfare, i.e., sum of players' utilities.

Power of Signaling (PoS). Our goal is to formally quantify the relative power of different types of signaling schemes as they become less constrained. In particular, we consider four types of signaling schemes with increasing power: full information (FI), optimal public signaling (Pub), optimal private signaling (Pri), and optimal *ex-ante* private signaling (exP). FI is a natural *benchmark* without any strategic use of information whereas Pub, Pri, exP are arguably the three most widely studied schemes in previous literature.[1] The power of signaling of

---

[1] Public and private signaling has been extensively studied in previous works. Several recent works study ex-ante private signaling with motivations from recommender systems [4,6,32].

scheme B over A for *any* A preceding B in the list {FI, Pub, Pri, exP}—termed PoS(B:A)—is defined as the ratio of the sender's utilities from scheme A and B. This ratio is *at least* 1 for cost-minimization games and *at most* 1 for payoff-maximization games (the same as the range of the price of anarchy). Moreover, the further it is from 1, the more powerful scheme B is than scheme A.

Characterizations of PoS. Our main result is a clean and tight characterization about the power of signaling. Concretely, for any cost-minimization game with a random state, we prove that all the aforementioned PoS ratios are upper bounded by the maximum PoA of its corresponding realized games. Moreover, all these upper bounds are tight in the following sense: for any ratio $r \geq 1$ and any scheme A preceding B in the list {FI, Pub, Pri, exP}, there exists a Bayesian cost-minimization game where all of its realized games have PoA $= r$ and moreover PoS(B:A) $= r$ as well. Invoking known PoA results, this implies the power of various types of signaling schemes in well studied games such as atomic or non-atomic games with linear or polynomial latency functions (each class has its own PoA bound which now all easily translate to the PoS bounds.) Next, we show that exactly the same results hold for payoff-maximization games—the PoSs are similarly bounded by PoA and all the bounds are tight.[2]

Our results reveal the intrinsic connections between the power of signaling and price of anarchy. Prior to this work, it was not clear that these two concepts are inherently related—PoA characterizes the worst-case equilibrium welfare whereas PoS characterizes how much information can be strategically used to improve welfare. To our knowledge, recent work [25] is the only one to observe this connection but only for the power of public signaling over full information in non-atomic routing with affine latency functions. Our results are systematic and much more general.

2   Preliminaries

2.1   Cost-Minimization/Payoff-Maximization Games

A cost-minimization game G is a standard strategic game where each player minimizes a non-negative cost function. Let n denotes the number of players in the game. Each player $i \in [n] = \{1, \cdots, n\}$ has action space S_i. Let $S = S_1 \times S_2 \cdots S_n$ denote the space of action profiles and $s \in S$ is a generic action profile. Player i is attempting to minimize $c_i(s) \geq 0$. A (randomized) mixed strategy for player i is a distribution $\mathbf{x}_i$ over S_i where $x_i(s_i)$ is the probability of taking action s_i. By convention, $\mathbf{x}$ denotes the profile of mixed strategies for all players, and $\mathbf{x}_{-i}$ denotes all the mixed strategies excluding i's. With slight abuse of notation, let $c_i(\mathbf{x}) = \mathbf{E}_{s_i \sim \mathbf{x}_i, \forall i} c_i(s)$ denote the expected utility of player i under mixed strategy $\mathbf{x}$. There is also a *global objective* which is simply to minimize the sum of the total costs $C(\mathbf{x}) = \sum_i c_i(\mathbf{x})$.

---

[2] Instead of FI, another natural benchmark scheme is to reveal *no* information. The PoSs compared to this benchmark turn out to be unbounded, which we show in Appendix C.

We adopt the standard mixed-strategy *Nash equilibrium* (NE) as the solution concept. A strategy profile $\mathbf{x}^*$ is a NE if for each player i, $c_i(\mathbf{x}^*) \leq c_i(\mathbf{x}_i, \mathbf{x}^*_{-i})$ for any $\mathbf{x}_i \in \Delta(S_i)$. Let X^* denote the set of all NEs. The well-studied concept of the price of anarchy (for mixed equilibria) for a cost-minimization game is defined as follows [15, 21, 22, 30]

$$\mathsf{POA} = \frac{\max_{\mathbf{x}^* \in X^*} C(\mathbf{x}^*)}{\min_{s \in S} C(s)} \in [1, \infty) \tag{1}$$

In other words, the POA is the ratio between the worst Nash equilibrium and the optimal social outcome.

Remark 1. *The PoA can also be defined with respect to pure Nash equilibrium in which case X^* consists of all pure equilibria. Since not every game admits a pure Nash equilibrium, in striving for generality, we choose to analyze the version w.r.t. to mixed equilibria since they always exist in finite games as well as in many infinite games. However, all our results—both upper and lower bound proofs—hold for pure equilibria as well, so long as they exist.*

Payoff-maximization games are defined similarly; here each player i maximizes expected payoff $u_i(\mathbf{x}) \geq 0$. The global objective is to maximize $U(\mathbf{x}) = \sum_i u_i(\mathbf{x})$. The price of anarchy here is defined similarly as $\mathsf{POA} = \frac{\min_{\mathbf{x}^* \in X^*} U(\mathbf{x}^*)}{\max_{s \in S} U(s)}$, which now lies in $[0, 1]$.

This paper concerns games with uncertainty. Specifically, players' payoffs depend also on a *random* state of nature θ drawn from support Θ with distribution λ. We use $c_i^\theta(s)/u_i^\theta(s)$ to denote the cost/payoff function at state θ. Such a *Bayesian game* is denoted by $\{G^\theta\}_{\theta \sim \lambda}$. As is standard in information design, the prior distribution λ is publicly known to every player. We assume Θ to be finite for ease of notation, and use $\lambda(\theta)$ to denote the probability of state θ. However, all our results hold for infinite state space.

2.2 Signaling Schemes and Equilibrium Concepts

This paper adopts the perspective of an informationally advantaged *sender* who has privileged access to the *realized* state θ and would like to strategically signal this information to players in order to influence their actions. The sender is equipped with the natural objective of optimizing the sum of the players' utilities, i.e., the global objective $C(\mathbf{x})$ or $U(\mathbf{x})$, at equilibrium. We consider three natural types of signaling schemes with increasing generality.

Public Signaling. At a high level, a *public signalling scheme* constructs a random variable σ from support Σ—called the *signal*—that is correlated with the state of nature θ. The scheme then sends the sampled signal σ publicly to all players, which carries information about the state θ due to their correlation. Such a public scheme φ can be fully described by variables $\{\varphi(\sigma; \theta)\}_{\sigma \in \Sigma, \theta \in \Theta}$ where $\varphi(\sigma; \theta)$ is the probability of sending signal σ *conditioned* on state of nature θ. Adopting the standard information design assumption [18, 19], the sender commits to the signaling scheme before state θ is realized. Therefore, φ is publicly known to all players. The probability of sending signal σ equals $\mathbf{Pr}(\sigma) = \sum_\theta \lambda(\theta)\varphi(\sigma;\theta)$. Upon receiving signal σ, all players perform a standard Bayesian update and infer the following posterior probability about the state θ: $\mathbf{Pr}(\theta|\sigma) = \lambda(\theta)\varphi(\sigma;\theta)/P(\sigma)$.

Since all players receive the same information, the game will be played according to the expected cost $\bar{c}_i(s;\sigma) = \sum_\theta \mathbf{Pr}(\theta|\sigma)c_i^\theta(s)$ or $\bar{u}_i(s;\sigma) = \sum_\theta \mathbf{Pr}(\theta|\sigma)u_i^\theta(s)$ for all i and signal σ. We assume that players will reach a NE of this average game. Let $C(\sigma)$ denote the sender's expected cost at equilibrium under signal σ and $C(\varphi) = \sum_\sigma \mathbf{Pr}(\sigma)C(\sigma)$ denote the expected sender cost under signaling scheme φ. Like the PoA literature, when there are multiple Nash equilibria, we always adopt the *worst* one in our analysis. Notations for payoff maximization are defined similarly.

Private Signaling. Private signaling relaxes public signaling by allowing the sender to send different, and possibly correlated, signals to different players. Specifically, let Σ_i denote the set of possible signals to player i and $\Sigma = \Sigma_1 \times ... \times \Sigma_n$ denote the set of all possible signal profiles. With slight abuse of notation, a private signaling scheme can be similarly captured by variables $\{\varphi(\sigma;\theta)\}_{\theta\in\Theta,\sigma\in\Sigma}$. When signal profile σ is restricted to have the same signal to all players, this degenerates to public signaling. Private signaling leads to a truly Bayesian game where each player holds different information about the state of nature. The standard solution concept in this case is the *Bayes correlated equilibrium* (BCE) introduced by Bergemann and Morris [1], which consists of all outcomes that can possibly arise at Bayes Nash equilibrium under all possible signaling schemes. Standard revelation-principle type argument shows that signals of private signaling schemes in a BCE can be interpreted as *obedient action recommendations* [1,12,19]. That is, Σ_i can W.L.O.G. be S_i and $\Sigma = S$. An action recommendation s_i to player i is *obedient* if following this recommended action is indeed a best response for i, or formally, for any $s_i, s_i' \in S_i$ we have

$$\sum_{\theta\in\Theta, s_{-i}\in S_{-i}} \varphi(s_i, s_{-i}; \theta)\lambda(\theta)c_i^\theta(s_i, s_{-i}) \geq \sum_{\theta\in\Theta, s_{-i}\in S_{-i}} \varphi(s_i, s_{-i}; \theta)\lambda(\theta)c_i^\theta(s_i', s_{-i})$$

$$(2)$$

Ex-ante Private Signaling. Motivated by recommender system applications, recent works [4,6,32] relax the obedience constraints (2) of BCE to a *coarse correlated equilibrium* type of obedience constraints, described as follows:

$$\sum_{\theta\in\Theta, s\in S} \varphi(s_i, s_{-i}; \theta)\lambda(\theta)c_i^\theta(s_i, s_{-i}) \geq \sum_{\theta\in\Theta, s\in S} \varphi(s_i, s_{-i}; \theta)\lambda(\theta)c_i^\theta(s_i', s_{-i}), \forall s_i' \in S_i. \quad (3)$$

That is, for any player i, following the recommendation is better than opting out of the signaling scheme and acting just according to his prior belief. A signaling scheme satisfying Constraint (3) is dubbed an *ex-ante* private scheme [4,6].

3 The Power of Signaling (PoS)

We now formalize the *Power of Signaling* (PoS) in cost-minimization and payoff-maximization games. Intuitively, the PoS characterizes how much additional power a class of signaling schemes has over another. Formally, let Φ^a and Φ^b be two classes of signaling schemes (e.g., public and private schemes). We say Φ^b is *less restricted* than Φ^a, conveniently denoted as $\Phi^a \subseteq \Phi^b$, if $\varphi \in \Phi^b$ whenever $\varphi \in \Phi^a$.

Definition 1. (PoS of Φ^b over Φ^a). *For any two classes of signaling schemes Φ^a, Φ^b where Φ^b is less restricted than Φ^a (i.e., $\Phi^a \subseteq \Phi^b$), the power of signaling of Φ^b over Φ^a, or $PoS(\Phi^b : \Phi^a)$ for short, is defined as*

$$PoS(\Phi^b : \Phi^a) = \frac{\min_{\varphi \in \Phi^a} C(\varphi)}{\min_{\varphi \in \Phi^b} C(\varphi)} \left(or \; \frac{\max_{\varphi \in \Phi^a} U(\varphi)}{\max_{\varphi \in \Phi^b} U(\varphi)} \right),$$

for cost-minimization (or payoff-maximization) games.

In other words, PoS is the ratio between the objectives of the optimal scheme from signaling class Φ^a and that from a less restricted class Φ^b. Similar to the PoA ratio, $PoS(\Phi^b : \Phi^a) \geq 1$ for cost-minimization games, and the larger this ratio is, the more powerful Φ^b is over Φ^a. In contrast, $PoS(\Phi^b : \Phi^a) \leq 1$ for payoff-maximization games, and the smaller this ratio is, the more powerful Φ^b is over Φ^a. If both the numerator and denominator are 0, we say the PoS is 1; if only the denominator is 0, the PoS is $+\infty$.

Though PoS is well-defined for any two classes of signaling schemes, in this paper we primarily consider the following well-studied classes of signaling schemes:

– Φ^1 or FI: full information ;
– Φ^2 or Pub: public signaling schemes;
– Φ^3 or Pri: private signaling schemes;
– Φ^4 or exP: ex-ante private signaling schemes.

The full information class FI only contains a single signaling scheme, i.e., fully revealing the state θ. This serves as a benchmark scheme where information is not strategically signaled. Another natural benchmark scheme is to reveal *no* information. We show in Appendix C that the PoS compared to this benchmark turns out to be unbounded.

4 PoS in Cost-Minimization Games

The main result of this section is the following tight characterization about the PoS ratios in cost-minimization games.

Theorem 1. *For any Bayesian cost-minimization game $\{G^\theta\}_{\theta \sim \lambda}$, let $PoA_{\max} = \max_\theta PoA(G^\theta)$ denote the worst PoA ratio among game $G^\theta s$. We have*

$$PoS(\Phi^j : \Phi^i) \leq PoA_{\max}, \quad \forall 1 \leq i < j \leq 4. \tag{4}$$

Moreover, these upper bounds are all tight in the following sense: for any $r \geq 1$ and $1 \leq i < j \leq 4$, there exits a Bayesian cost-minimization game $\{G^\theta\}_{\theta \sim \lambda}$ with $PoA(G^\theta) = r$ for any θ and $PoS(\Phi^j : \Phi^i) = r$ as well.

The remainder of this section is devoted to the proof of Theorem 1. The following simple observation follows from Definition 1 of the PoS, and will be useful for proving the tightness of the bounds in Inequality (4).

Fact 1. *For any* $1 \leq i < j < j' \leq 4$, *we have*

$$PoS(\Phi^j : \Phi^i) \leq PoS(\Phi^{j'} : \Phi^i).$$

As a consequence of Fact 1, if we prove the tightness of $PoS(\Phi^2 : \Phi^1)$, i.e., PoS(Pub:FI) by constructing an example with PoS(Pub:FI) = PoA$_{\max}$, the example must satisfy PoS(Pri:FI) = PoS(exP:FI) = PoA$_{\max}$ as well, implying their tightness. Therefore, to prove the tightness of Inequality (4), we only need to prove the tightness of PoS(Pub:FI), PoS(Pri:Pub) and PoS(exP:Pri).

4.1   Proof of the PoS Upper Bounds

We first prove all the upper bounds in Inequality (4) through a unified result, summarized in the following proposition.

Proposition 1. *For any Bayesian cost-minimization game* $\{G^\theta\}_{\theta \sim \lambda}$, *we have* $PoS(\Phi^b : \Phi^a) \leq \max_\theta PoA(G^\theta)$ *for any two classes of signaling schemes* Φ^a, Φ^b *satisfying* $\Phi^a \subseteq \Phi^b$ *and that the full information scheme is contained in* Φ^a.

Proof. Let PoA$_{\max} = \max_\theta PoA(G^\theta)$ be the worst (i.e., the maximum) price of anarchy ratio among game G^θs. Denote by $C^*(G_\theta) = \min_{s \in S} C(S)$ the minimum total social cost among all outcomes (not necessarily an equilibrium) for game G_θ; let $s^{\theta*}$ be an strategy profile that achieves $C^*(G_\theta)$. φ^0 denotes the full information revelation scheme.

Observe that for any signaling scheme φ we must have $C(\varphi) \geq \sum_\theta \lambda(\theta) C^*(G^\theta)$ because regardless of how players act in the scheme φ, its expected total cost can never be less than the minimum possible total cost $\sum_\theta \lambda(\theta) C^*(G^\theta)$. Now since Φ^a contains φ^0, we thus have

$$\min_{\varphi' \in \Phi^a} C(\varphi') \leq C(\varphi^0)$$

$$= \sum_{\theta \in \Theta} \lambda(\theta) C(G_\theta)$$

$$\leq \sum_{\theta \in \Theta} \lambda(\theta) \cdot r C^*(G_\theta)$$

$$\leq r C(\varphi), \text{ for any scheme } \varphi$$

where $C(G_\theta)$ is the worst (i.e., maximum) equilibrium cost of game G^θ and the second inequality is by the definition of the price of anarchy. As a result, $PoS(\Phi^b : \Phi^a) = \frac{\min_{\varphi' \in \Phi^a} C(\varphi')}{\min_{\varphi \in \Phi^b} C(\varphi)} \leq r$, as desired.

4.2 Tightness of the Upper-Bound for PoS

Non-atomic Routing. It turns out that all the PoS bounds in Theorem 1 are tight in a special and well-studied class of cost-minimization games, i.e., non-atomic routing. The game takes place on a directed graph $G = (V, E)$ with V as the vertex set and E as the edge set. There is a continuum of players, each controlling a negligible amount of flow characterized by a pair of nodes (s, t) where $s \in V$ is the starting node of the flow and $t \in V$ is its destination. In non-atomic routing with incomplete information, each edge $e \in E$ can be described by a congestion function $c_e^\theta(x)$ which depends on the total amount of flow x on edge e as well as a random state of nature $\theta \in \Theta$. Each player (s, t) optimizes her own utility by taking a minimum-cost directed path from s to t. The sender minimizes overall congestion cost. There is an essentially unique pure Nash equilibrium for non-atomic routing under a public scheme. Therefore, equilibrium selection is not an issue in non-atomic routing.

Tightness of PoS(Pub:FI). We now show the tightness of PoS(Pub:FI) for any ratio $r \geq 1$ via a non-atomic routing game example, which implies the tightness of PoS(Pri:FI) and PoS(exP:FI) by Fact 1. Consider a variant of Pigou's example [27], as depicted in Fig. 1, where cost functions are described on each edge. The traffic demand from s_1 to t is set as $d(\alpha) = \left(\frac{1}{\alpha+1}\right)^{\frac{1}{\alpha}}$ whereas the demand from s_2 to t is $1 - d(\alpha)$. Each state of nature occurs with probability 0.5.

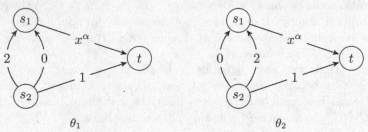

Fig. 1. A tight example for PoS(Pub:FI)

First, we compute the price of anarchy (PoA) of each game, as a function of α. Clearly, at equilibrium no flow will pass through the edge with cost 2 since deviating to the edge with cost 0 is strictly better. It is easy to see that at equilibrium all flow will go through the edge with cost x^α, leading to total congestion 1 at equilibrium. The optimal flow, however, is that all flow at s_1 goes through edge (s_1, t) and all flow at s_2 goes through (s_2, t), leading to minimum total congestion $\left(\frac{1}{\alpha+1}\right)^{\frac{(\alpha+1)}{\alpha}} + 1 - \left(\frac{1}{\alpha+1}\right)^{\frac{1}{\alpha}}$. Therefore, the PoA as a function of α in this instance is

$$\text{PoA} = \frac{1}{\left(\frac{1}{\alpha+1}\right)^{\frac{(\alpha+1)}{\alpha}} + 1 - \left(\frac{1}{\alpha+1}\right)^{\frac{1}{\alpha}}}, \tag{5}$$

which is a continuous function of $\alpha > 0$. Standard analysis shows that this function tends to ∞ as $\alpha \to \infty$ and tends to 1 as $\alpha \to 0^+$. For the special case of $\alpha = 0$, it can be directly verified that the PoA ratio is 1. Therefore, this PoA ratio can take any value $r \geq 1$ with a proper choice of α.

We now consider the cost of the full information scheme FI. In this case, all flow will always go through the edge with cost 1 at equilibrium, leading to total cost 1. The optimal public scheme in this example happens to be revealing no information. Without being able to distinguish the zero-cost edge from the edge of cost 2, all flow at s_2 will take the (s_2, t) path. This achieves the minimum total cost, rendering the PoS(Pub:FI) ratio equal the PoA for any n.

Tightness of PoS(Pri:Pub) and PoS(exP:Pri). Due to space limit, we give the instance constructions for the tightness of PoS(Pri:Pub) in Fig. 2 and tightness of PoS(exP:Pri) in Fig. 3. The formal computation about the PoS bound is relatively standard and is deferred to Appendix A.

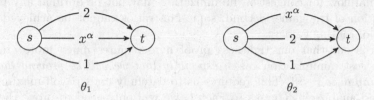

Fig. 2. A tight example for PoS(Pri:Pub)

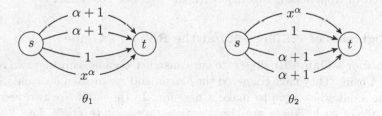

Fig. 3. A tight example for PoS(exP:Pri)

5   PoS in Payoff-Maximization Games

In this section, we show that a similar tight characterization as in Theorem 1 holds for payoff-maximization games as well.

Theorem 2. *For any Bayesian payoff-maximization game $\{G^\theta\}_{\theta \sim \lambda}$, let $PoA_{\min} = \min_\theta PoA(G^\theta)$ denote the worst-case PoA ratio among game G^θs. We have*

$$PoS(\Phi^j : \Phi^i) \geq PoA_{\min}, \quad \forall 1 \leq i < j \leq 4. \tag{6}$$

Moreover, these lower bounds are all tight in the following sense: for any $r \in (0, 1]$ and any $1 \leq i < j \leq 4$, there exits a Bayesian payoff-maximization game $\{G^\theta\}_{\theta \sim \lambda}$ with $PoA(G^\theta) = r$ for any θ and $PoS(\Phi^j : \Phi^i) = r$ as well.

Remark 2. *In payoff-maximization games, the smaller PoS is, the more powerful signaling is. Therefore, Inequality* (6) *is a lower bound for the PoS ratio but an upper bound for the power of signaling. Similar discrepancies also arise in the definition of the PoA for payoff-maximization game* [30].

The proof of the PoS lower bounds in Inequality (6) follow an analogous argument Proposition 1, and thus is omitted here due to space limit. We only prove their tightness. One might wonder whether the tightness proof here can be adapted from that for cost-minimization games by simply reversing minimizing cost functions to be maximizing their negations (plus a large constant to make it positive). The answer turns out to be *no*. We illustrate the detailed reasons in the Appendix D, but at a high level there are at least two reasons. First, the optimal flow for congestion minimization may not be optimal any more in the negation of the game. Second, some PoA ratios cannot be achieved in the negation of routing games.

It turns out that our tightness proof here requires much more creativity than the cost minimization cases, *especially that we will be proving the tightness at any ratio $r < 1$*. This requires us to carefully craft payoff-maximization games and analyze it. Thanks to Fact 1, we will only need to prove the tightness of PoS(Pub:FI), PoS(Pri:Pub) and PoS(exP:Pri). The tightness proof for PoS(Pri:Pub) turns out to be the most basic and involved, which we show next. The proof of PoS(exp:Pri) is based on a variant to the *Robber's game*, which is deferred to Appendix B together with the proof of PoS(Pub:FI).

5.1   Tightness of PoS(Pri:Pub) at the Robber's Game

The Robber's Game. To illustrate our constructed game, consider two robbers robbing a bank. They have triggered the alarm, and are pressed for time. As they enter the vault, they have to make a decision. In the vault are two *safes*, and a huge pile of cash. There are three options: attempt to crack Safe 1 (action S_1), Safe 2 (action S_2), or simply take as much cash as they can carry (action C), worth α. In order to protect the bank's valuable items, one of the safes is a decoy safe, and is *empty*. Inside the other safe are two objects: a gold bar worth $\alpha + \epsilon$, where $1 \geq \alpha > \epsilon > 0$, and an extremely·delicate but valuable crystal worth 1. The two robbers (players) have very different skill sets. Player 1 (P1) is a lock picking expert, and Player 2 (P2) is a demolitions expert. The players' payoffs equal whatever they individually steal from the vault. Only P2 is capable of carrying the crystal without breaking it. If P1 cracks the safe using his lockpicking, P2 can take the crystal on the way out, leaving P1 with the gold bar. However, if P2 cracks the safe, the crystal will be destroyed due to his explosions, leaving him only the gold bar. In addition, if both players crack the same safe, they get in each others' way and are forced to leave with nothing.

	P2 S_1	C	S_2
S_1	$(0,0)$	$(0,\alpha)$	$(0,\alpha+\epsilon)$
P1 C	$(\alpha,0)$	(α,α)	$(\alpha,\alpha+\epsilon)$
S_2	$(\alpha+\epsilon,1)$	$(\alpha+\epsilon,1)$	$(0,0)$

θ_1

	P2 S_1	C	S_2
S_1	$(0,0)$	$(\alpha+\epsilon,1)$	$(\alpha+\epsilon,1)$
P1 C	$(\alpha,\alpha+\epsilon)$	(α,α)	$(\alpha,0)$
S_2	$(0,\alpha+\epsilon)$	$(0,\alpha)$	$(0,0)$

θ_2

Fig. 4. Payoffs of the robber's game achieving tight PoS(Pri:Pub); each state occurs with probability 0.5.

Concretely, the payoff matrix for the aforementioned game is in the above tables. At θ_1, safe 1 is empty, and at θ_2, safe 2 is empty. Note that θ_2 simply exchanges the payoffs of strategies S_1 and S_2 symmetrically for both players. A-priori, both robbers do not know the state and share the common uniform random prior. The sender is a heist leader and knows which safe is which. The sender gets to pocket a cut of the total haul, so naturally she is interested in maximizing the total utility. In public signaling, both robbers use the same radio to communicate with the sender, but in private signaling each player has his own communication radio.

We first calculate the PoA for the game at each state. W.l.o.g., we consider equilibria for state θ_1 as θ_2 is symmetric. Since action C dominates action S_1 (i.e., cracking the empty safe) for both players, without loss of generality we will assume both players do not play action S_1 in our following analysis. It is easy to see that (S_2, C) and (C, S_2) are the only two pure Nash equilibria after excluding action S_1. We now consider mixed equilibrium in this game. Note that since both players have only two actions, both players must randomize in any mixed equilibrium. Let $\lambda_1, \lambda_2 \in (0,1)$ denote the probability that player 1 and 2 play C, respectively. We have $\lambda_2(\alpha + \epsilon) = \alpha$ since player 1's both actions must be equally good, and similarly $\lambda_1\alpha + (1 - \lambda_1) = (\alpha + \epsilon)\lambda_1$. This implies $\lambda_1 = 1/(1 + \epsilon)$ and $\lambda_2 = \alpha/(\alpha + \epsilon)$. The total expected payoff of this mixed strategy equilibrium is $\alpha + (\alpha + \epsilon)/(1 + \epsilon)$, which is the smallest equilibrium total payoff. The largest social payoff is however $\alpha + \epsilon + 1$. Therefore, the PoA of this game is $\frac{2\alpha+\epsilon+\alpha\epsilon}{(1+\alpha+\epsilon)(1+\epsilon)}$.

We now show that the optimal public signalling scheme is to reveal full information, assuming worst-case equilibrium selection. We prove this by arguing that for any public signal with posterior probability $p \in [0,1]$ of state θ_1, the sender's utility is at most $U_0 = \alpha + (\alpha+\epsilon)/(1+\epsilon)$ in the worst equilibrium, which is the sender utility of full information revelation. This follows a case analysis:

1. When $p(\alpha \mid c) < \alpha$ and $(\alpha + \epsilon)(1 - p) < \alpha$, i.e., $\frac{\epsilon}{\alpha+\epsilon} < p < \frac{\alpha}{\alpha+\epsilon}$ (recall that our parameter choice satisfies $\alpha > \epsilon$). In this case, for P1, the utility α of the safer action C is strictly larger than the best possible expected utility $(\alpha + \epsilon)(1 - p)$ of taking action S_1 and larger than the best possible expected utility $(\alpha + \epsilon)p$ of taking action S_2 as well. Given that P1 will always take C, P2 strictly prefers C as well as his utility $(\alpha + \epsilon)(1 - p)$ for S_1 and utility $(\alpha+\epsilon)p$ for S_2 are both smaller. Therefore, both players taking action C is the

unique equilibrium, leading to sender utility 2α which is less than U_0 since $U_0 > \alpha + (\alpha + \epsilon\alpha)/(1 + \epsilon) = 2\alpha$.

2. When $p(\alpha + \epsilon) \geq \alpha$. In this case, state θ_1 is very likely and action S_1 is strictly *dominated* by action C for P1 and thus will not be taken by P1. We show that there exists a mixed strategy that has sender utility worse than U_0. In particular, consider the following mixed strategies: (1) P1 chooses action C with probability $p_C = \frac{p + \alpha(1-p)}{\epsilon p + p}$ and action S_2 with remaining probability $1 - p_C$; (2) P2 chooses action C with probability $q_C = \frac{\alpha}{(\alpha + \epsilon)p}$ and action S_2 with remaining probability $1 - q_C$. We claim that this is a mixed strategy equilibrium. In particular, both action C and S_2 have expected utility α to P1 and both action C and S_2 have expected utility $p_C(\alpha + \epsilon)p$. Therefore, the sender's utility at the worst mixed equilibrium is at most

$$\alpha + \frac{p + \alpha(1-p)}{\epsilon p + p}(\alpha + \epsilon)p = \alpha + \frac{\alpha + \epsilon}{\epsilon + 1}[p + \alpha(1-p)]$$

$$\leq \alpha + \frac{\alpha + \epsilon}{\epsilon + 1} = U_0.$$

3. When $(1 - p)(\alpha + \epsilon) \geq \alpha$, i.e. $p \leq \frac{\epsilon}{\alpha + \epsilon}$. This case is symmetric to Case 2, and thus has the same conclusion.

Finally, we argue that the optimal private scheme results in the optimal social outcome. Consider the private scheme that reveals no information to Player 2 but full information to Player 1. Given no information, Player 2 has a strictly dominant action C since $\epsilon < \alpha$. With full information, Player 1 will always open the non-empty safe. This leads to the optimal outcome and sender utility $1 + \alpha + \epsilon$. Therefore, the `PoS(Pri:Pub)` for this game equals the `PoA` $= \frac{2\alpha + \epsilon + \alpha\epsilon}{(1 + \alpha + \epsilon)(1 + \epsilon)}$. We see that when $\alpha = 1$ and $\epsilon \to 0$, the `PoA` tends to 1; when $\alpha \to 0$ and $\epsilon \to 0$, the `PoA` tends to 0. When $\alpha = 1$ and $\epsilon = 0$, we have a trivial case, with `PoA` and `PoS` 1. Thus its ratio takes any value within $(0, 1]$.

6   Discussions and Future Work

In this paper, we initiate and formalize the concept of the power of signaling (`PoS`). In the general classes of cost-minimization and payoff-maximization games, we show that the `PoS` is inherently related to, in fact precisely characterized by, the *price of anarchy* (`PoA`).

There are many possibilities for future research. In our analysis, we use the full information scheme (`FI`) as the benchmark, since the no information scheme (`NI`) will lead to infinite bound. However, another natural and stronger benchmark is the *better* of these two schemes `FI`, `NI`. One interesting question is whether we will have strictly less power of signaling when compared to this strong benchmark. Another direction is to consider weaker class of signaling schemes. For example, we can consider the power of signaling for restricted classes of schemes, such as schemes with limited communication power [11] or schemes with costly communication [16]? For these classes of schemes, the power of signaling will also decrease. It is interesting to study how much these restrictions limit the power of signaling.

A   Omitted Proofs for Cost-Minimization PoS Bounds

A.1   Tightness of the Upper-Bound for PoS(Pri:Pub)

We show the tightness of PoS(Pri:Pub) for any ratio $r \geq 1$, which implies the tightness of PoS(exP:Pub) by Fact 1. We construct a Bayesian non-atomic routing game as depicted in Fig. 2, which can be viewed as another variant of Pigou's example.[3] There is a 1 unit of flow demand from s to t. Each state has equal probability 0.5.

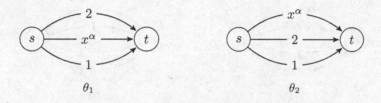

Fig. 5. A tight example for PoS(Pri:Pub)

Similarly to the calculation for the Example in Figure 1, the PoA for each game here also equals that as described in Eq. (5). We now argue that the expected cost of any public signaling scheme will equal 1 in this example i.e., all public schemes are equally bad and will not be able to reduce any congestion. Any public signal gives the same information about the game state to all players. Let $\lambda \in [0, 1]$ denote the posterior probability of θ_1 given any public signal. W.l.o.g., consider the case $\lambda \geq 0.5$ since the other case is symmetric. The top edge will have expected cost $2\lambda + x^\alpha(1 - \lambda) > 1$ for any $x > 0$, therefore this edge will never be taken since the bottom edge is a strictly better choice. Consequently, players will be choosing between the middle edge, with expected cost function $f(x) =: x^\alpha \lambda + 2(1 - \lambda)$, and the bottom edge with fixed cost 1. Note that $f(0) = 2(1 - \lambda) \leq 1$ and $f(1) = 1 - \lambda \geq 1$. Therefore, at the unique equilibrium, the amount of flow through the middle edge will be exactly the x^* such that $f(x^*) = 1$ whereas the remaining flow will be through the bottom edge. The expected total cost at this equilibrium is 1.

Finally, we show that the optimal private signaling will be able to induce the optimal flow, concluding our tightness proof. Consider the following private signaling scheme: revealing full information to a randomly selected $x^* =$

---

[3] This example generalizes an earlier example observed by Cheng, Dughmi and Xu [10]. It also avoided their use of (impractical) ∞ flow cost, and relies on a more involved analysis due to the finite edge cost.

$(1/(\alpha+1))^{1/\alpha}$ fraction of the players, and revealing no information to the remaining players. The x^* fraction of players given the full information has a dominant action of taking the edge with cost x^α at the told state. For the remaining players with no information, their cost of taking either the top or the middle edge will be at least $2 \times (1/2) + (x^*)^\alpha \times (1/2) > 1$. Therefore, their optimal response will be taking the bottom edge. This leads to exactly the optimal flow for each state, as desired.

A.2   Tightness of the Upper-Bound for PoS(exP:Pri)

Finally, we prove the tightness of PoS(exP:Pri). Consider the non-atomic routing game depicted in Fig. 3. There is one unit of flow from s to t and the two states θ_1, θ_2 occurs with equal probability 0.5.

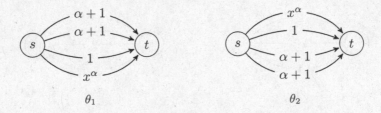

Fig. 6. A tight example for PoS(exP:Pri)

Similar to the analysis for previous examples, the PoA for each game equals also the function described in Eq. (5), which takes value in $[1, \infty)$ as we vary the parameter α.

Next we argue that the optimal private signaling scheme is full-revelation, with cost 1. Recall from the preliminary section, any private scheme can be viewed as obedient action recommendations. Numbering the edges from the top to the bottom as edge $1, 2, 3, 4$, we claim that any obedient action recommendation should never recommend edge 2 and 3. This is because if a non-zero amount of players are recommended, e.g., to edge 2, switching to edge 1 or 4 will be strictly better. In particular, if the players are certain that they are at state θ_1, they will prefer to switch to edge 4. Otherwise, there is non-zero probability that they are at state θ_2. In this case, switching to edge 1 is strictly better. Consequently, the optimal private scheme can be captured by two variables: (1) x: the amount of flow recommended to edge 4 at state θ_1 (thus edge 1 consumes the remaining $1 - x$ amount); (2) y: the amount of flow recommended to edge 1 at θ_2. It can be shown that the parameterized total expected cost $[x^{\alpha+1} + (\alpha+1)(1-x) + y^{\alpha+1} + (\alpha+1)(1-y)]/2$ is minimized at $x = 1, y = 1$, i..e, the full information scheme.

Finally, we show that the optimal ex-ante private scheme achieves the minimum possible total social cost, concluding the tightness proof of PoS(exP:Pri). In particular, consider the ex-ante private scheme that induces the optimal flow by recommending a randomly chosen $(1/(\alpha + 1))^{1/\alpha}$ amount of the flow to the edge with cost x^α at any state and the remaining flow to the edge with cost 1. This is indeed obedient in the ex-ante sense because opting out from this scheme and taking any path will lead to cost at least $(\alpha + 1 + \frac{1}{\alpha+1})/2$, which is at least 1 and thus is larger than its expected utility in the scheme (less than 1).

B   Omitted Proofs for Payoff-Maximization PoS Bounds

B.1   Tightness of the Lower-Bound for PoS(Pub:FI)

Consider the following game played by two players P1, P2 where α, ϵ are parameters satisfying $\alpha - 1 > 2\epsilon > 0$. Each player has two actions, conveniently denoted as A, B. There are two states θ_1, θ_2 with equal probability 0.5. The only difference of the two states is the payoffs for action profile (B, A), which is $(1, \alpha)$ at state θ_1 and $(\alpha, 1)$ at state θ_2.

		P2	
		A	B
P1	A	$(1+\epsilon, 1)$	$(1-\epsilon, 1-\epsilon)$
	B	$(1, \alpha)$	$(1, 1+\epsilon)$

θ_1

		P2	
		A	B
P1	A	$(1+\epsilon, 1)$	$(1-\epsilon, 1-\epsilon)$
	B	$(\alpha, 1)$	$(1, 1+\epsilon)$

θ_2

We first consider full information (FI). At state θ_1, action A is a strictly dominant action for Player 2. This implies that both players choosing action A is the only Nash equilibrium, resulting in sender utility $2 + \epsilon$. However, the optimal outcome at state θ_1 is that Player 1 chooses B and Player 2 chooses A, leading to total payoff $\alpha + 1$ ($> 2 + 2\epsilon$ since $\alpha - 1 > 2\epsilon$). State θ_2 is symmetric. The expected utility of full information is $2 + \epsilon$, and the PoA of each game is $\frac{2+\epsilon}{\alpha+1}$.

We now show that the optimal public signaling scheme can maximize total payoff. Consider the scheme which reveals no information at all. The only change is that the expected payoffs of action profile (B, A) becomes $\frac{\alpha+1}{2}$ for both players. We see then, that A remains a strictly dominant strategy for Player 2 and B becomes a strictly dominant action for Player 1. The only equilibrium is the action profile (B, A), resulting in expected sender utility $\alpha + 1$. Therefore, PoS(Pub:FI) $= \frac{2+\epsilon}{\alpha+1}$, equaling the price of anarchy. If $\alpha \to 1$, the PoS ratio tends to 1, and $\alpha \to \infty$ with a fixed ϵ makes the PoS ratio continuously tend to 0. Setting $\alpha = 1 + \epsilon$ gives a trivial case where the PoS and PoA are both 1. Thus, the ratio achieves all possible values within $(0, 1]$.

B.2   Tightness of the Lower-Bound for PoS(exP:Pri)

Finally, for any $r \in (0, 1]$, we show the tightness of PoS(exP:Pri). Consider the same two robbers, robbing the same bank. However, the bank has now upgraded its anti-theft countermeasures. Instead of a decoy safe, there is now an entire decoy vault. Naturally, any robber who goes into it will leave empty-handed. In the real vault, there is again a large pile of cash, but now (only) one safe. A robber can choose to take cash for a guaranteed payoff ($\alpha + \epsilon$ for Player 1, α for Player 2). The safe contains the same fragile, valuable crystal as in the previous section, but this time, no gold bar. If both players attempt to crack the safe, they get in each others' way and will fail. After cracking the safe, each player has enough time to take cash instead of the contents of the safe if they choose. We assume $1 \geq \alpha > \epsilon \geq 0$ and $\alpha + \epsilon < 1$. If Player 2 attempts to take the cash, and Player 1 cracks the safe, Player 2 can take the crystal on the way out, *instead of* the cash. However, if Player 2 does not go into the correct vault, he will leave with nothing. The detailed payoff is as in the following table. Here, θ_1 corresponds to the first vault being the decoy one, whereas θ_2 corresponds to the second vault being empty. Each state occurs with equal probability 0.5. The sender as the heist leader knows which vault is empty.

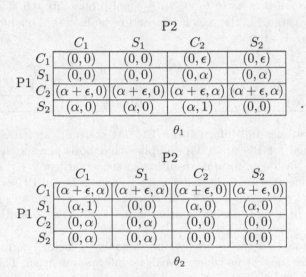

$$\theta_1$$

		C_1	S_1	C_2	S_2
	C_1	$(0,0)$	$(0,0)$	$(0,\epsilon)$	$(0,\epsilon)$
P1	S_1	$(0,0)$	$(0,0)$	$(0,\alpha)$	$(0,\alpha)$
	C_2	$(\alpha+\epsilon,0)$	$(\alpha+\epsilon,0)$	$(\alpha+\epsilon,\alpha)$	$(\alpha+\epsilon,\alpha)$
	S_2	$(\alpha,0)$	$(\alpha,0)$	$(\alpha,1)$	$(0,0)$

P2

		C_1	S_1	C_2	S_2
	C_1	$(\alpha+\epsilon,\alpha)$	$(\alpha+\epsilon,\alpha)$	$(\alpha+\epsilon,0)$	$(\alpha+\epsilon,0)$
P1	S_1	$(\alpha,1)$	$(0,0)$	$(\alpha,0)$	$(\alpha,0)$
	C_2	$(0,\alpha)$	$(0,\alpha)$	$(0,0)$	$(0,0)$
	S_2	$(0,\alpha)$	$(0,\alpha)$	$(0,0)$	$(0,0)$

$$\theta_2$$

Fig. 7. Payoffs of the robber's game variant with tight `PoS(exP:Pri)`; each state occurs with probability 0.5.

We first calculate the `PoA` of each game. For the game at state θ_1, it is easy to see that action C_2 strictly dominates all other actions for Player 1. This leads to (C_2, C_2) and (C_2, S_2) be the two unique NEs and the total payoff of any equilibrium is $(\alpha + \epsilon) + \alpha = 2\alpha + \epsilon$. The maximum total payoff however is $\alpha + 1$, under action profile (S_2, C_2). Similar analysis holds for state θ_2. The `PoA` for each game is thus $(2\alpha + \epsilon)/(1 + \alpha)$ (recall $\alpha + \epsilon < 1$ in our game).

We now argue that an optimal private scheme is to reveal full information. Crucially, Player 1's strictly dominant action is to take C_1 or C_2, whichever is more likely to get the $\alpha + \epsilon$ cash amount in the posterior distribution of his private signal. Given this, Player 2's optimal action is to choose C_1/S_1 or C_2/S_2, whichever is more likely to get the α payoff (from cash or from opening the safe) in the posterior distribution of his private signal. Consequently, any partial information will lead to Player 1 utility at most $\alpha + \epsilon$ and Player 2 utility at most α. This renders full information revelation optimal, leading to total player utility $2\alpha + \epsilon$.

Finally, we show that an ex-ante signaling scheme induces the optimal outcome. The scheme simply recommends (S_2, C_2) at state θ_1 and (S_1, C_1) at state θ_2. This satisfies the ex-ante obedience constraint (3) because: (1) if Player 1 opts out and acts according to his prior belief, he gets expected utility at most $\frac{1}{2}(\alpha + \epsilon)$, which is strictly less than his utility α in the scheme; (2) Player 2 gets utility 1 in the scheme and certainly does not want to opt out. Therefore, the PoS(exP:Pri) ratio in this game equals precisely the PoA of each game $\frac{2\alpha+\epsilon}{1+\alpha}$. This ratio tends to 1 as $\alpha + \epsilon \rightarrow 1$ and tends to 0 as $\alpha, \epsilon \rightarrow 0^+$. If $\alpha + \epsilon = 1$, the PoA and PoS(exP:Pri) are trivially equal to 1. Therefore, the ratio takes any value $r \in (0, 1]$.

C   PoS w.r.t the No Information (NI) Benchmark

In the main body of our paper, we choose full information (FI) as our benchmark scheme. One might wonder what happens if the no information scheme is used instead. It turns out that no information (NI) may lead to very bad social welfare, and examples are fairly easy to construct. Massicot and Langbort gave a Bayesian cost-minimization game with PoA = 4/3 for each game—more concretely, a non-atomic routing game example with affine latency function—such that PoS(Pub:NI) $\rightarrow \infty$ [25].

We now exhibit a simple reward-maximization game with PoA = 1 but PoS(Pub:NI) $\rightarrow 0$. Consider a (trivial) game with n actions $A_1, A_2, A_3, ..., A_n$ and a single player. There are n equally likely states of nature, with each state of nature θ_i gives utility 1 to action A_i and 0 utility to all other actions. The price of anarchy of this game is trivially 1, since there is only one reward-maximizing player, and full information as the optimal public scheme achieves optimal welfare 1. However, in the case of no information, the player can only get expected utility $\frac{1}{n}$. As $n \rightarrow \infty$, PoS(Pub:NI) $\rightarrow 0$.

D   Non-tightness of PoS in "Reverse" Routing

When proving the tightness for *payoff-maximization* games, a very natural first attempt is, perhaps, to convert the previously constructed cost-minimization routing games into payoff-maximization games by flipping the sign of cost functions and adding a large constant to make it positive. One example by reversing our game constructed for the tightness of PoS(Pri:Pub) is depicted in Fig. 8. That is, any edge with cost function $c_e(x)$ in our original construction can be changed to instead having payoff $N - c_e(x)$ for large positive constant N. Clearly, this is a valid payoff-maximization game, which we term "reverse" routing game for convenience.

We argue this natural adaptation of our previous routing game constructions in this way does not produce an example with, e.g., tight PoS(Pri:Pub) ratio.

Fig. 8. A reverse routing example

This is why we must turn to new constructions of payoff-maximization games. There are two reasons. First, this adaption cannot lead to any price of anarchy ratio within $(0, 1)$. In particular, it can be verified that the PoA of the game in Fig. 8 is at least $1/2$ since $N \geq 2$. The second major reason is that optimal routes in the standard cost-minimization routing game may not be optimal any more in its natural adaption to the reward-maximization situation. For example, in cost minimization, one never wants to route through a cycle but in its reward maximization variant, we would like to route through a cycle as much as possible to collect rewards (such examples are fairly easy to construct).

References

1. Bergemann, D., Morris, S.: Bayes correlated equilibrium and the comparison of information structures in games. Theor. Econ. **11**(2), 487–522 (2016)
2. Bhaskar, U., Cheng, Y., Kun Ko, Y., Swamy, C.: Conference on economics and computation, EC 2016, pp. 479–496. Association for Computing Machinery, New York (2016)
3. Miltersen, P.B., Sheffet, O.: Send mixed signals: Earn more, work less. In: Proceedings of the 13th ACM Conference on Electronic Commerce, EC 2012, pp. 234–247, ACM, New York (2012)
4. Castiglioni, M., Celli, A., Marchesi, A., Gatti, A.: Signaling in Bayesian network congestion games: the subtle power of symmetry. ArXiv, abs/2002.05190 (2020)
5. Castiglioni, M., Celli, A., Gatti, N.: Persuading voters: it's easy to whisper, it's hard to speak loud. In: Thirty-Forth AAAI Conference on Artificial Intelligence (2020)
6. Celli, A., Coniglio, S., Gatti, N.: Private Bayesian persuasion with sequential games. In: Proceedings of the AAAI Conference on Artificial Intelligence, vol. 34, pp. 1886–1893, April 2020
7. Chawla, S., Sivan, B.: Bayesian algorithmic mechanism design. ACM SIGecom. Exchanges **13**(1), 5–49 (2014)
8. Conitzer, V., Freeman, R., Shah, N.: Fair public decision making. In: Proceedings of the 2017 ACM Conference on Economics and Computation, pp. 629–646 (2017)
9. Das, S., Kamenica, E., Mirka, R.: Reducing congestion through information design. In: 2017 55th Annual Allerton Conference on Communication, Control, and Computing (Allerton), pp. 1279–1284. IEEE (2017)
10. Dughmi, S.: Algorithmic information structure design: a survey. ACM SIGecom. Exchang. **15**(2), 2–24 (2017)
11. Dughmi, S., Kempe, D., Qiang. R.: Persuasion with limited communication. In: Proceedings of the 2016 ACM Conference on Economics and Computation, pp. 663–680 (2016)

12. Dughmi, S., Xu, H.: Algorithmic Bayesian persuasion. In: Proceedings of the Forty-Eighth Annual ACM Symposium on Theory of Computing, STOC 2016, pp. 412–425. ACM (2016)

13. Dughmi, S., Xu, H.: Algorithmic persuasion with no externalities. In: Proceedings of the 2017 ACM Conference on Economics and Computation, pp. 351–368. ACM (2017)

14. Emek, Y., Feldman, M., Gamzu, I., Paes Leme, R., Tennenholtz, M.: Signaling schemes for revenue maximization. In: Proceedings of the 13th ACM Conference on Electronic Commerce, EC 2012, pp. 514–531. ACM (2012)

15. Feldman, M., Immorlica, N., Lucier, B., Roughgarden, T., Syrgkanis, V.: The price of anarchy in large games. In: Proceedings of the Forty-Eighth Annual ACM symposium on Theory of Computing, pp. 963–976 (2016)

16. Gentzkow, M., Kamenica, E.: Costly persuasion. Am. Econ. Rev. **104**(5), 457–62 (2014)

17. Immorlica, N., Ligett, K., Ziani, J.: Access to population-level signaling as a source of inequality. In: Proceedings of the Conference on Fairness, Accountability, and Transparency, pp. 249–258 (2019)

18. Kamenica, E.: Bayesian persuasion and information design. Ann. Rev. Econ. **11**(1), 249–272 (2019)

19. Kamenica, E., Gentzkow, M.: Bayesian persuasion. Am. Econ. Rev. **101**(6), 2590–2615 (2011)

20. Keren, S., Haifeng, X., Kwapong, K., Parkes, D., Grosz, B.: Information shaping for enhanced goal recognition of partially-informed agents. In: Proceedings of the AAAI Conference on Artificial Intelligence, vol. 34, pp. 9908–9915 (2020)

21. Koutsoupias, E., Papadimitriou, C.: Worst-case equilibria. In: Meinel, C., Tison, S. (eds.) STACS 1999. LNCS, vol. 1563, pp. 404–413. Springer, Heidelberg (1999). https://doi.org/10.1007/3-540-49116-3_38

22. Kulkarni, J., Mirrokni, V.: Robust price of anarchy bounds via lP and fenchel duality. In: Proceedings of the Twenty-Sixth Annual ACM-SIAM Symposium on Discrete Algorithms, pp. 1030–1049. SIAM (2014)

23. Li, Z., Das, S.: Revenue enhancement via asymmetric signaling in interdependent-value auctions. In: Proceedings of the AAAI Conference on Artificial Intelligence, pp. 2093–2100 (2019)

24. Mansour, Y., Slivkins, A., Syrgkanis, V., Wu, Z.S.: Bayesian exploration: incentivizing exploration in Bayesian games. In: Proceedings of the 2016 ACM Conference on Economics and Computation (2016)

25. Massicot, O., Langbort, C.: On the comparative performance of information provision policies in network routing games. In: 2018 52nd Asilomar Conference on Signals, Systems, and Computers, pp. 1434–1438 (2018)

26. Nisan, N., Ronen, A.: Algorithmic mechanism design. Games Econ. Behav. **35**(1–2), 166–196 (2001)

27. Pigou, A.C.: The Economics of Welfare. Palgrave Macmillan, London (2013)

28. Procaccia, A.D., Tennenholtz, M.: Approximate mechanism design without money. ACM Tran. Econ. Comput. (TEAC) **1**(4), 1–26 (2013)

29. Rabinovich, Z., Jiang, A.X., Jain, M., Xu, H.: Information disclosure as a means to security. In: Proceedings of the 14th International Conference on Autonomous Agents and Multiagent Systems (AAMAS) (2015)

30. Roughgarden, T.: Intrinsic robustness of the price of anarchy. J. ACM (JACM) **62**(5), 32 (2015)

31. Vetta, A.: Nash equilibria in competitive societies, with applications to facility location, traffic routing and auctions. In: Proceedings of the 43rd Annual IEEE Symposium on Foundations of Computer Science, 2002, pp. 416–425. IEEE (2002)
32. Xu, H.: On the tractability of public persuasion with no externalities. In: Proceedings of the 2020 ACM-SIAM Symposium on Discrete Algorithms (2020)
33. Xu, H., Rabinovich, Z., Dughmi, S., Tambe, M.: Exploring information asymmetry in two-stage security games. In: Proceedings of the Twenty-Ninth AAAI Conference on Artificial Intelligence, pp. 1057–1063. AAAI Press (2015)
34. Xu, H., Wang, K., Vayanos, P., Tambe, M.: Strategic coordination of human patrollers and mobile sensors with signaling for security games. In: Thirty-Second AAAI Conference on Artificial Intelligence (2018)
35. Yan, C., Xu, H., Vorobeychik, Y., Li, B., Fabbri, D., Malin, B.A.: To warn or not to warn: online signaling in audit games. In: 2020 IEEE 36th International Conference on Data Engineering (ICDE), pp. 481–492. IEEE (2020)

Uncertainty-Aware Low-Rank Q-Matrix Estimation for Deep Reinforcement Learning

Tong Sang, Hongyao Tang, Jianye Hao$^{(\boxtimes)}$, Yan Zheng$^{(\boxtimes)}$, and Zhaopeng Meng

College of Intelligence and Computing, Tianjin University, Tianjin, China
{2019218044,bluecontra,jianye.hao,yanzheng,mengzp}@tju.edu.cn

Abstract. Value estimation is one key problem in Reinforcement Learning. Albeit many successes have been achieved by Deep Reinforcement Learning (DRL) in different fields, the underlying structure and learning dynamics of value function, especially with complex function approximation, are not fully understood. In this paper, we report that decreasing rank of Q-matrix widely exists during learning process across a series of continuous control tasks for different popular algorithms. We hypothesize that the low-rank phenomenon indicates the common learning dynamics of Q-matrix from stochastic high dimensional space to smooth low dimensional space. Moreover, we reveal a positive correlation between value matrix rank and value estimation uncertainty. Inspired by above evidence, we propose a novel **U**ncertainty-**A**ware **L**ow-rank Q-matrix **E**stimation (**UA-LQE**) algorithm as a general framework to facilitate the learning of value function. Through quantifying the uncertainty of state-action value estimation, we selectively erase the entries of highly uncertain values in state-action value matrix and conduct low-rank matrix reconstruction for them to recover their values. Such a reconstruction exploits the underlying structure of value matrix to improve the value approximation, thus leading to a more efficient learning process of value function. In the experiments, we evaluate the efficacy of UA-LQE in several representative OpenAI MuJoCo continuous control tasks.

Keywords: Reinforcement learning · Value estimation · Uncertainty · Low rank

1 Introduction

In recent years, Deep Reinforcement Learning (DRL) has been demonstrated to be a promising approach to solve complex sequential decision-making problems in different domains [6,11,16,21,22,26]. Albeit the progress made in RL, the intriguing learning dynamics is still not well known, especially when with complex function approximation, e.g., deep neural network (DNN). The little understanding of how RL performs and proceeds prevents the deep improvement and utilization of RL algorithms.

© Springer Nature Switzerland AG 2022
J. Chen et al. (Eds.): DAI 2021, LNAI 13170, pp. 21–37, 2022.
https://doi.org/10.1007/978-3-030-94662-3_2

To understand the learning dynamics of RL, value function is one of the most important functions to consider. Value function defines the expected cumulative rewards (i.e., returns) of a policy, indicating how a state or taking an action under a state could be beneficial when performing the policy. It plays a vital role in RL for both value-based methods (e.g., DQN [16]) and policy-based methods (e.g., DDPG [11]). The learning dynamics and underlying structure of value function can be easily investigated in tabular RL by classical RL algorithms [24], where typically the action-value function Q of finite states and actions is maintained as a table. Towards larger state-action space, linear function approximation is widely as a representative option used for analysis [8,13,20]. Nevertheless, with complex non-linear function approximation as DRL, the learning dynamics and underlying structure are not well understood. A few works provide some insights on this with different empirical discoveries [7,10,12,28].

One notable work is Structured Value-based RL (SVRL) [28]. Yang et al. [28] identifies the existence of low-rank Q-functions in both classical control with Dynamic Programming [17] and Atari games with DQN [16]. In specific, the approximate rank of Q-matrix spanned by (sampled) states and actions decreases during the learning process. This indicates an underlying low-rank structure of the optimal value function and the learning dynamics of rank decrease. By leveraging this low-rank structure of Q-function, SVRL performs random drop of Q-matrix entries and then matrix estimation (or matrix completion). The intuition behind SVRL is the utilization of global structure of Q-function to reduce the value estimation errors thus facilitates the value function learning. Despite of the promising results achieved by SVRL, the low-rank structure of value function has not been investigated in continuous control. In addition, the selection of Q-matrix entries is aimless, in other words, it does not take into account the quality of estimate values, thus the underlying information from the low-rank structure is not fully and accurately utilized.

In this paper, we aim at addressing the above two problems and propose a novel algorithm called **U**ncertainty-**A**ware **L**ow-rank Q-matrix **E**stimation (**UA-LQE**), as a general framework to facilitate the learning of value function of DRL agents. We first examine the low-rank structure and learning dynamics of Q-functions in continuous control task. A quick view of the conclusion is shown in Fig. 1, the rank of sampled Q-matrix of representative continuous control DRL algorithms (DDPG [11], TD3 [4] and SAC [5]) decreases during the learning process. Our empirical evidence completes the SVRL's discovery in DRL, indicating the generality of low-rank structure of Q-function across different algorithms and environments. Thus, we can expect that an effective method that makes use of the low-rank structure to improve RL will be beneficial in RL community broadly. Moreover, we take a step further and empirically reveal a positive correlation between value matrix rank and value estimation uncertainty, i.e., a value matrix of high rank often has an high overall value estimation uncertainty. Thus, we hypothesize that value estimation uncertainty can be used as the indicator of the target entries in Q-value matrix. This is because that the value estimate with high uncertainty often has a large estimation error that may severely violate the underlying structure of Q-value matrix, making themselves be the major 'sinners' to the induced high rank of value matrix. To this end, rather than erasing the entries and performing matrix estimation to reconstruct the value aimlessly as in SVRL, we quantify the uncertainty of state-action value estimates, then reconstruct the entries of high uncertainty in Q-value matrix. We expect the uncertainty-aware reconstruction to better reduce the value estimation errors and facilitate the emergence of the low-rank structured value function. In specific, we adopt deep ensemble-based and count-based uncertainty quantification and conduct value matrix reconstruction on both online value estimates (from evaluate network) and target value estimates (from target network). In the experiments, we evaluate the effectiveness of UA-LQE with DDPG [11] as the base algroithm in OpenAI MuJoCo continuous control tasks.

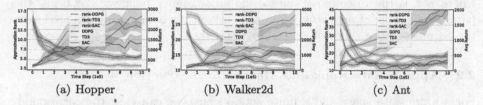

| (a) Hopper | (b) Walker2d | (c) Ant |

Fig. 1. Low-rank structure of Q-matrix in MuJoCo continuous control tasks. The left and right vertical axes denote the approximate rank of Q-matrix and average return. For all tasks and all algorithms, the rank decreases along the learning process. The results are averaged over 3 runs and the shaded region denotes a standard deviation.

We summarize our contributions as follows: 1) We empirically demonstrate the existence of low-rank structure of Q-function for representative DRL algorithms in widely-adopted continuous control tasks. 2) We empirically reveals a positive correlation between value matrix rank and value estimation uncertainty. 3) We propose a novel algorithm, i.e., Uncertainty-Aware Low-rank Q-matrix Estimation (UA-LQE). We demonstrate the effectiveness of UA-LQE in improving learning performance in our experiments.

2   Background

2.1   Reinforcement Learning

Consider a Markov Decision Process (MDP) defined by the tuple $\langle S, A, P, R, \gamma, T \rangle$ with state space $S \in \mathbb{R}^{d_s}$, action space $A \in \mathbb{R}^{d_a}$, transition function $P : S \times A \times S \rightarrow [0,1]$, reward function $R : S \times A \rightarrow \mathbb{R}$, discount factor $\gamma \in [0,1)$ and horizon T. The agent interacts with the MDP by performing its policy $\pi : S \rightarrow A$. An RL agent aims at optimizing its policy to maximize the expected discounted cumulative reward $J(\pi) = \mathbb{E}_\pi[\sum_{t=0}^{T} \gamma^t r_t]$, where $s_0 \sim \rho_0(s_0)$ the initial state distribution, $a_t \sim \pi(s_t)$, $s_{t+1} \sim \mathcal{P}(s_{t+1} \mid s_t, a_t)$ and $r_t = R(s_t, a_t)$. The state-action value function Q^π is defined as the expected cumulative discounted reward: $Q^\pi(s,a) = \mathbb{E}_\pi \left[\sum_{t=0}^T \gamma^t r_t \mid s_0 = s, a_0 = a \right]$ for all $s, a \in S \times A$.

With function approximation, the agent maintains a parameterized policy π_ϕ, while Q^{π_ϕ} can be approximated by Q_θ with parameter θ typically through minimizing Temporal Difference loss [24]:

$$L(\theta) = \mathbb{E}_{s,a,r,s' \sim D} \left[Q_\theta(s,a) - \mathbb{E}_{a' \sim \pi_{\bar\phi}(s')} (r + \gamma Q_{\bar\theta}(s',a')) \right]^2. \tag{1}$$

The policy π_ϕ can be updated by taking the gradient of the objective $\nabla_\phi J(\pi_\phi)$, e.g., with the deterministic policy gradient (DPG) [23]:

$$\nabla_\phi J(\pi_\phi) = \mathbb{E}_{s \sim \rho^{\pi_\phi}} \left[\nabla_\phi \pi_\phi(s) \nabla_a Q^\pi(s,a)|_{a=\pi_\phi(s)} \right], \tag{2}$$

where ρ^{π_ϕ} is the discounted state distribution under policy π_ϕ. $\bar\phi, \bar\theta$ are the parameters of *target networks*, which are periodically or smoothly updated from the parameters (i.e., ϕ, θ) of *evaluation networks*.

2.2 Approximate Rank and Matrix Reconstruction

For a finite state space and a finite action space, we can view the corresponding Q-function as a Q-value matrix (shortly Q-matrix) of shape $|S| \times |A|$. The approximate rank [10,28] for a threshold δ is defined as arank $_\delta(Q) = \min\{k : \sum_{i=1}^k \sigma_i(Q) \geq (1 - \delta) \sum_{i=1}^d \sigma_i(Q)\}$, where $\{\sigma_i(Q)\}$ are the singular values of Q-value matrix in decreasing order, i.e., $\sigma_1 \geq \cdots \geq \sigma_d \geq 0$. In this paper, we use $\delta = 0.01$ and omit the subscript for clarity. In other words, approximate rank is the first k singular values that capture more than 99% variance of all singular values. Further, for continuous (infinite) state and action space, we can define that arank $(Q) = \frac{1}{N} \sum_{j=1}^N$ arank $(\tilde{Q}_j)$, i.e., the empirical average approximate rank of sampled Q-value submatrix $\{\tilde{Q}_i\}$, each of which corresponds to sampled state and action subspaces S_i, A_i.

If arank$(Q) \ll \min\{|S|, |A|\}$, we say a Q-value matrix (or Q-function) is *low-rank*. This means that the matrix can be *redundant* in the sense that the whole matrix can be reconstructed or completed with only some entries are known. In this paper, we consider Soft-Impute algorithm [15] for matrix reconstruction and more background about matrix estimation is provided in Appendix A. In the context of RL, we also view this as the underlying structure of Q-value matrix, of which we make use to improve value function learning in RL.

3 Low-Rank Q-Matrix in DRL

In this section, we first empirically investigate the rank of Q-matrix for representative DRL agents in continuous control tasks (Sect. 3.1), which complements prior studies in discrete-action control tasks. Then, we introduce two ways to leverage the low-rank structure for DRL continuous control (Sect. 3.2).

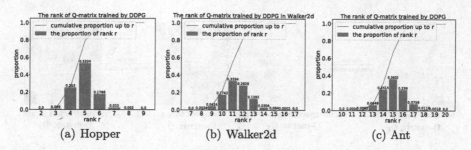

(a) Hopper	(b) Walker2d	(c) Ant

Fig. 2. Distribution of approximate rank of 10k sampled Q-matrix learned by DDPG agent after 1 million training on four MuJoCo environments.

3.1 Empirical Study of Low-Rank Q-Matrix in MuJoCo

To investigate the low-rank structure of Q-matrix of DRL agents in continuous control, we choose three representative DRL agents: DDPG [11], TD3 [4], SAC [5], and we use MuJoCo environments as benchmarks. We train each algorithm in each environment for 1 million interaction steps; for each evaluation time point, we record the average return and the approximate rank by calculating the average of 10k sampled Q-matrix (i.e., $\{\tilde{Q}_i\}_{i=1}^{10000}$ with $|S_i| = |A_i| = 64$ for all i, sampled from replay buffer). In Fig. 1, we plot the curves of average return and approximate rank of learned Q network (i.e., $\mathrm{arank}(Q_\theta)$.[1] The results show that the rank of sampled Q-matrix of all three algorithms decreases during the learning process in all the environments. This reveals a general but intriguing underlying dynamics of value learning. Moreover, there shows no clear correlation between the rank and the average return by comparing across different algorithms. Intuitively, we hypothesize that such phenomena are caused by the integration of function approximation and TD backup, where the induced generalization and bootstrapping smooth the value estimates.

In Fig. 2, we statistic the distribution of the approximate rank of 10k sampled Q-matrix learned by DDPG. We can find that most Q-matrices concentrate at a low rank with the value about 2 to 3 times of the action dimenionality controlled. This shows that the (near) optimal Q-matrix can be low-rank. In addition, consider the extreme case that Q-values are independently influenced by each dimension of action and linear to action, the rank of Q-matrix should be the action dimensionality (i.e., $\dim(A)$) since each action's value under some state can be represented by the values of $\dim(A)$ actions. The results indicate that the (near) optimal Q-matrix learned possesses a similarly simple relation between value and action as the extreme case.

---

[1] For TD3 and SAC which adopt double critics $Q_{\theta_1}, Q_{\theta_2}$, we calculate the approximate rank for Q_{θ_1} since the two are equivalent.

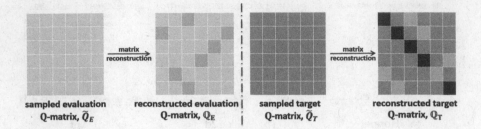

sampled evaluation   **reconstructed evaluation**   **sampled target**   **reconstructed target**
Q-matrix, $\tilde{Q}_E$ Q-matrix, $\mathbb{Q}_E$ Q-matrix, $\tilde{Q}_T$ Q-matrix, $\mathbb{Q}_T$

Fig. 3. Illustration of two ways of Q-matrix reconstruction. Consider a batch of transitions $\{(s,a,r,s')_i\}$ of size N sampled from replay buffer. *(Left)* For the sampled *evaluation Q-matrix* (green) $\tilde{Q}_E = \{Q_\theta(s,a)\}_{i,j=1,1}^{N,N}$, the entries (orange) selected (e.g., at random) are erased and then matrix estimation algorithm is performed to obtain the reconstructed $\mathbb{Q}_E$, which is used to regularize the evaluation network. *(Right)* For the sampled *target Q-matrix* (blue) $\tilde{Q}_T = \{Q_{\bar{\theta}}(s',a')\}_{i,j=1,1}^{N,N}$ where each $a' = \pi_{\bar{\phi}}(s')$, the entries are removed and re-estimated in the same way and the diagonal (red) is used as modified target values to approximate. (Color figure online)

3.2 Q-Matrix Reconstruction for DRL

The empirical results above show the dynamics of the approximate rank of Q-matrix, i.e., gradually decreasing during the learning process to a low rank. This reveals the smoothing nature of learning process and the underlying low-rank structure of the desired Q function. Based on similar empirical results obtained with discrete action space, Yang et al. [28] propose SVRL algorithm to improve conventional learning process, by conducting matrix reconstruction for target value matrix when performing TD learning of DQN [16]. In the following, we first make a simple extension of the matrix reconstruction for target value matrix to the continuous-action setting.

With discrete (and usually finite) action space, for each training of value function (recall Eq. 1), a mini-batch of N transitions $\{(s,a,r,s')_i\}$ are sampled from the replay buffer. SVRL [28] uses the N next states and all discrete actions to form the sampled *target Q-matrix*. While it is infeasible to use all actions from a continuous action space, we make use of the N next actions sampled by target policy $\pi_{\bar{\phi}}$ under corresponding N next states $\{(s',a')_i\}$ and form an $N \times N$ matrix, denoted by $\tilde{Q}_T = \{Q_{\bar{\theta}}(s',a')\}_{i,j}$. As in the left of Fig. 3, by randomly removing a portion of entries in $\tilde{Q}_T$ and then performing Soft-Impute matrix estimation, we can obtain the reconstructed Q-matrix denoted by $\mathbb{Q}_T$, in which the removed entries are re-estimated by leveraging the underlying rank structure of Q-matrix. Then, we can obtained the modified TD value function learning with target values from $\mathbb{Q}_T$ as shown by the loss function below:

$$L_T(\theta) = \mathbb{E}_{s,a,r,s'\sim D}\left[Q_\theta(s,a) - \mathbb{E}_{a'\sim\pi_{\bar{\phi}}(s')}\left(r + \gamma\mathbb{Q}_T(s',a')\right)\right]^2. \qquad (3)$$

As proposed in SVRL [28], the reconstructed value estimates are expected and empirically demonstrated to be better and thus benefit the learning process.

However, we can observe that only the diagonal value estimates in $\mathbb{Q}_T$ is used during updates; the reconstructed value estimates at randomly removed entries may not be used at all, thus being inefficient in leveraging the information of the underlying structure of Q-matrix.

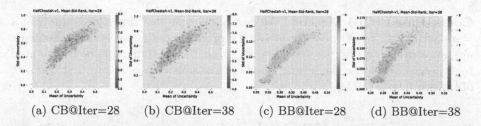

(a) CB@Iter=28   (b) CB@Iter=38   (c) BB@Iter=28   (d) BB@Iter=38

Fig. 4. Illustration of the positive correlation between approximate rank and value estimate uncertainty. We use DDPG agent on HalfCheetah for demonstration. Each point, corresponding to a sampled Q-matrix, is colored by its approximate rank with purple and red for the lowest and highest; and is placed on the 2D plane with the x-axis for the mean of its uncertainty matrix $U(\tilde{Q}_i)$ and the y-axis for the std. From (a)–(d), we plot the results in two different iterations for both count-based (CB) and bootstrapped-based (BB) uncertainty quantification methods.

To this end, we additionally propose matrix reconstruction on evaluate value matrix for the purpose of making more use of reconstructed value estimates. As in the right of Fig. 3, for a mini-batch of N transitions, we use the N state-action pairs to form the sampled *evaluation Q-matrix*, denoted by $\tilde{Q}_E = \{Q_\theta(s,a)\}_{i,j}$. Then we can obtain the reconstructed $\mathbb{Q}_E$ and regularize Q network as follows:

$$L_E(\theta) = \mathbb{E}_{s,a\sim D} \left[\tilde{Q}_E(s,a) - \mathbb{Q}_E(s,a) \right]^2. \tag{4}$$

The above two ways improve value learning process differently. The former way provides superior target values to approximate by re-estimating values based on rank information involved in matrix reconstruction on target Q-matrix; while the latter works more like consistency regularization according to underlying rank structure.

4   DRL with Uncertainty-Aware Q-Matrix Reconstruction

After the complementary study to prior work on empirical rank structure and reconstructed Q-matrix estimates in previous section, in this section, we take a further step to reveal the correlation between the approximate rank of Q-matrix and value estimation uncertainty. Based on this discovery, we propose an uncertainty-aware Q-matrix reconstruction method for DRL agents rather than the random removal and reconstruction.

4.1   Connection Between Rank and Uncertainty

To investigate the relation between approximate rank of Q-matrix and value estimation uncertainty, we first introduce two techniques for uncertainty quantification: **count-based method** [1] and **bootstrapped-based method** [18].

Count-Based (CB) Method. We denote the sampled times for training of a state-action pair by $N(s, a)$. For feasibility, original continuous state-action pairs are converted into discrete *Hash* codes [25] and thus similar pairs are counted together. We then use the inverse number of sampled times as our first measure of uncertainty [27]: $U_{\mathrm{CB}}(s, a) = \frac{1}{N(s,a)}$.

Bootstrapped-Based (BB) Method. Following previous works [3,18], we establish a bootstrapped ensemble of K independent value function estimators $\{Q_i\}_{i=1}^{K}$. For any state-action pair s, a, we obtain the uncertainty $U_{\mathrm{BB}}(s, a)$ by calculating the standard deviation of the value estimates from the ensemble: $U_{\mathrm{BB}}(s, a) = \sqrt{\frac{1}{N} \sum_{i=1}^{N} [Q_i(s, a) - \bar{Q}(s, a)]^2}$ where $\bar{Q}$ is the mean of the ensemble. This is also widely known as *epistemic uncertainty* in the literature of Bayesian exploration.

Next, with the above two uncertainty quantification of value estimates, we investigate the connection between the approximate rank of sampled Q-matrix and value estimate uncertainty. First, for some time point among the learning process, we sample 100 evaluation Q-matrix (similarly as done in Figs. 1 and 2). For each sampled Q-matrix $\tilde{Q}_i$, we obtain its approximate rank; we also quantify the value estimate uncertainty of each entry in $\tilde{Q}_i$, resulting in an uncertainty matrix $U(\tilde{Q}_i)$, and then calculate the mean and standard deviation (i.e., std for short) of $U(\tilde{Q}_i)$. In Fig. 4, we draw the scatter plots with the x-axis for the mean of each $U(\tilde{Q}_i)$ and the y-axis for the std. Each point in Fig. 4 denotes a sampled Q-matrix and is colored by its approximate rank (from purple to red). Similar results can also be observed for target Q-matrix and for other MuJoCo environments. Figure 4 shows a positive correlation between the approximate rank of sampled Q-matrix and value estimate uncertainty. There also shows a slight trend that this correlation is more obvious as the training marches.

Intuitively, when the mean and std are both large, it means that there are many entries with high uncertainty in the Q-matrix. The highly inaccurate value estimates deviate from the underlying structure of true value matrix, thus leading to a high approximate rank. On the contrary, when the uncertainty is low, in other words the value estimates are relatively certain, the approximate rank is often low since the underlying structure is well approximated. Now, we reach the important insight that value estimate uncertainty that reflects value approximation error, can be the signifier in finding the target entries which deviate from the desired learning process and prevent the emergence of low-rank structure. Therefore, rather than performing at random, uncertainty-aware removal and reconstruction is aimless and thus can be more efficient in reducing the value approximation error and maintaining the underlying structure of Q-matrix, finally improving the learning process.

4.2   Uncertainty-Aware Q-Matrix Reconstruction for DRL

Based on the insight revealed in previous paragraph, in the following we proposed Uncertainty-Aware Low-rank Q-matrix Estimation (**UA-LQE**) algorithm.

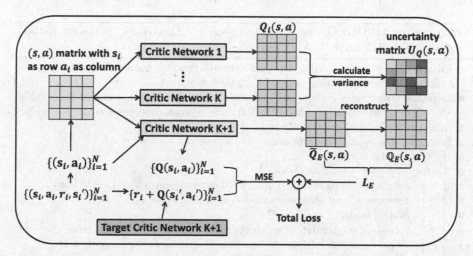

Fig. 5. Uncertainty-aware Q-matrix reconstruction based on bootstrapped-based uncertainty quantification. The bootstrapped ensemble of first K critic networks is used to calculate uncertainty, which is used for the selection of re-estimated entries in the Q-matrix. The reconstructed matrix is then used for the learning of the $K+1$ critic network, i.e., the one for primal value estimation and policy learning. Only matrix reconstruction on sampled evaluation Q-matrix $\mathbb{Q}_E$ is plot for demonstration.

The main idea of UA-LQE is to quantify the uncertainty of Q-value estimates during the learning process, and then perform Q-matrix reconstruction as introduced in Sect. 3.2 with re-estimated entries that are selected according to the quantified uncertainty. Let us take UA-LQE for *evaluationQ-matrix* reconstruction with bootstrapped-based uncertainty quantification for demonstration. Figure 5 illustrates the process. We maintain an ensemble of K independent critic networks $\{Q_i\}_{i=1}^{K}$ (i.e., Q-function approximator) parallel to the major critic network Q_{K+1}. The critic networks in the ensemble are trained in convention according to Eq. 1 during the learning process. For each time of update for the major critic network, we can obtain the uncertainty matrix of sampled evaluation Q-matrix through the ensemble, i.e., $U_{\text{BB}}(\tilde{Q}_E)$. According to $U_{\text{BB}}(\tilde{Q}_E)$, we select the entries of **top p-percent highest uncertainty** for each row (corresponding to each sampled state) of $\tilde{Q}_E$ to remove; and then we obtain reconstructed evaluation Q-matrix $\mathbb{Q}_E$ and conduct the training as Eq. 4. The processes for target Q-matrix reconstruction and for count-based uncertainty quantification are similar which are omitted in Fig. 3 for clarity.

Next, we propose a practical implementation of DRL algorithm for continuous control by integrating DDPG as a base algorithm and UA-LQE. Note that UA-LQE is compatible with most off-the-shelf DRL algorithms. We use DDPG for an representative implementation and leave the other potential combination for future work. A detailed pseudo-code of DDPG with UA-LQE, with algorithmic options of sampled Q-matrix reconstruction and uncertainty quantification methods, is provided in Algorithm 1.

Algorithm 1: DDPG with Uncertainty-aware Low-Rank Q-Matrix Estimation (**DDPG with UA-LQE**)

---

1 Initialize actor π_ϕ and critic networks Q_θ with random parameters ϕ, θ, and
   target actor $\pi_{\bar{\phi}}$ and critic networks $Q_{\bar{\theta}}$ with $\bar{\theta} = \theta$, $\bar{\phi} = \phi$
2 Initialize uncertainty estimator $U_\omega(\cdot)$ with parameter ω (i.e., random Hash table
   for CB and ensemble of critics for BB) and re-estimation percentage p
3 Initialize replay buffer $\mathcal{D}$, loss weight β
4 **for** *Episode $e \leftarrow 1$ to Max Episode Number* **do**
5   Obtain the initial state s_0
6   **for** *Timestep $t \leftarrow 1$ to T* **do**
7   Select action $a_t \sim \pi_\phi(s_t) + \epsilon_e$, with exploration noise $\epsilon_e \sim \mathcal{N}(0, \sigma)$
8   Execute a_t and observe reward r_t and new state s_{t+1}
9   Store $\{s_t, a_t, r_t, s_{t+1}\}$ in $\mathcal{D}$
10   Sample a mini-batch of experience $\Omega = \{s_i, a_i, r_i, s_{i+1}\}_{i=1}^{N}$ from $\mathcal{D}$
11   // perform UA-LQE with algorithmic options (refer to Sect. 3.2)
12   **if** *using Target Q-matrix Reconstruction:* **then**
13   Calculate conventional TD loss $L(\theta)$ as critic loss $\triangleright$ see Eq. 1
14   **else**
15   Establish evaluation Q-matrix $\tilde{Q}_T$, remove p-percentage entries
             based on $U_\omega(\tilde{Q}_T)$ and obtain the reconstructed Q-matrix $\mathbb{Q}_T$
16   Calculate $L_T(\theta)$ as critic loss $\triangleright$ see Eq. 3
17   **if** *using Evaluation Q-matrix Reconstruction:* **then**
18   Establish evaluation Q-matrix $\tilde{Q}_E$, remove p-percentage entries
             based on $U_\omega(\tilde{Q}_E)$ and obtain the reconstructed Q-matrix $\mathbb{Q}_E$
19   Calculate $L_E(\theta)$, weight it by β and add to critic loss $\triangleright$ see Eq. 4
20   Update Q_θ according to the total critic loss calculated above
21   Update π_ϕ with deterministic policy gradient $\triangleright$ see Eq. 2
22   // update uncertainty estimator (refer to Sect. 4.1)
23   Update ω accordingly, i.e., update the counts in Hash table for CB and
        update each critic in the ensemble with $L(\theta_i)$ for BB $\triangleright$ see Eq. 1

---

5 Experiment

In the experiment, we evaluate the efficacy of UA-LQE in several representative MuJoCo continuous control environments.

5.1   Experiment Setup

We use four MuJoCo environments: *Hopper, Walker2d, Ant, HalfCheetah*, for our experimental evaluation. We run each algorithm (introduced in the next subsection) for 1 million steps for all four environments, and evaluate the performance of trained agents every 5000 steps. For each configuration, we run three independent trials with random environmental seeds and network initialization.

Baselines and Variants. We provide an overview of the properties of the baseline and variants we considered in our experiments in Table 1. We use DDPG [11] as our base algorithm for the implementation of UA-LQE. We also extend prior algorithm SVRL [28] also based on DDPG for continuous control. With SVRL, we can evaluate the efficacy of uncertainty-aware Q-matrix reconstruction compared with at random. We also evaluate several UA-LQE variants of different algorithmic options from the perspectives of Q-matrix reconstruction losses (i.e., **E** or **T**) and uncertainty quantification methods (i.e., **CB** or **BB**).

Table 1. Properties of baselines and variants evaluated in our experiments. All the algorithms share the DDPG base. N.A. is short for 'not applicable'.

Alg./Prop.	Reconstruction	Rec. Loss	Selection	Uncertainty ($U_\omega(\cdot)$)
DDPG	✗	N.A.	N.A.	N.A.
SVRL-E	✓	$L_E(\theta)$	Random	N.A.
UA-LQE-E-CB	✓	$L_E(\theta)$	Uncertainty-aware	Count-based
UA-LQE-E-BB	✓	$L_E(\theta)$	Uncertainty-aware	Bootstrapped-based
SVRL-T	✓	$L_T(\theta)$	Random	N.A.
UA-LQE-T-CB	✓	$L_T(\theta)$	Uncertainty-aware	Count-based
UA-LQE-T-BB	✓	$L_T(\theta)$	Uncertainty-aware	Bootstrapped-based

Structure and Hyperparameters. For all algorithms, we use two-layer fully connected networks with 200 nodes and ReLU activation (not including input and output layers) for both actor and critic networks of DDPG base structure. The learning rate is 0.0001 for actor and 0.001 for critic. We use a discount factor $\gamma = 0.99$. The parameters of target networks $(\bar{\theta}, \bar{\phi})$ is softly replaced with the ratio 0.001. After each episode ends, the actor and critic are trained for T times where T here denotes the number of time steps of the episode. The batch size is set to 64 for both actor and critic mini-batch training and sampled Q-matrix (i.e., with the size of 64×64). The uncertainty-aware re-estimation percentage is set to $p = 20$. For count-based (CB) method, we use 8-byte Hash codes [25] for each state-action pair. For bootstrapped-based (BB) method, we maintain an ensemble of $K = 10$ critic networks. For UA-LQE on evaluation Q-matrix, we weight $L_E(\theta)$ by $\beta = 0.1$ when it is added to the total loss.

5.2   Results and Analysis

The evaluation results of algorithms are shown in Table 2. Complete learning curves are provided in Appendix C. We conclude the results from different angles

Table 2. Evaluation results of each algorithm (refer to Table 1) in four MuJoCo environments The results are means and stds of average return after 1 million step over 3 trials. Top-1 results in each environment are marked in bold.

Alg./ Env.	Walker2d	HalfCheetah	Ant	Hopper
DDPG	1192 ± 225	7967 ± 439	715 ± 220	1348 ± 339
SVRL-E	1458 ± 238	8039 ± 398	443 ± 221	1690 ± 468
UA-LQE-E-CB	1835 ± 348	5429 ± 1020	421 ± 140	622 ± 129
UA-LQE-E-BB	1716 ± 503	9555 ± 262	707 ± 203	**1751 ± 480**
SVRL-T	1863 ± 296	**9819 ± 392**	340 ± 102	1721 ± 422
UA-LQE-T-CB	2430 ± 506	6764 ± 346	400 ± 194	1029 ± 202
UA-LQE-T-BB	**2477 ± 549**	8727 ± 697	**1182 ± 568**	823 ± 221

of comparison: **1)** Compared with DDPG baseline, all variants with Q-matrix reconstruction (both SVRL and UA-LQE) show an overall better performance (except for some variants in specific environments), demonstrating the efficacy of Q-matrix reconstruction in improving value estimation and thus the entire learning process. **2)** UA-LQE variants outperforms SVRL baselines significantly in Walker2d and Ant and show comparable results in HalfCheetah and Hopper. This shows the effectiveness of uncertainty-aware target entry selection during the process of Q-matrix reconstruction. This reflects the intuition derived from the empirical positive correlation revealed in Sect. 4.1: it is more proper to re-estimate the value entries of high uncertainty based on the other ones which are relatively more reliable. **3)** The family of evaluation Q-matrix reconstruction (**-E**) performs comparably to the family of target Q-matrix reconstruction (**-T**): the former performs better in HalfCheetah and Hopper, while the latter is better in Walker2d and Ant. We hypothesize that this is because two families influence the learning process in different ways (i.e., self-consistency regularization v.s. error reduction in approximation target, as discussed in Sect. 3.2). There remains the possibility to combine of these two orthogonal ways for better results, and we leave the study on such algorithms in the future. **4)** For two kinds of uncertainty quantification methods, bootstrapped-based (**BB**) variants show superiority over count-based method (**CB**) in an overall view. This is consistent to the popularity of bootstrapped-based uncertainty quantification in RL exploration researches [18,19]. Notably, among all variants, UA-LQE-T-BB achieves significant improvement in Walker2d and Ant, which are of more complex kinematics among the four environments. This reveals the potential of UA-LQE in improving the performance of DRL agents.

Finally, it remains much space for improving and developing our proposed algorithm UA-LQE. Our results in MuJoCo is preliminary and obtained without much hyperparameter tuning. A fine-grained hyperparameter tuning is expected to further render the effectiveness of UA-LQE. Moreover, it remains space to improve uncertainty quantification with more accurate model and matrix reconstruction with more efficient algorithm. We leave these aspects for future work.

6   Conclusion

In this paper, we demonstrate the existence of low-rank structure of value function of DRL agents in continuous control problem and introduce two ways of performing Q-matrix reconstruction for value function learning with continuous action space. Moreover, we are the first to empirically reveal a positive correlation between the rank of value matrix and value estimation uncertainty. Based on this, we further propose a novel Uncertainty-aware Low-rank Q-matrix Estimation (UA-LQE) algorithm, whose effectiveness is demonstrated by DDPG-based implementation in several MuJoCo environments. Our work provides some first-step exploration on leveraging underlying structure in value function to improve DRL, which can be further studied, e.g., with more advanced and efficient matrix estimation techniques and other re-estimation schemes in the future.

Acknowledgement. The work is supported by the National Natural Science Foundation of China (Grant No. 62106172)

Appendix

A   More Background on Matrix Estimation

Matrix estimation is mainly divided into two types, matrix completion and matrix restoration. In this paper, we focus on the former. We use matrix estimation and matrix completion alternatively. Consider a low-rank matrix M, in which the values of some entries are missing or unknown. The goal of matrix completion problem is to recover the matrix by leveraging the low-rank structure. We use Ω to denote the set of subscripts of known entries and the use $\hat{M}$ to denote the recovered (also called reconstructed in this paper) matrix. Formally, the matrix completion problem is formalized as follows:

$$
\begin{aligned}
\min \; & \|\hat{M}\|_*, \\
\text{subject to } & \hat{M}_{i,j} = M_{i,j}, \quad \forall (i,j) \in \Omega
\end{aligned}
\tag{5}
$$

Here $\| \cdot \|_*$ is nuclear norm. The purpose of the above optimization problem is to fill the original matrix as much as possible while keeping the rank of the filled matrix $\hat{M}$ low.

Some representative methods to solve this optimization problem are Fix Point Continuation (FPC) algorithm [14], Singular Value Thresholding (SVT) algorithm [2], OptSpace algorithm [9], Soft Impute algorithm [15] and so on.

In this paper, we use Soft-Impute algorithm for matrix completion, i.e., Q-matrix reconstruction. A pseudocode for Soft-Impute is shown in Algorithm 2. Define function P_Ω as below:

$$
P_\Omega(M)(i,j) = \begin{cases} M_{ij}, & if \ (i,j) \in \Omega \\ 0, & if \ (i,j) \notin \Omega \end{cases}
\tag{6}
$$

Algorithm 2: Soft-Impute Algorithm

---

1 Input: Raw matrix M, the set of the subscripts of known entries Ω and the set of the subscripts of unknown (or missing) entries $\bar{\Omega}$

2 Reset the value of the entries in Ω of M with 0 and initialize $M_{old} = M$

3 Initialize singular filtering divider ζ and iteration termination threshold ϵ

4 **for** $t \leftarrow 1$ to Max Iteration Number **do**

5   // matrix decomposition, singular filtering and matrix reconstruction

6      $U \times S \times V = \mathrm{SVD}(P_\Omega(M_{old}) + P_{\bar{\Omega}}(M))$

7   // S is the diagonal matrix with singular values $\{s_i\}$ on the diagonal

8      $\lambda = \max_i s_i / \zeta$

9      $S_\lambda = S - \lambda * E$, where E is the identity matrix

10     $M_{\mathrm{new}} = U \times S_\lambda \times V$

11   // check the termination condition

12   **if** $\frac{|M_{new} - M_{old}|_F^2}{|M_{old}|_F^2} > \epsilon$: **then**

13   Set $M_{\mathrm{old}} = M_{\mathrm{new}}$

14   **else**

15   Set Set $\hat{M} = M_{\mathrm{new}}$

16   break;

17 Output $\hat{M}$

Similar, we define function $P_{\bar{\Omega}}$. Soft-Impute algorithm completes the missing entries in $\bar{\Omega}$ in an iterative reconstruction fashion. During each iteration, the singular matrix S is obtained by SVD decomposition, and then a filtering is performed through subtracting the value λ controlled by a filtering divider hyperparameter ζ; next, the new matrix M_{new} is reconstructed with the subtracted singular matrix S_λ. Afterwards, an iteration termination check is performed by comparing whether the difference between the matrices before and after the iteration exceeds the termination threshold ϵ. When the termination threshold is met or the max iteration number is reached, the algorithm finally outputs the completed matrix $\hat{M}$.

B   Additional Experimental Details

All regular hyperparametyers and implementation details are described in the main body of the paper.

For code-level details, our codes are implemented with Python 3.6.9 and Torch 1.3.1. All experiments were run on a single NVIDIA GeForce GTX 1660Ti GPU. Our Q-matrix reconstruction algorithm is implemented with reference to https://github.com/YyzHarry/SV-RL.

For matrix reconstruction algorithm, i.e., SoftImpute, we set the filtering divider $\zeta = 50$, termination threshold $\epsilon = 0.0001$ and Max Iteration Number to be 100.

C   Complete Learning Curves of Table 2

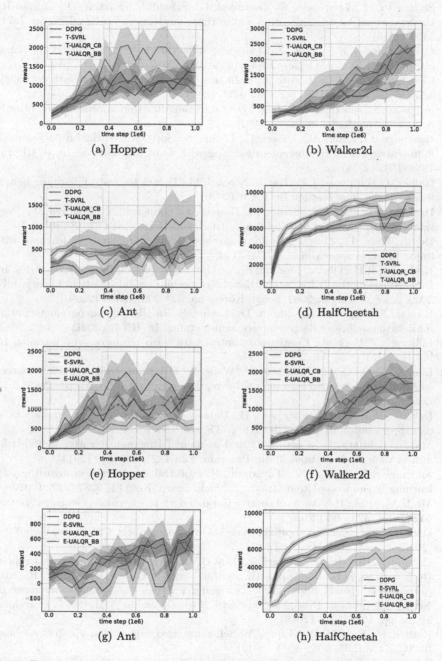

Fig. 6. For the reward curve, the first four are the results of reconstruction on the target Q-matrix, and the last four are the results of reconstruction on the current Q-matrix. Results are the means and stds over 3 trials.

References

1. Bellemare, M., Srinivasan, S., Ostrovski, G., Schaul, T., Saxton, D., Munos, R.: Unifying count-based exploration and intrinsic motivation. In: NeurIPS, pp. 1471–1479 (2016)
2. Cai, J., Candès, E.J., Shen, Z.: A singular value thresholding algorithm for matrix completion. SIAM J. Optim. **20**(4), 1956–1982 (2010)
3. Ciosek, K., Vuong, Q., Loftin, R., Hofmann, K.: Better exploration with optimistic actor critic. In: NeurIPS, pp. 1785–1796 (2019)
4. Fujimoto, S., van Hoof, H., Meger, D.: Addressing function approximation error in actor-critic methods. In: ICML (2018)
5. Haarnoja, T., Zhou, A., Abbeel, P., Levine, S.: Soft actor-critic: off-policy maximum entropy deep reinforcement learning with a stochastic actor. In: ICML, pp. 1856–1865 (2018)
6. Hafner, D., Lillicrap, T.P., Ba, J., Norouzi, M.: Dream to control: learning behaviors by latent imagination. In: ICLR (2020)
7. Hasselt, H., Doron, Y., Strub, F., Hessel, M., Sonnerat, N., Modayil, J.: Deep reinforcement learning and the deadly triad. CoRR abs/1812.02648 (2018)
8. He, J., Zhou, D., Gu, Q.: Uniform-pac bounds for reinforcement learning with linear function approximation. CoRR abs/2106.11612 (2021)
9. Keshavan, R.H., Oh, S., Montanari, A.: Matrix completion from a few entries. In: Proceedings of the IEEE International Symposium on Information Theory, ISIT 2009, June 28–July 3, 2009, Seoul, Korea, pp. 324–328. IEEE (2009)
10. Kumar, A., Agarwal, R., Ghosh, D., Levine, S.: Implicit under-parameterization inhibits data-efficient deep reinforcement learning. In: ICLR (2021)
11. Lillicrap, T.P., et al.: Continuous control with deep reinforcement learning. In: ICLR (2015)
12. Luo, X., Meng, Q., He, D., Chen, W., Wang, Y.: I4R: promoting deep reinforcement learning by the indicator for expressive representations. In: IJCAI, pp. 2669–2675. ijcai.org (2020)
13. Lyle, C., Rowland, M., Ostrovski, G., Dabney, W.: On the effect of auxiliary tasks on representation dynamics. In: AISTATS. vol. 130, pp. 1–9 (2021)
14. Ma, S., Goldfarb, D., Chen, L.: Fixed point and Bregman iterative methods for matrix rank minimization. Math. Program. **128**(1–2), 321–353 (2011)
15. Mazumder, R., Hastie, T., Tibshirani, R.: Spectral regularization algorithms for learning large incomplete matrices. J. Mach. Learn. Res. **11**, 2287–2322 (2010)
16. Mnih, V., et al.: Human-level control through deep reinforcement learning. Nature **518**(7540), 529–533 (2015)
17. Ong, H.: Value function approximation via low-rank models. CoRR abs/1509.00061 (2015)
18. Osband, I., Blundell, C., Pritzel, A., Roy, B.V.: Deep exploration via bootstrapped DQN. In: Lee, D.D., Sugiyama, M., von Luxburg, U., Guyon, I., Garnett, R. (eds.) Advances in Neural Information Processing Systems 29: Annual Conference on Neural Information Processing Systems 2016, December 5–10, 2016, Barcelona, Spain, pp. 4026–4034 (2016)
19. Pathak, D., Gandhi, D., Gupta, A.: Self-supervised exploration via disagreement. In: ICML, vol. 97, 5062–5071 (2019)
20. Scherrer, B., Ghavamzadeh, M., Gabillon, V., Lesner, B., Geist, M.: Approximate modified policy iteration and its application to the game of Tetris. J. Mach. Learn. Res. **16**, 1629–1676 (2015)

21. Schreck, J.S., Coley, C.W., Bishop, K.J.: Learning retrosynthetic planning through simulated experience. ACS Central Sci. **5**(6), 970–981 (2019)
22. Silver, D., et al.: Mastering the game of go with deep neural networks and tree search. Nature **529**(7587), 484–489 (2016)
23. Silver, D., Lever, G., Heess, N., Degris, T., Wierstra, D., Riedmiller, M.A.: Deterministic policy gradient algorithms. In: ICML, pp. 387–395 (2014)
24. Sutton, R.S., Barto, A.G.: Reinforcement learning: an introduction. IEEE Trans. Neural Netw. **16**, 285–286 (1988)
25. Tang, H., et al.: #Exploration: a study of count-based exploration for deep reinforcement learning. In: Guyon, I., et al. (eds.) Advances in Neural Information Processing Systems 30: Annual Conference on Neural Information Processing Systems 2017, 4–9 December 2017, pp. 2753–2762. Long Beach, CA, USA (2017)
26. Vinyals, O., et al.: Grandmaster level in StarCraft ii using multi-agent reinforcement learning. Nature **575**(7782), 350–354 (2019)
27. Yang, T., et al.: Exploration in deep reinforcement learning: a comprehensive survey. CoRR abs/2109.06668 (2021)
28. Yang, Y., Zhang, G., Xu, Z., Katabi, D.: Harnessing structures for value-based planning and reinforcement learning. In: ICLR (2020)

SEIHAI: A Sample-Efficient Hierarchical AI for the MineRL Competition

Hangyu Mao[1], Chao Wang[1], Xiaotian Hao[2], Yihuan Mao[3], Yiming Lu[3], Chengjie Wu[3], Jianye Hao[1,2(✉)], Dong Li[1], and Pingzhong Tang[3]

[1] Huawei Noah's Ark Lab, Beijing, China
{maohangyu1,wangchao358,haojianye,lidong106}@huawei.com
[2] Tianjin University, Tianjin, China
{xiaotianhao,jianye.hao}@tju.edu.cn
[3] IIIS, Tsinghua University, Beijing, China
{maoyh20,luym19,wucj19}@mails.tsinghua.edu.cn, kenshin@tsinghua.edu.cn

Abstract. The MineRL competition is designed for the development of reinforcement learning and imitation learning algorithms that can efficiently leverage human demonstrations to drastically reduce the number of environment interactions needed to solve the complex *ObtainDiamond* task with sparse rewards. To address the challenge, in this paper, we present **SEIHAI**, a **S**ample-**e**fficient **H**ierarchical **AI**, that fully takes advantage of the human demonstrations and the task structure. Specifically, we split the task into several sequentially dependent subtasks, and train a suitable agent for each subtask using reinforcement learning and imitation learning. We further design a scheduler to select different agents for different subtasks automatically. SEIHAI takes the first place in the preliminary and final of the NeurIPS-2020 MineRL competition.

Keywords: MineRL competition · Reinforcement learning · Imitation learning · Sample efficiency

1 Introduction

Reinforcement learning has achieved tremendous breakthroughs in games [26], robotic manipulations [4], network configurations [21,22], and multi-agent systems [20,23,24]. However, the state-of-the-art reinforcement learning algorithms usually require a large number of environment interactions, which can be expensive or even infeasible in real-world applications. In addition, most methods can hardly be applied directly to tasks with very sparse rewards.

Recently, the MineRL competition [5–9,25] has been introduced in order to promote researches in sample-efficient reinforcement learning and imitation learning that leverage human demonstrations to solve the complex *ObtainDiamond* task in the Minecraft game. Specifically, an agent starts off from a random position on a randomly-generated Minecraft map with the goal to obtain

---

H. Mao, C. Wang, X. Hao, Y. Mao, Y. Lu, C. Wu—These authors contribute equally to this work.

© Springer Nature Switzerland AG 2022
J. Chen et al. (Eds.): DAI 2021, LNAI 13170, pp. 38–51, 2022.
https://doi.org/10.1007/978-3-030-94662-3_3

a Diamond, which can only be accomplished by mining materials and crafting necessary tools. Moreover, mining or crafting some items requires crafting other prerequisite items. For instance, a wooden pickaxe is needed for collecting stones, which is further required by other more advanced tools. Besides, an agent would be rewarded if and only if it successfully obtains the right item at the right time. In other word, the environment is with sparse rewards. In summary, the *ObtainDiamond* task requires the completion of a series of sequentially dependent subtasks with sparse rewards, which usually takes thousands of steps. The randomness of generated worlds, the visual observations with low signal-to-noise ratios, and the temporal dependence of underlying subtasks and the sparse-reward setting constitute the challenges for efficient learning.

In this paper, we leverage the hierarchical dependency property of different items, and propose a hierarchical solver to address the challenge of the MineRL competition. Specifically, we split the *ObtainDiamond* task into several subtasks according to the natural sequential dependency among the items, and train a suitable reinforcement learning or imitation learning agent to tackle each subtask. To further alleviate the sparse-reward problem and improve the sample efficiency, the task structure knowledge and the human demonstrations are employed. Moreover, to make our methods fully automatic and generalizable to other domains[1], we design a scheduler to select different agent in different game states. The proposed hierarchical agent SEIHAI is optimized based on limited human demonstrations and environment interactions as the competition required[2]. Our final model wins the first place in both the preliminary and final of the MineRL competition at NeurIPS-2020, among 90+ teams and around 500 submissions. The contributions are summarized as follows.

1) We proposed a fully automatic hierarchical method to address the *Obtain-Diamond* task with sparse rewards and limited human demonstrations in the MineRL competition.

2) We introduced a few learning-based methods to extract critical actions from the large continuous and domain-agnostic action spaces, which enhances the sample efficiency of our methods.

3) We evaluated our methods in the MineRL competition, and won the first place in the preliminary and final of NeurIPS-2020 MineRL competition. The results demonstrate the effectiveness of our methods.

2   Background

2.1   The MineRL Competition

The MineRL competition is designed to promote the development of sample-efficient, domain-agnostic, and robust algorithms for solving complex tasks with

---

[1] The competition requires that the submitted methods should not include any meta-actions or rule-based heuristics.

[2] Specifically, over 60 million frames of human demonstrations and 8 million online interactions with the environment.

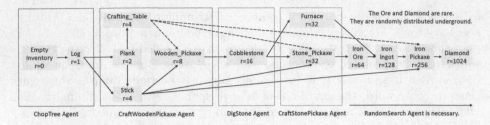

Fig. 1. The background about the MineRL competition: the grey rectangles represent different items, e.g., Log, Plank, Crafting_Table, Stick, Wooden_Pickaxe, and so on; the arrows demonstrate the typical item hierarchy for obtaining a Diamond; the red texts denote the reward when the agent mines or crafts an item for the first time. We split the *ObtainDiamond* task into five subtasks according to the natural hierarchy among items, and train a suitable agent for each subtask. (Color figure online)

sparse rewards using human demonstrations [5]. The competition consists of the following challenges.

Limited Human Demonstrations and Environment Interactions. The competition provides 60 million frames of human demonstrations [8], but only 211 trajectories are collected in the *ObtainDiamond* environment[3]. Besides, participants are restricted to use at most 8 million frames of environment interactions. Therefore, it is necessary to use imitation learning techniques to fully leverage the demonstrations.

Domain Agnosticism. The original states and discrete actions have been obfuscated into continuous feature vectors by an unknown autoencoder. This prevents participants from using domain-specific actions or rule-based heuristics.

2.2   The ObtainDiamond Task

The goal of the MineRL competition is to address the *ObtainDiamond* task. This task is featured with the following challenges:

Item Dependency. In order to obtain a Diamond, the agent needs to craft or obtain different items in a certain order. The typical item dependency (or item hierarchy) for obtaining a Diamond is shown in Fig. 1. To finish the task, the agent must follow the item hierarchy, otherwise it will not get any rewards. For example, it is impossible for the agent to dig Cobblestones before crafting the Wooden_Pickaxe.

Sparse Reward. The agent receives a reward *the first time* it crafts or obtains an item, and the different reward value are depicted in Fig. 1. Note that obtaining the same item several times does not incur additional rewards, although it is necessary for the agent to collect enough base items in order to craft all the tools it needs.

---

[3] Most human demonstrations are scattered among several small environments, like *Navigation, Treechop*.

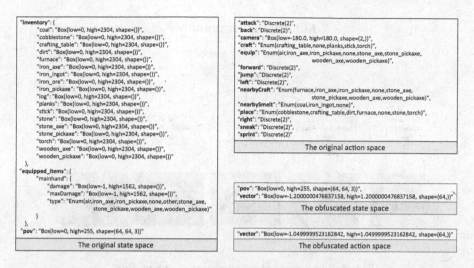

"inventory": {
    "coal": "Box(low=0, high=2304, shape=())",
    "cobblestone": "Box(low=0, high=2304, shape=())",
    "crafting_table": "Box(low=0, high=2304, shape=())",
    "dirt": "Box(low=0, high=2304, shape=())",
    "furnace": "Box(low=0, high=2304, shape=())",
    "iron_axe": "Box(low=0, high=2304, shape=())",
    "iron_ingot": "Box(low=0, high=2304, shape=())",
    "iron_ore": "Box(low=0, high=2304, shape=())",
    "iron_pickaxe": "Box(low=0, high=2304, shape=())",
    "log": "Box(low=0, high=2304, shape=())",
    "planks": "Box(low=0, high=2304, shape=())",
    "stick": "Box(low=0, high=2304, shape=())",
    "stone": "Box(low=0, high=2304, shape=())",
    "stone_axe": "Box(low=0, high=2304, shape=())",
    "stone_pickaxe": "Box(low=0, high=2304, shape=())",
    "torch": "Box(low=0, high=2304, shape=())",
    "wooden_axe": "Box(low=0, high=2304, shape=())",
    "wooden_pickaxe": "Box(low=0, high=2304, shape=())"
},
"equipped_items": {
    "mainhand": {
        "damage": "Box(low=-1, high=1562, shape=())",
        "maxDamage": "Box(low=-1, high=1562, shape=())",
        "type": "Enum(air,iron_axe,iron_pickaxe,none,other,stone_axe,
                      stone_pickaxe,wooden_axe,wooden_pickaxe)"
    }
},
"pov": "Box(low=0, high=255, shape=(64, 64, 3))"

The original state space

"attack": "Discrete(2)",
"back": "Discrete(2)",
"camera": "Box(low=-180.0, high=180.0, shape=(2,))",
"craft": "Enum(crafting_table,none,planks,stick,torch)",
"equip": "Enum(air,iron_axe,iron_pickaxe,none,stone_axe,stone_pickaxe,
               wooden_axe,wooden_pickaxe)",
"forward": "Discrete(2)",
"jump": "Discrete(2)",
"left": "Discrete(2)",
"nearbyCraft": "Enum(furnace,iron_axe,iron_pickaxe,none,stone_axe,
                     stone_pickaxe,wooden_axe,wooden_pickaxe)",
"nearbySmelt": "Enum(coal,iron_ingot,none)",
"place": "Enum(cobblestone,crafting_table,dirt,furnace,none,stone,torch)",
"right": "Discrete(2)",
"sneak": "Discrete(2)",
"sprint": "Discrete(2)"

The original action space

"pov": "Box(low=0, high=255, shape=(64, 64, 3))",
"vector": "Box(low=-1.2000000476837158, high=1.2000000476837158, shape=(64,))"

The obfuscated state space

"vector": "Box(low=-1.0499999523162842, high=1.0499999523162842, shape=(64,))"

The obfuscated action space

Fig. 2. The state and action spaces in the *ObtainDiamond* task. The obfuscated states and actions are more complex since we do not know their semantic meanings.

Long Episode Length. As can be seen from Fig. 1, mining or crafting each item takes lots of steps. According to the human demonstrations, it usually takes more than thousands of steps, for a skilled human to obtain the Diamond. As a result, it aggravates the sparse-reward situation.

2.3   The ObtainDiamond MDP

The *ObtainDiamond* environment can be formulated as a Markov Decision Process (MDP), which can be formally defined by the tuple $\langle S, A, T, R, \gamma \rangle$, where S is the set of possible states s; A represents the set of possible actions a; $T(s'|s,a) : S \times A \times S \rightarrow [0,1]$ denotes the state transition function (or the environment dynamics); $R(s,a) : S \times A \rightarrow \mathbb{R}$ is the reward function; $\gamma \in [0,1]$ is the discount factor.

We use s_t, a_t and $r_t = R(s_t, a_t)$ to denote the state, action and reward at time step t, respectively. In the *ObtainDiamond* task, the agent tries to learn a policy $\pi(a_t|s_t)$ that can maximize $\mathbb{E}[G]$ using limited human demonstrations and environment interactions, where $G = \Sigma_{t=0}^{H}\gamma^t r_t$ is the return, and H is the time horizon. Reinforcement learning [32] and imitation learning [27] are very general approaches to solve the *ObtainDiamond* MDP problem.

In the *ObtainDiamond* task, the state and action spaces are very complex. As can be seen from Fig. 2, the original state space consists of two parts: the inventory state space and the image observation space. An inventory state includes a number of different items (i.e., the inventory in Fig. 2) and the state of the equipped items (i.e., the equipped_items in Fig. 2). An image observation is a game screen shot (i.e., the 64 * 64 * 3-dimensional pov in Fig. 2). The original action space includes moving left/right/forward/back, attacking, crafting tables,

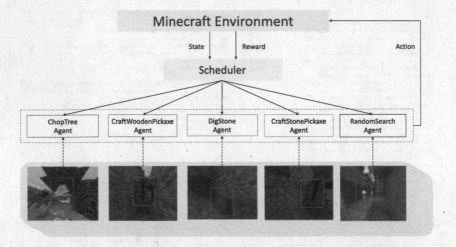

Fig. 3. The proposed architecture of our method.

placing tables, rotating the camera, nearby crafting, nearby smelting and other necessary actions. To prevent participants from directly using these states and actions which have clear semantic meanings, both the inventory state space and the action space are encoded into 64-dimensional vector spaces, which further increases the difficulty of the competition, as shown in Fig. 2.

3   Method

3.1   The Overall Framework

As can be seen from the item hierarchy shown in Fig. 1, mining a Diamond depends on the completion of a series of different subtasks, each requires mastering different skills. For example, at the beginning of an episode, the agent should search for the Log; after that, the agent should craft the Wooden_Pickaxe; when this is done, the agent should dig down and search for Cobblestone and Iron_Ore under the ground. It is hard to imagine how a single policy could learn all these skills (without forgetting other skills). Therefore, we propose to split the *ObtainDiamond* task into five subtasks according to the item dependency, and train a reinforcement learning agent or imitation learning agent (i.e., subpolicies) suitable for each subtask. Specifically, there are five agents: the ChopTree, CraftWoodenPickaxe, DigStone, CraftStonePickaxe, and RandomSearch agent.

A scheduler is needed for choosing which agent to use in different scenario. One way to implement the scheduler is to switch over different agents based on the reward obtained during execution, as obtaining a reward often means the completion of a subtask. However, this method is sometimes problematic and does not generalize well. For example, to craft the Wooden_Pickaxe, the ChopTree agent should collect at least 3 Logs. Consequently, switching to the

CraftWoodenPickaxe agent after the ChopTree agent getting a reward $r = 1$ would cause failure of the entire task. Second, it will violate the rules because the competition requires that the submitted method should not include any meta-actions or rule-based heuristics. To make our method totally automatic and general, we propose to implement a learning-based scheduler to select different agents in different states. Specifically, the scheduler is trained using human demonstrations and the supervised learning technique. In the inference phase, it takes as input the current state and predicts which agent to activate. The overall framework of our hierarchical method SEIHAI is illustrated in Fig. 3.

In the following subsections, we give a detailed introduction of each agent and the scheduler.

3.2   Action Discretization

Because of the domain agnosticism requirement, we can only get the obfuscated states and actions. However, the obfuscated action space is very large, which is a 64-dimensional vector space and each dimension takes a continuous value ranging from -1.049 to 1.049, as shown in Fig. 2. As far as we know, current reinforcement learning or imitation learning algorithms can hardly handle this challenge. For example, we conduct preliminary experiments using Deep Deterministic Policy Gradient (DDPG) [17,30], an algorithm that is widely used to handle continuous action space, but the performance is unsatisfactory.

To deal with the continuous action space, we transform the MDP with a continuous action space into a MDP with a discrete action space, which can be efficiently solved by algorithms specifically designed for MDPs with discrete action spaces such as Deep Q-network (DQN) [26].

We adopt different techniques to discretize actions in different phases. For the ChopTree agent, the DigStone agent and the RandomSearch agent, we first categorize their corresponding actions into two sets depending on whether an action results in inventory state changes, since we observe that the two sets of actions play different roles: generally, actions that do not incur inventory state changes are responsible for agent movement, such as moving forward and backward; by contrast, actions that contribute to inventory state changes are those for chopping trees and digging stones. We leverage the K-Means [16,18,19] algorithm to cluster each action set into 30 classes. As a result, we transform the continuous action space into a discrete action set with 60 actions in total. For the CraftWoodenPickaxe agent and the CraftStonepickaxe agent, we only extract the actions that contribute to inventory state changes and cluster them into several categories using the Non-parametric Bayesian method [2]. We neglect actions that do not incur inventory state changes since crafting different items only requires the inventory information.

3.3   The ChopTree Agent

At the beginning of the *ObtainDiamond* task, the agent is initialized at a random place on a randomly-generated Minecraft map, with nothing in its inventory.

According to the item hierarchy shown in Fig. 1, the primary goal of SEIHAI is to chop trees and collect enough Logs.

We implement the ChopTree agent using a deep neural network. The network consists of three 2D convolutional layers and two fully connected layers, with a dueling architecture at the last layer to generate the Q-values for every discrete actions. Although this network is very simple, it can balance the training efficiency and testing performance very well[4].

We train the agent using the Soft Q Imitation Learning (SQIL) algorithm [28]. It encourages the agent to follow the trajectories in human demonstrations by giving a penalty (i.e., negative reward) when the agent generates actions that lead to the out-of-distribution states. It has been shown that SQIL can be interpreted as a regularized variant of behavioral cloning that uses a sparse prior to encourage imitation. Therefore, the agent can easily obtain the Logs as human does after training[5].

Specifically, the training process consists of two-stage. In the first stage, we train the agent using an additional *TreeChop* environment provided by the competition organizer, so that the agent could focus on the chopping tree task. While in the second stage, we fine-tune the agent using the *ObtainDiamond* environment. In the fine-tune phase, we only use the truncated human demonstrations which contain treechop-related interactions only. The trick is crucial to final performance since demonstrations of digging or crafting can mislead the training of the ChopTree agent. To be specific, we only use demonstrations before the reward of crafting a plank is obtained.

In order to deal with the obfuscated high-dimensional observation space, we add an auxiliary task to help with feature extraction. Ideally, an efficient strategy should learn the segmentation of different parts in observations, as well as recognizing their semantic meanings. For instance, it should recognize the position of trees. Therefore, we add an reconstruction loss to the original reinforcement learning objective.

3.4   The CraftWoodenPickaxe Agent

In this phase, the agent's goal is to craft a Wooden_Pickaxe that can be used to obtain other items in the following phase. To this end, the agent needs to first craft Planks using Logs, then make a Crafting_Table and some Sticks using Planks. Finally, the agent needs to place the Crafting_Table in an empty area, stand nearby the Crafting_Table and then craft a Wooden_Pickaxe. It is worth noting that the Sticks and the Crafting_Table can be crafted in a random order and the Crafting_Table can be placed before or after Sticks are crafted.

Since the behaviors in this phase show a clear pattern, we implement the CraftWoodenPickaxe agent using supervised learning, also known as behavior

---

[4] For example, we test deep ResNet [10], but the performance has not improved much, while the training time has increased a lot.

[5] We also test the Generative Adversarial Imitation Learning (GAIL) algorithm [12], but it is not as stable as SQIL.

cloning in the sequential decision making setting. Specifically, we first extract those critical actions $\{a_i\}_{i=0}^N$ resulting in inventory state changes and their corresponding inventory states $\{s_i\}_{i=0}^N$ from the demonstrations. Note that the image states are not included since crafting those items only requires inventory information. The extracted actions $\{a_i\}_{i=0}^N$ are further clustered into K classes using the non-parametric Bayesian clustering method mentioned in Sect. 3.2. We build a training set using the inventory states and the corresponding discretized actions and use a fully connected network with two hidden layers each with 64 neurons as the function approximator for the classifier.

3.5   The DigStone Agent

The CraftWoodenPickaxe agent is followed by the DigStone agent, which aims at digging enough Cobblestones.

The DigStone agent is implemented using a deep neural network that is exactly the same as that of the ChopTree agent. It can balance the training efficiency and executing performance as mentioned above.

We train the agent based on a variant of imitation learning method, which we call Large Margin Imitation (LarMI). Specifically, LarMI imitates the actions in the demonstrations with a large margin constraint as follows:

$$J(Q) = \max_{a \in A, a \neq a_D} [Q(s, a) + T] - Q(s, a_D) \qquad (1)$$

where $Q(s, a)$ represents the "relative value"[6] of action a in state s, and a_D is the action that the demonstrator took in state s, and T is the margin threshold. Equation (1) forces the relative values of the other actions to be at least a margin lower than the relative value of the demonstrator's action. This avoids the overestimation of the unseen actions, and the greedy policy induced by $a^* \leftarrow \arg\max_a Q(s, a)$ will easily imitate the demonstrations. Note that our method is very similar to the pretraining stage of the Deep Q-learning from Demonstration (DQfD) algorithm [11].

The reasons that we do not use DQfD or SQIL are two-fold. First, DQfD and SQIL require interacting with the environment, so they will reduce the sample efficiency. Second, since digging stone is an intermediate subtask, we cannot control the environment to switch to the exact states in the demonstrations, which further prevents us from applying DQfD and SQIL.

3.6   The CraftStonePickaxe Agent

In this phase, the agent needs to craft a Stone Pickaxe and a Furnace for the purpose of mining Iron_Ore and crafting Iron_Ingots.

Similar to the CraftWoodenPickaxe phase, we first extract critical actions related to this subtask and then cluster them into several classes for action discretization. Different from the CraftWoodenPickaxe, the Stone_Pickaxe and the

---

[6] This is different from the Q-value in reinforcement learning, since here $Q(s, a)$ does not estimate the expected cumulative rewards.

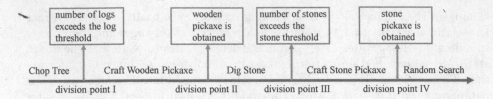

Fig. 4. The division points of an episode.

Furnace can be crafted in a random order, so we implement the CraftStonePickaxe agent with a policy that randomly selects actions from the clustered action set. The Stone_Pickaxe and the Furnace can be obtained as long as the critical actions are executed for enough times. The online evaluation results demonstrate that this policy works quite well.

3.7   The RandomSearch Agent

After obtaining the Stone_Pickaxe, the agent should continue to mine Iron_Ore, which is the raw material for crafting Iron_Ingot and Iron_Pickaxe. The agent can further use the Iron_Pickaxe to obtain the Diamond. However, the Iron_Ore may locate at any places underground, which is totally unpredictable. Consequently, the agent must search around for Iron_Ore. To this end, we combine the DigStone agent and a fully random search agent as the RandomSearch agent to obtain Iron_Ore. Specifically, at each time step, the RandomSearch agent executes an action either proposed by the DigStone agent or randomly selected from the action set. The randomness of the RandomSearch agent is controlled by an exploration threshold that ranges from zero to one and gradually increases over time. The larger the exploration threshold is, the more random the policy is.

The Iron_Ingot, the Iron_Pickaxe and the Diamond can be obtained by designing agents similar to these already introduced agents. However, we find out that the success rate of obtaining the Iron_Ore has dropped to around 7% (illustrated in Table 2), which leaves little room for optimization and improvement of subsequent agents. Therefore, we employ the RandomSearch agent to address the remaining tasks by setting the exploration threshold to one, in the hope that it can obtain any left items by luck.

3.8   The Scheduler

The scheduler plays the role of determining which agent to use in different states. The sequential dependency of each agents is shown in Fig. 3. To finish the *Obtain-Diamond* task, one must finish each subtask sequentially. Deviating from the agent dependency (or task dependency) will cause failure of the task.

Given the simple sequential dependency of different agents, training the scheduler boils down to a classification task. To build the training set, we split each episode into five parts, which is detailed in Fig. 4. We set the log threshold as the minimum number[7] of Logs required to finish crafting the Stone_Pickaxe, and the stone threshold as the minimum number of Cobble_Stones required to finish crafting the Stone_Pickaxe and the Furnace. Moreover, we build the training set for the scheduler using the dense demonstrations, since the agent can obtain a certain amount of rewards every time it chops a tree or digs a stone.

4 Experiments

We present the official results[8] reported by the MineRL competition organizers. There are 90+ teams and around 500 submissions in total. The official evaluation consists of two rounds. In round 1, the submitted model is evaluated. In round 2, only codes are submitted, and the model is retrained by the competition organizer from scratch. The final score is averaged over 200 episodes. For each episode, the score is computed as the sum of the rewards (shown in Fig. 1) achieved by the agent.

4.1 Overall Evaluation

Table 1 shows the scores of the best-performing submissions from both rounds. As can be seen, our methods rank first in the preliminary and final of the NeurIPS-2020 competition. Besides, scores obtained by our methods surpass those of the second-place team by a large margin, demonstrating the effectiveness of our methods. It is also worth noting that our method can be further improved: our round 1 submission only consists of the ChopTree, the Craft-WoodenPickaxe, and the DigStone agents, and the corresponding score is 19.84; in contrast, we further add the CraftStonePickaxe and the RandomSearch agents in round 2, and the score becomes 39.55. If we design specific agents to craft Iron_Ingot and Iron_Pickaxe, we believe the score can be further improved. We leave it as our future work.

4.2 Agent-Level Evaluation

Table 2 demonstrates the detailed performance of our final submission, which is evaluated for 200 episodes. We can see that there are 72 episodes with zero rewards, meaning that the ChopTree agent fails to collect any Logs in the online evaluation environment. The reason is that in the ObtainDiamond environment, the agent often starts off from some deserted places (e.g., in water, in the desert or on the cliff). As a result, the ChopTree agent must search around for trees, which may be risky or take a lot of time. We believe that the ChopTree agent can be further improved by developing techniques handling such extreme cases.

---

[7] Note that the minimum number is learned from the demonstrations, and this is allowed by the competition organizers.

[8] https://www.aicrowd.com/challenges/neurips-2020-minerl-competition/leaderboards.

Table 1. The final scores of the official baselines (left) and the best-performing submissions from Round 1 (middle) and Round 2 (right). Note that only six teams achieved a non-zero score in round 2. Our team is HelloWorld.

Baselines		Round 1 (Preliminary)		Round 2 (Final)	
Name	Score	Team name	Score	Team name	Score
SQIL	2.94	**HelloWorld**	**19.84**	**HelloWorld**	**39.55**
DQfD	2.39	NoActionWasted	16.48	michal_opanowicz	13.29
Rainbow	0.42	michal_opanowicz	9.29	NoActionWasted	12.79
PDDDQN	0.11	CU-SF	6.47	Rabbits	5.16
		NuclearWeapon	4.34	MajiManji	2.49
		murarinCraft	3.61	BeepBoop	1.97

Table 2. Details of our final submission. The "# Episode" column shows the total number of episodes end up with the corresponding score. The "Cond. Rate" column depicts the conditional success rate of each agent given the success of its previous agent. The "Rate" column shows the success rate of SEIHAI in different phases.

Agent	Item	Score	# Episode	Cond. Rate	Rate
ChopTree	None	0	72	64.0%	64.0%
	Log	1	11		
CraftWoodenPickaxe	Plank	3	15	78.6%	46.0%
	Stick	7	3		
	Crafting_Table	11	6		
	Wooden_Pickaxe	19	20		
DigStone	Cobblestone	35	9	78.3%	36.0%
CraftStonePickaxe	Stone_Pickaxe	67	2	84.7%	31.5%
	Furnace	99	47		
RandomSearch	Iron_Ore	163	14	23.0%	7.0%
	Iron_Ignots	291	0		
	Iron_Pickaxe	547	0		
	Diamond	1571	0		

Fortunately, the CraftWoodenPickaxe agent, the DigStone agent and the CraftStonePickaxe agent perform well, achieving a conditional success rate of more than 78.0%. However, as different agents are called in tandem, the success rate of SEIHAI in a certain phase is the multiplication of the conditional success rates of all the previously activated agents and would degrades exponentially as more agents are called. Evaluation results show that the success rate of SEIHAI at the crafting Stone_Pickaxe phase has dropped to 31.5%, which is very low.

The RandomSearch agent is responsible for the remaining tasks after obtaining the Stone_Pickaxe. Results show that this policy can obtain the Iron_Ore with a conditional success rate of 23.0%, and the success rate of SEIHAI in this phase drops to 7.0%. Since we do not design specific agents to attack these tasks due to the complexity of the underground environment and the difficulty of obtaining effective actions, we believe that well-designed agents for obtaining Iron_Ore and subsequent items can further improve the overall performance.

5   Related Work

The MineRL competition [5–9,25] has become a popular competition promoting sample-efficient reinforcement learning and imitation learning algorithms. There are lots of relevant studies. For example, HDQfD [31] utilizes the hierarchical structure of expert trajectories, presenting a structured task-dependent replay buffer and an adaptive prioritizing technique to gradually erase poor-quality expert data from the buffer. HDQfD won the first place in 2019. The runner-up method [1] highlights the influence of network architecture, loss function, and data augmentation. In contrast, the third winner [29] proposes a training procedure where policy networks are first trained with human demonstrations and later fine-tuned by reinforcement learning; additionally, they propose a policy exploitation mechanism, an experience replay scheme and an additional loss regularizer to prevent catastrophic forgetting of previously learned skills. However, these methods include many hand-crafted meta-actions or rules, which cannot generalize to new environment and is the main reason that the MineRL 2020 introduces the domain agnosticism requirement. Different from these methods, our method is domain-agnostic and generalizes well in new online training and evaluation environments.

Learning from demonstration is an effective way to achieving data efficiency for the MineRL competition. The simplest way is behaviour cloning, but it cannot handle complex situations. Recent studies such as DQfD [11], DDPGfD [33], POfD [13], SQIL [28] and GIAL [12] are more effective, but they can hardly handle the challenging MineRL environment with very sparse rewards. Consequently, we only adopt some of them to attack subtasks in our work. We notice that offline reinforcement learning algorithms like BCQ [3], BEAR [14] and CQL [15] are also promising approaches, and we leave the exploration of these methods as our future work.

6   Conclusion

In this work we present a sample-efficient hierarchical method to handle the *ObtainDiamond* task with limited human demonstrations, sparse rewards but an explicit task structure. The task is divided into several sequentially dependent subtasks (e.g., Treechop task, CraftWoodenPickaxe task, and DigCobblestone task) based on human priors. Furthermore, we design an imitation learning or a reinforcement learning agent suitable for each subtask. Besides, an imitation learning based scheduler is developed to determine which agent to use in different states. We win the first place in the preliminary and final of the NeurIPS-2020 MineRL competition, which demonstrates the efficiency of our hierarchical method, SEIHAI. We also identify several directions for further improvement based on detailed analysis of the online evaluation results. We believe that developing methods that properly combine human priors and sample-efficient learning-based techniques is a competitive way to solve complex tasks with limited demonstrations, sparse rewards but an explicit task structure.

Acknowledgement. The authors would like to thank Mengchen Zhao, Weixun Wang, Rundong Wang, Shixun Wu, Zhanbo Feng and the anonymous reviewers for their comments.

References

1. Amiranashvili, A., Dorka, N., Burgard, W., Koltun, V., Brox, T.: Scaling imitation learning in minecraft. arXiv preprint arXiv:2007.02701 (2020)
2. Blei, D.M., Griffiths, T.L., Jordan, M.I.: The nested Chinese restaurant process and Bayesian nonparametric inference of topic hierarchies. J. ACM (JACM) **57**(2), 1–30 (2010)
3. Fujimoto, S., Meger, D., Precup, D.: Off-policy deep reinforcement learning without exploration. In: International Conference on Machine Learning, pp. 2052–2062. PMLR (2019)
4. Gu, S., Holly, E., Lillicrap, T., Levine, S.: Deep reinforcement learning for robotic manipulation with asynchronous off-policy updates. In: 2017 IEEE International Conference on Robotics and Automation (ICRA), pp. 3389–3396. IEEE (2017)
5. Guss, W.H., et al.: The MineRL 2020 competition on sample efficient reinforcement learning using human priors. arXiv preprint arXiv:2101.11071 (2021)
6. Guss, W.H., et al.: NeurIPS 2019 competition: the MineRL competition on sample efficient reinforcement learning using human priors. arXiv preprint arXiv:1904.10079 (2019)
7. Guss, W.H., et al.: The MineRL competition on sample efficient reinforcement learning using human priors. arXiv e-prints (2019)
8. Guss, W.H., et al.: MineRL: a large-scale dataset of minecraft demonstrations. arXiv preprint arXiv:1907.13440 (2019)
9. Guss, W.H., et al.: Towards robust and domain agnostic reinforcement learning competitions: MineRL 2020. In: NeurIPS 2020 Competition and Demonstration Track, pp. 233–252. PMLR (2021)
10. He, K., Zhang, X., Ren, S., Sun, J.: Identity mappings in deep residual networks. In: Leibe, B., Matas, J., Sebe, N., Welling, M. (eds.) ECCV 2016. LNCS, vol. 9908, pp. 630–645. Springer, Cham (2016). https://doi.org/10.1007/978-3-319-46493-0_38
11. Hester, T., et al.: Deep Q-learning from demonstrations. In: Thirty-Second AAAI Conference on Artificial Intelligence (2018)
12. Ho, J., Ermon, S.: Generative adversarial imitation learning. Adv. Neural Inf. Process. Syst. **29**, 4565–4573 (2016)
13. Kang, B., Jie, Z., Feng, J.: Policy optimization with demonstrations. In: International Conference on Machine Learning, pp. 2469–2478. PMLR (2018)
14. Kumar, A., Fu, J., Tucker, G., Levine, S.: Stabilizing off-policy Q-learning via bootstrapping error reduction. arXiv preprint arXiv:1906.00949 (2019)

15. Kumar, A., Zhou, A., Tucker, G., Levine, S.: Conservative Q-learning for offline reinforcement learning. arXiv preprint arXiv:2006.04779 (2020)
16. Likas, A., Vlassis, N., Verbeek, J.J.: The global k-means clustering algorithm. Pattern Recogn. **36**(2), 451–461 (2003)
17. Lillicrap, T.P., et al.: Continuous control with deep reinforcement learning. In: ICLR (2016)
18. Lloyd, S.: Least squares quantization in PCM. IEEE Trans. Inf. Theor. **28**(2), 129–137 (1982)
19. MacQueen, J., et al.: Some methods for classification and analysis of multivariate observations. In: Proceedings of the Fifth Berkeley Symposium on Mathematical Statistics and Probability, vol. 1, pp. 281–297. Oakland, CA, USA (1967)
20. Mao, H., Gong, Z., Ni, Y., Xiao, Z.: ACCNet: actor-coordinator-critic net for "learning-to-communicate" with deep multi-agent reinforcement learning. arXiv preprint arXiv:1706.03235 (2017)
21. Mao, H., et al.: Neighborhood cognition consistent multi-agent reinforcement learning. In: Proceedings of the AAAI Conference on Artificial Intelligence, vol. 34, pp. 7219–7226 (2020)
22. Mao, H., Zhang, Z., Xiao, Z., Gong, Z.: Modelling the dynamic joint policy of teammates with attention multi-agent DDPG. In: Proceedings of the 18th International Conference on Autonomous Agents and MultiAgent Systems (2019)
23. Mao, H., Zhang, Z., Xiao, Z., Gong, Z., Ni, Y.: Learning agent communication under limited bandwidth by message pruning. In: Proceedings of the AAAI Conference on Artificial Intelligence, vol. 34, pp. 5142–5149 (2020)
24. Mao, H., Zhang, Z., Xiao, Z., Gong, Z., Ni, Y.: Learning multi-agent communication with double attentional deep reinforcement learning. Auton. Agents Multi Agent Syst. **34**(1), 1–34 (2020). https://doi.org/10.1007/s10458-020-09455-w
25. Milani, S., et al.: Retrospective analysis of the 2019 MineRL competition on sample efficient reinforcement learning. In: NeurIPS 2019 Competition and Demonstration Track, pp. 203–214. PMLR (2020)
26. Mnih, V., et al.: Human-level control through deep reinforcement learning. Nature **518**(7540), 529–533 (2015)
27. Osa, T., Pajarinen, J., Neumann, G., Bagnell, J.A., Abbeel, P., Peters, J., et al.: An algorithmic perspective on imitation learning. Found. Trends Rob. **7**(1–2), 1–179 (2018)
28. Reddy, S., Dragan, A.D., Levine, S.: SQIL: imitation learning via reinforcement learning with sparse rewards. In: ICLR (2019)
29. Scheller, C., Schraner, Y., Vogel, M.: Sample efficient reinforcement learning through learning from demonstrations in minecraft. In: NeurIPS 2019 Competition and Demonstration Track, pp. 67–76. PMLR (2020)
30. Silver, D., Lever, G., Heess, N., Degris, T., Wierstra, D., Riedmiller, M.: Deterministic policy gradient algorithms. In: ICML, pp. 387–395. PMLR (2014)
31. Skrynnik, A., Staroverov, A., Aitygulov, E., Aksenov, K., Davydov, V., Panov, A.I.: Hierarchical deep q-network from imperfect demonstrations in minecraft. Cogn. Syst. Res. **65**, 74–78 (2021)
32. Sutton, R.S., Barto, A.G.: Reinforcement Learning: An Introduction. MIT Press, Cambridge (2018)
33. Vecerik, M., et al.: Leveraging demonstrations for deep reinforcement learning on robotics problems with sparse rewards. arXiv preprint arXiv:1707.08817 (2017)

BGC: Multi-agent Group Belief with Graph Clustering

Tianze Zhou[1], Fubiao Zhang[1(✉)], Pan Tang[1], and Chenfei Wang[2]

[1] Beijing Institute of Technology, Beijing, China
{tianzezhou, fubiao.zhang}@bit.edu.cn
[2] Boston University, Boston, USA
wang1029@bu.edu

Abstract. Recent advances have witnessed that value decomposed-based multi-agent reinforcement learning methods make an efficient performance in coordination tasks. Most current methods assume that agents can communicate to assist decisions, which is impractical in some real situations. In this paper, we propose an observation-to-cognition method to enable agents to realize high efficient coordination without communication. Inspired by the neighborhood cognitive consistency (NCC), we introduce the group concept to help agents learn a belief, a type of consensus, to realize that adjacent agents tend to accomplish similar subtasks to achieve cooperation. We propose a novel agent structure named Belief in Graph Clustering (BGC) via Graph Attention Network (GAT) to generate agent group belief. In this module, we further utilize an MLP-based module to characterize special agent features to express the unique characteristics of each agent. Besides, to overcome the consistent agent problem of NCC, a split loss is introduced to distinguish different agents and reduce the number of groups. Results reveal that the proposed method makes excellent coordination and achieves a significant improvement in the SMAC benchmark. Due to the group concept, our approach maintains excellent performance with an increase in the number of agents.

Keywords: Multi-agent reinforcement learning · Graph attention network · Group concept

1 Introduction

The multi-agent system is widely applied in many real-life applications, including sensor networks, aircrafts formation flight, multi-robot cooperative control [8], and networked autonomous vehicles. In these systems, multiple agents can make coordination to improve the actual performance [1]. Most methods assume that agents can exchange information to communicate, which is impractical in some real situations, such as the heavy interference scenario. Even if communication is available, the integrity of the communication data is unreliable. In this situation, the solution will be sub-optimal.

In nature, animals tend to work in groups to execute tasks, such as ants and geese, which hints that the agents in the same group can make less communication or even

J. Chen et al. (Eds.): DAI 2021, LNAI 13170, pp. 52–63, 2022.
https://doi.org/10.1007/978-3-030-94662-3_4

Fig. 1. The process in SMAC via the Graph Clustering method. In this 8m_vs_9m map, agents are divided into three groups to complete the corresponding sub-tasks.

no communication to accomplish the specific task. The intention behind this situation is that agents in the same group tend to accomplish similar tasks due to similar cognition. Intuitively, adjacent agents are more likely to become the same group, and similar cognitions are more likely to be generated in neighbors. Moreover, viewing agents into groups reduces the coordination complexity by decomposing all agents' relationships into the relationship between groups and the inner group agents. Inspired by this, the group concept is introduced to multi-agent reinforcement learning (MARL) to handle cooperative tasks without communication. Expressly, we assume that each agent has a belief to assist in decision-making, and agents in the same group tend to generate similar beliefs.

In the large-scale multi-agents scenario, it is challenging to model all agents in a limited time efficiently. A simple way is to use a bidirectional recurrent neural network to represent all agents as a group explicitly. However, it limits by the agent order and ignores agents' relationship. In this paper, we propose a straightforward approach to map agents. Inspired by the neighborhood cognitive consistency [13], we propose to use the k-Nearest Neighbor (kNN) method to map agents via agent-relative position to tend adjacent agents to make similar decisions. However, the neighborhood cognitive consistency-based method may lead to the consistent agent problem and tend all agents to get close to each other. It causes the kNN method to divide n agents into n groups. Therefore, it is necessary to separate agents and prevent agent consistency.

To overcome the above problems, we make four contributions. (1) We design a novel modular agent structure that can achieve the Centralised Training and Decentralised Execution (CTDE) to realize no communication in the execution phase, compared to the communication-based method. (2) We propose using the agent belief to represent the approximate group feature and take an MLP-based module to preserve the specificity of each agent. (3) We propose an explicit method to map agents to achieve similar beliefs in adjacent agents via the GAT module. (4) To overcome the consistent problem brought by NCC, we introduce a split loss to distinguish agents.

The proposed group-based method is evaluated on several unit micromanagement tasks based on StarCraft II [22]. The results show that our approach outperforms traditional methods in the test-win results. Besides, the performance of the method scales well with the number of agents. To interpret the generated agent group belief, we utilize the t-SNE method [10] to express the group belief, and the result reveals that the features of agents in adjacent positions are also adjacent.

2   Related Work

Over the past years, deep multi-agent reinforcement learning has made a considerable breakthrough, widely used in games, traffic control, and other fields [2,5,23]. This research concentrates on cooperative MARL with a value function-based method. All teams receive a global team reward in this setting, and it is essential to divide the global team reward into individual rewards. VDN decomposes the joint Q value into the individual Q value of each agent [18]. QMIX [16] presents the constraint in which the joint Q value and the individual Q value have a monotonous setting based on the VDN algorithm. The suboptimal problem and decentralization in multi-agents are balanced by QTRAN via an L2-penalty term [17].

Among current multi-agent algorithms [3,6], communication is a crucial point. The communication-based approaches assume that multiple agents can make essential information interactions to assist in deciding. RIAL-DIAL applies a deep feedforward neural network to generate communication vectors for agents' communication [6]. BIC-NET [15] utilizes a bidirectional recurrent neural network for communication between multiple agents. LIIC [3] proposes a selectable point-to-point communication method adopted to determine whether agents communicate with each other by constructing a belief vector. Unlike these methods, GCRL [9] employs graph convolution network and multi-head dot-product attention [20] to aggregate agent features. Because agents generate their actions based on other agents' beliefs or relative features, these methods always need centralized execution that follows the centralized training with centralized execution (CTCE) framework. However, when the actual scenario restricts some agents from being difficult to communicate, the agent may obtain inaccurate beliefs or wrong beliefs, leading to sub-optimal actions.

Current researches on agent mapping always focus on the attention mechanism [20]. MAGA [11] utilizes two-stage attention to build an adjacent matrix. Hard attention is devoted to deleting related weak edges, and soft attention generates the weight coefficients of the retained graph structure. EPC-MADDPG [12] uses the scalar dot product attention to fuse the variable entity features to achieve a global mapping. However, simply using the attention mechanism to capture the association between agents for mapping will cause the algorithm to fall into a locally optimal solution. If the initial attention mechanism considers that two far apart agents are similar and tend to similar tasks, all agents will behave the same. Then all agents will be consistent and lose the meaning of grouping.

3   Background

In the present research, the problem is regarded as a fully cooperative multi-agent task viewed as a Dec-POMDP [14] comprising of a tuple $\langle I, S, U, Z, P, R, O, n, \gamma \rangle$, where $s \in S$ depicts the global state of the environment. At any time, each agent $i \in I \equiv$

$\{1, ..., n\}$ interacts with the environment by generating corresponding actions $u_i \in U$ through the observation vector $z_i \in Z$ according to the observation function $O(s, i)$. Agents learn to maximize the reward R for environmental feedback. This process is based on a state transition function $P(s' \mid s, a)$. Moreover, n signifies the number of agents, and γ represents a discount factor.

Interaction-based MARL algorithms are generally implemented via the CTCE framework. Agent's strategy $\pi_i(u_i|\tau_i)$ is generated based on the observation sequence τ, the global state s, and the interactive features of other agents. Via introducing the modular mechanism, our framework is designed as the distributed and applying a regularization tool to generate representative group features without interaction.

3.1 Graph Attention Network

Graph Attention Network (GAT) can correlate similar features between agents using masked self-attentional layers. By introducing the attention mechanism, the neural network can focus on the most relevant parts of the input, which helps learn the correlation features adaptively between nodes [21]. Besides, GAT can capture relative features between disconnect nodes by stacking GAT layers. Traditional GAT has two attention mechanism types, Global Graph Attention and Masked Graph Attention. The former builds the attention operation between nodes, while the latter only performs the same operation on neighboring nodes.

Distant agents may bring a negative effect on the current agent. Also, it is very computationally expensive to calculate the relationship between all agents, which causes the Global Graph Attention-based method to be unreasonable in large-scale scenarios. While Masked Graph Attention-based method only captures the relationship between the adjacent agents, which leads neighbor agents to hold similar actions. In the masked Graph Attention-based method, agents can capture the unconnected agent features via stacking the Masked Graph Attention layers and broadening the receptive field of agents.

4 Method

In this section, we introduce the group concept into the multi-agent algorithm and propose the Belief in Graph Clustering (BGC) method. The overall framework of the algorithm is shown in Fig. 2. To obtain a distributed framework, a modular approach is applied to build an algorithm network and generate group beliefs and individual features of agents separately. Due to the idea of modular construction [24], we utilize a new group belief network to represent agent group beliefs via minimizing the difference of original group beliefs without agent intersection. We utilize the group beliefs to replace the communication to implement the distributed execution.

In the overall pipeline, each agent generates an individual Q value via its local observation and then passes it into a mixed network (such as QMIX) for generating the global Q value. Specifically, each agent encodes its local observation sequence via a GRU network to produce agent group beliefs and individual features. The group beliefs are

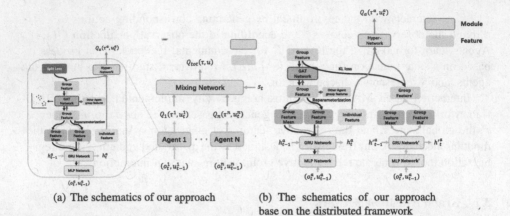

(a) The schematics of our approach   (b) The schematics of our approach base on the distributed framework

Fig. 2. (a) The Encoder generates individual features and Gaussian distribution's group features. A hyper-network module merges the group features and the individual features into the individual Q value to generate the total Q value via QMIX. The black line indicates the route of gradient propagation of the split loss. (b) In the inference stage, we utilize a new group feature network instead of the original group module to achieve distributed execution. The new group feature network is independent of other agents. This new network is optimized by minimizing the KL divergence between the group feature network's output and the original group network's output.

obtained via exercising the multivariate Gaussian distribution conditioned on the observation sequence. The reparameterization trick [4] is practiced for sampling to ensure the continuity of the gradient. Then, the GAT network is adopted to cluster group beliefs and obtain relationship features from adjacency agents. Group and individual features are fed into a hyper network [7] to generate individual Q Value. Besides, we introduce a split loss to distinguish non-adjacent agents to prevent all group beliefs from converging together.

4.1 Adjacent Matrix via kNN

In this subsection, we introduce the k-Nearest Neighbor (kNN) method to construct agent relationships on topology and generate the adjacency matrix for the masked attention-based GAT network. Due to the centralized training setting, it is available to get all agents' position information from agent observation. Agent adjacency graph is constructed by defining the agent with the k nearest agents as the same type. Then we take the Laplace transform and regularization of adjacent graphs to produce the agent adjacency matrix.

In our setting, the hyperparameter k in kNN is two. In this setting, all agents will be divided into several groups, as shown in Fig. 4(d). The non-information exchange between the group level leads different groups to achieve different sub-tasks to realize group diversity. Besides, when all agents are adjacent, the group-level information transfer, which helps accomplish the finished task via group coordination.

4.2   Belief in Graph Clustering

In this part, we introduce the details on how to generate agent group belief via graph clustering. First, we embed agent observation via the MLP network. To capture the relevant features on time series, we introduce the GRU module. To add the uncertainty to agent group features and enhance the exploration ability, we use a Gaussian Sampling module, which takes the condition on the embedding features of the GRU module. This module uses MLP networks to generate the mean and variance (take the exponential function on the logarithmic variance term) to construct the independent Gaussian distribution. Besides, we introduce the reparameterization trick to sample the actual group features for the gradient's continuity.

$$(\mu_{g_i}, \sigma_{g_i}, s_i) = f(\tau_i; \theta_i) \tag{1a}$$

$$g_i = \mu_{g_i} + \sigma_{g_i} \odot \varepsilon_i \quad \varepsilon_i \sim N(0,1) \tag{1b}$$

where θ_i represents parameters of network, μ_{g_i} and σ_{g_i} signify the mean and standard deviation of group features, s_i represents individual features, g_i signifies the actual group beliefs, and ε_i is noise.

In centralized training, agents can utilize the adjacent agent features to assist in deciding. After getting the adjacency matrix via the kNN method, we utilize the GAT module to cluster the relevant features into group beliefs. Under this setup, the agent interacts with agents in the adjacent position via the attention mechanism and generates the correlation weight. This weight is used to calculate the agent relationship features for the current agent, and the agent fuses these features to produce final group beliefs.

$$g_i' = \sigma \left(\sum_{j \in \mathcal{N}_i} \alpha_{ij} g_j \right) \tag{2}$$

where $\sigma(\cdot)$ is the ReLU activation function, and α represents the clustering coefficient calculated via the function.

$$e_{ij} = a([Wg_i \| Wg_j]), j \in \mathcal{N}_i \tag{3a}$$

$$\alpha_{ij} = \frac{\exp(\text{LeakyReLU}(e_{ij}))}{\sum_{k \in \mathcal{N}_i} \exp(\text{LeakyReLU}(e_{ik}))} \tag{3b}$$

Finally, we use a hyper network module to fuse group and individual features to generate agent action. In this module, the group features are utilized to generate the MLP network weight, multiplying individual features to get the final action.

4.3   Split Loss

Although mask attention-based GAT network can prevent all agents from forming a single group, current agents' group belief will still extend to all other agents due to the time perspective and lead all agents to be consistent. We propose a split loss to alleviate this issue. We use Kullback-Leibler (KL) divergence [19] to measure differences between agents' group features and keep the non-adjacent agents at a fixed distance.

$$\max_{\pi \in \Pi} \quad \mathbb{E}_{\tau \sim \rho_\pi} \left[\sum_{t=0}^{T} r\left(\mathbf{s}_t, \mathbf{a}_t\right) \right] \tag{4a}$$

$$s.t. \quad \mathbf{KL}(g_i \| g_j) \geq \delta, \forall\, edge(i, j) = \phi \tag{4b}$$

where g_i represents agent group features and $edge(i, j)$ determines whether there is a connection between agent i and agent j.

Therefore, the split loss is introduced to separate the agent:

$$L_{split} = -\sum_{i}^{N} \sum_{j}^{N} min(\mathbf{KL}(g_i \| g_j) - \delta, 0),$$

$$\forall\, edge(i, j) = \phi, i \neq j \tag{5}$$

The hyperparameter δ is employed to keep agents at the δ distance, changing with the scenario.

4.4 Decentralization Execution

To realize the complete decentralized execution without agent interaction, we apply a new sub-network to learn the new group features and use the KL divergence to minimize the distance between the new group belief and the original group belief learned by the GAT network, as shown in Fig. 2(b). Then, the gradient will flow through the new sub-network and train the new sub-network. This frame can be viewed as a teacher-student structure.

5   Experiment

5.1   Starcraft II

We evaluate the proposed algorithm in the SMAC benchmark. SMAC is a Starcraft II environment that focuses on unit micromanagement. It leverages the natural multi-agent microstructure by proposing a modified version of the problem designed specifically for decentralized control. We run our algorithm in 6 to 10 million time steps in 4 random seeds to test the algorithm's robustness. Besides, to estimate the performance of the method, we experiment in five scenarios, three standard maps of 5m_vs_6m (very hard), 8m_vs_9m (very hard), and 10m_vs_11m (very hard), and two custom maps of 12m_vs_14m (very hard) and 15m_vs_17m (very hard).

Figure 3 illustrates the performance difference between the proposed algorithm and other benchmark algorithms in five scenarios. We first compare our method with the origin QMIX algorithm. The result shows that BGC-QMIX outperforms the origin QMIX on the final test win rate and the performance variance, as shown in Fig. 3 (BGC+QMIX vs. QMIX). These results indicate that agents can realize high efficient coordination to achieve a high win rate.

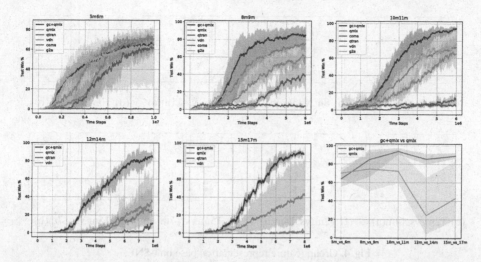

Fig. 3. Result of starcraft II. The comparison between the group-based method and other algorithms with agents increased from 5 to 15. Empirical results show that as the number of agents increases, our algorithm shows more superiority than the baseline algorithm. Due to the limitation of graphics card memory (RTX2080Ti), we did not test the actual effect of G2ANet in the 12m_vs_14m map and 15m_vs_17m map.

Except for the scenario with a small number of $5m\_vs\_6m$(very hard), our algorithm demonstrates a significant improvement compared to the baseline methods. In the $5m\_vs\_6m$ scenario, the proposed algorithm has excellent advantages over other convergence speeds. In this scenario set, the small number of agents causes consistent agents' group features, leading to less than ideal performance. To check the influence of the number of agents, we further perform our method in other scenarios. Figure 3 (bottom right) shows that it can maintain an excellent performance with the number of agents increasing while QMIX does not. The results show that the mask attention-based GAT method and the split loss performance.

Furthermore, our algorithm is compared with G2ANet, which is an attention communication method. G2ANet practices hard attention to determine the communication target and soft attention to complete the communication. G2ANet overlooks agent with topology information's association using a bi-LSTM network to construct the hard attention operation. Further, the G2ANet is tested based on the QMIX framework, and the empirical result shows that our graph clustering-based algorithm performs better. (As the tests were under the hardware condition of RTX 2080Ti, $12m\_vs\_14m$ and $15m\_vs\_17m$ scenarios can cause Cuda memory exhausted. Thus, only three standard scenes are tested.)

5.2 Representation

To demonstrate the interpretability of the proposed algorithm, the t-SNE algorithm is used to reduce the group beliefs' dimension for visually illustrating the effect of agent grouping. Pictures show the representation after compressing the group belief of the

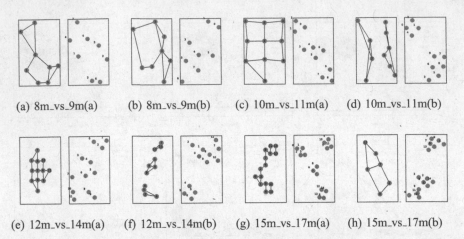

(a) 8m_vs_9m(a)   (b) 8m_vs_9m(b)   (c) 10m_vs_11m(a)   (d) 10m_vs_11m(b)

(e) 12m_vs_14m(a)   (f) 12m_vs_14m(b)   (g) 15m_vs_17m(a)   (h) 15m_vs_17m(b)

Fig. 4. Group feature representation base on t-SNE.

agent to two dimensions in multiple map scenarios. Pictures include various scenarios from the beginning of the game, the middle of the game, and the agent's death later. Figure 4(c) shows that BGC-QMIX first divides the agents into three belief groups at the beginning. Agents' group beliefs are divided into two groups without any information interaction between groups with the game running. Figure 4(h) indicates that group beliefs of the dead agents can be divided into a specific group that demonstrates the rationality of the algorithm. Moreover, we find that the agent mapping becomes more straightforward and representative as the number of agents increases.

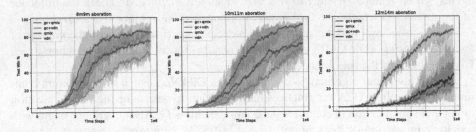

Fig. 5. Ablation test. We test the our graph clustering-based method base on the VDN framework. The result shows its superiority.

5.3 Ablation

To check the robustness of our proposed algorithm, an ablation experiment is conducted. The VDN network is used instead of the QMIX network to aggregate the independent Q values of the agent. Experimental results in 8m_vs_9m map, 10m_vs_11m map, and 12m_vs_14m map are compared. In two scenarios, where the number of agents is small, the group clustering-based method has improved significantly compared with the VDN and the QMIX algorithm. In the more significant number 12m_vs_14m scenario, the VDN algorithm is better than QMIX; the VDN method based on graph clustering is still better than these two.

Further, we verify the influence of different parameters of kNN on the method structure and the split loss. We find that a massive value of hyperparameter k will cause the agent to appear consistent, leading to sub-optimal agent cooperation strategies, and affecting the final winning rate. As shown in Fig. 6, the split loss can significantly improve the final performance when the agent population is huge.

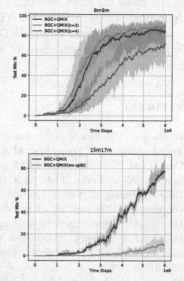

Fig. 6. Ablation test in kNN hyperparameter (above) and split loss (below).

5.4 Distributed Execution

The introduced algorithm is tested within the distributed framework. The KL divergence method is used for the agents trained based on the CTCE framework to minimize the new and group beliefs after graph clustering. The overall algorithm training is about 0.5M steps, and the original network error on the test_battle_win_mean index is within ±1%.

6 Conclusion

This paper introduces the group concept into multi-agent reinforcement learning to achieve excellent performance in the non-communication setup. We design a novel modular structure to realize this non-communication goal via an individual agent module to generate individual features and an agent group belief module to produce group features. Apply a fusion module to fuse these features to generate agent actions. In the agent group belief module, we take the GAT network to merge the adjacent agent features and use a split loss to prevent all agents from being consistent.

Empirical results show that the proposed algorithm's effect is vastly improved compared to the current baseline algorithms, and this improvement increases with the increase in the number of agents. The t-SNE method is applied to verify that the group features of adjacent agents are more similar, indicating the agent group features representability under the graph attention network and the split loss. Additionally, the overall algorithm is integrated into the VDN network, and results show the performance improved compared to the original VDN network, even the QMIX method.

In future work, we will consider applying intrinsic rewards to group agents more efficiently and feasibly.

References

1. Busoniu, L., Babuska, R., De Schutter, B.: A comprehensive survey of multiagent reinforcement learning. IEEE Trans. Syst. Man Cybern. Part C (Appl. Rev.) **38**(2), 156–172 (2008)
2. Cui, J., Liu, Y., Nallanathan, A.: Multi-agent reinforcement learning-based resource allocation for UAV networks. IEEE Trans. Wirel. Commun. **19**(2), 729–743 (2019)
3. Ding, Z., Huang, T., Lu, Z.: Learning individually inferred communication for multi-agent cooperation. arXiv preprint arXiv:2006.06455 (2020)
4. Doersch, C.: Tutorial on variational autoencoders. arXiv preprint arXiv:1606.05908 (2016)
5. Foerster, J., Farquhar, G., Afouras, T., Nardelli, N., Whiteson, S.: Counterfactual multi-agent policy gradients. In: Proceedings of the AAAI Conference on Artificial Intelligence, vol. 32 (2018)
6. Foerster, J.N., Assael, Y.M., de Freitas, N., Whiteson, S.: Learning to communicate with deep multi-agent reinforcement learning. CoRR abs/1605.06676 (2016). http://arxiv.org/abs/1605.06676
7. Ha, D., Dai, A., Le, Q.V.: Hypernetworks. arXiv preprint arXiv:1609.09106 (2016)
8. Hüttenrauch, M., Sosic, A., Neumann, G.: Guided deep reinforcement learning for swarm systems. CoRR abs/1709.06011 (2017). http://arxiv.org/abs/1709.06011
9. Jiang, J., Dun, C., Lu, Z.: Graph convolutional reinforcement learning for multi-agent cooperation. CoRR abs/1810.09202 (2018). http://arxiv.org/abs/1810.09202
10. Laurens, V.D.M., Hinton, G.: Visualizing data using t-SNE. J. Mach. Learn. Res. **9**(2605), 2579–2605 (2008)
11. Liu, Y., Wang, W., Hu, Y., Hao, J., Chen, X., Gao, Y.: Multi-agent game abstraction via graph attention neural network. In: AAAI, pp. 7211–7218 (2020)
12. Long, Q., Zhou, Z., Gupta, A., Fang, F., Wu, Y., Wang, X.: Evolutionary population curriculum for scaling multi-agent reinforcement learning. arXiv preprint arXiv:2003.10423 (2020)
13. Mao, H., et al.: Neighborhood cognition consistent multi-agent reinforcement learning. arXiv preprint arXiv:1912.01160 (2019)
14. Oliehoek, F.A., Amato, C., et al.: A Concise Introduction to Decentralized POMDPs, vol. 1. Springer, New York (2016). https://doi.org/10.1007/978-3-319-28929-8
15. Peng, P., et al.: Multiagent bidirectionally-coordinated nets for learning to play StarCraft combat games. CoRR abs/1703.10069 (2017). http://arxiv.org/abs/1703.10069
16. Rashid, T., Samvelyan, M., de Witt, C.S., Farquhar, G., Foerster, J.N., Whiteson, S.: QMIX: monotonic value function factorisation for deep multi-agent reinforcement learning. CoRR abs/1803.11485 (2018). http://arxiv.org/abs/1803.11485
17. Son, K., Kim, D., Kang, W.J., Hostallero, D., Yi, Y.: QTRAN: learning to factorize with transformation for cooperative multi-agent reinforcement learning. CoRR abs/1905.05408 (2019). http://arxiv.org/abs/1905.05408
18. Sunehag, P., et al.: Value-decomposition networks for cooperative multi-agent learning. CoRR abs/1706.05296 (2017). http://arxiv.org/abs/1706.05296
19. Van Erven, T., Harremos, P.: Rényi divergence and kullback-leibler divergence. IEEE Trans. Inf. Theor. **60**(7), 3797–3820 (2014)

20. Vaswani, A., et al.: Attention is all you need (2017)
21. Veličković, P., Cucurull, G., Casanova, A., Romero, A., Lio, P., Bengio, Y.: Graph attention networks. arXiv preprint arXiv:1710.10903 (2017)
22. Vinyals, O., et al.: StarCraft II: a new challenge for reinforcement learning. CoRR abs/1708.04782 (2017). http://arxiv.org/abs/1708.04782
23. Wang, X., Ke, L., Qiao, Z., Chai, X.: Large-scale traffic signal control using a novel multi-agent reinforcement learning. IEEE Trans. Cybern. **51**(1), 174–187 (2020)
24. Watanabe, T.: A study on multi-agent reinforcement learning problem based on hierarchical modular fuzzy model. In: 2009 IEEE International Conference on Fuzzy Systems, pp. 2041–2046. IEEE (2009)

Incomplete Distributed Constraint Optimization Problems: Model, Algorithms, and Heuristics

Atena M. Tabakhi[1]([⊠]), William Yeoh[1], and Roie Zivan[2]

[1] Washington University in St. Louis, St. Louis, USA
{amtabakhi,wyeoh}@wustl.edu
[2] Ben Gurion University of the Negev, Beer Sheva, Israel
zivanr@bgu.ac.il

Abstract. The *Distributed Constraint Optimization Problem* (DCOP) formulation is a powerful tool to model cooperative multi-agent problems, especially when they are sparsely constrained with one another. A key assumption in this model is that all constraints are fully specified or known a priori, which may not hold in applications where constraints encode preferences of human users. In this paper, we extend the model to *Incomplete DCOPs* (I-DCOPs), where some constraints can be partially specified. User preferences for these partially-specified constraints can be elicited during the execution of I-DCOP algorithms, but they incur some elicitation costs. Additionally, we extend SyncBB, a complete DCOP algorithm, and ALS-MGM, an incomplete DCOP algorithm, to solve I-DCOPs. We also propose parameterized heuristics that those algorithms can utilize to trade off solution quality for faster runtime and fewer elicitation. They also provide theoretical quality guarantees when used by SyncBB when elicitations are free. Our model and heuristics thus extend the state-of-the-art in distributed constraint reasoning to better model and solve distributed agent-based applications with user preferences.

Keywords: Multi-agent problems · Distributed constraint optimization problems · Preference elicitation · Distributed problem solving

1 Introduction

The *Distributed Constraint Optimization Problem* (DCOP) [6,22,25] formulation is a powerful tool to model cooperative multi-agent problems. DCOPs are well-suited to model many problems that are distributed by nature and where agents need to coordinate their value assignments to minimize the aggregated constraint

This research is partially supported by BSF grant #2018081 and NSF grants #1812619 and #1838364.

© Springer Nature Switzerland AG 2022
J. Chen et al. (Eds.): DAI 2021, LNAI 13170, pp. 64–78, 2022.
https://doi.org/10.1007/978-3-030-94662-3_5

costs. This model is widely employed for representing distributed problems such as meeting scheduling [19], sensor and wireless networks [37], multi-robot teams coordination [41], smart grids [21], and smart homes [7,31].

The study and use of DCOPs have matured significantly over more than a decade since its inception [22]. DCOP researchers have proposed a wide variety of solution approaches, from complete approaches that use distributed search-based techniques [22,37] to distributed inference-based techniques [25]. There is also a significant body of work on incomplete methods that can be similarly categorized into local search-based methods [5,39], inference GDL-based techniques [35], and sampling-based methods [24].

One of the core limitations of all these approaches is that they assume that the constraint costs in a DCOP are specified or known a priori. In some applications, such as meeting scheduling problems, constraints encode the preferences of human users. As such, some of the constraint costs may be unspecified and must be elicited from human users.

To address this limitation, researchers have proposed the *preference elicitation problem for DCOPs* [29]. In this preference elicitation problem, some constraint costs are initially unknown, and they can be accurately elicited from human users. The goal is to identify which subset of constraints to elicit in order to minimize a specific form of expected error in solution quality. Unfortunately, it suffers from two limitations: First, it assumes that the cost of eliciting constraints is uniform across all constraints, which is unrealistic as providing the preferences for some constraints may require more cognitive effort than the preferences for other constraints. Second, it decouples the elicitation process from the DCOP solving process since the elicitation process must be completed before one solves the DCOP with elicited constraints. As both the elicitation and solving process are actually coupled, this two-phase decoupled approach prohibits the elicitation process from relying on the solving process.

Therefore, in this paper, we propose the *Incomplete DCOP* (I-DCOP) [30,36] model, which *integrates* both the elicitation and solving problems into a single integrated optimization problem. In an I-DCOP, some constraint costs are unknown and can be elicited. Elicitation of unknown constraint costs will incur elicitation costs, and the goal is to find a solution that minimizes the sum of constraint and elicitation costs incurred. To solve this problem, we adapt a complete algorithm – Synchronous Branch-and-Bounds (SyncBB) [14] – and an incomplete algorithm – an *Anytime Local Search* (ALS) [40] variant of *Maximum Gain Message* (MGM) [18], which we call ALS-MGM. We also introduce parameterized heuristics that can be used by SyncBB and ALS-MGM to trade off solution quality for faster runtimes and fewer elicitations, and provide quality guarantees for I-DCOPs without elicitation costs when the underlying DCOP algorithm is correct and complete.

2   Background

Distributed Constraint Optimization Problems (DCOPs): A DCOP [6,22,25] is defined by $\langle \mathcal{A}, \mathcal{X}, \mathcal{D}, \mathcal{F}, \alpha \rangle$, where $\mathcal{A} = \{a_i\}_{i=1}^{p}$ is a set of *agents*;

$\mathcal{X} = \{x_i\}_{i=1}^n$ is a set of decision *variables*; $\mathcal{D} = \{D_x\}_{x \in \mathcal{X}}$ is a set of finite *domains* and each variable $x \in \mathcal{X}$ takes values from the set D_x; $\mathcal{F} = \{f_i\}_{i=1}^m$ is a set of *constraints*, each defined over a set of decision variables: $f_i : \prod_{x \in \mathbf{x}^{f_i}} D_x \to \mathbb{R} \cup \{\infty\}$, where infeasible configurations have ∞ costs, $\mathbf{x}^{f_i} \subseteq \mathcal{X}$ is the *scope* of f_i; and $\alpha : \mathcal{X} \to \mathcal{A}$ is a *mapping function* that associates each decision variable to one agent.

A *solution* σ is a value assignment for a set $\mathbf{x}_\sigma \subseteq \mathcal{X}$ of variables that is consistent with their respective domains. The cost $\mathcal{F}(\mathbf{x}_\sigma) = \sum_{f \in \mathcal{F}, \mathbf{x}^f \subseteq \mathbf{x}_\sigma} f(\mathbf{x}_\sigma)$ is the sum of the costs across all the applicable constraints in $\mathbf{x}_\sigma$. A solution σ is a *complete solution* if $\mathbf{x}_\sigma = \mathcal{X}$ and is a *partial solution* otherwise. The goal is to find an optimal complete solution $\mathbf{x}^* = \text{argmin}_\mathbf{x} \mathcal{F}(\mathbf{x})$.

A *constraint graph* visualizes a DCOP, where nodes in the graph correspond to variables in the DCOP and edges connect pairs of variables appearing in the same constraint. A *pseudo-tree* arrangement has the same nodes as the constraint graph and includes all the edges of the constraint graph. The edges in the pseudo-tree are divided into *tree edges*, which connect parent-child nodes and all together form a rooted tree, and *backedges*, which connect a node with its *pseudo-parents* and *pseudo-children*. Finally, two variables that are constrained together in the constraint graph must appear in the same branch of the pseudo-tree. When the pseudo-tree has only a single branch, it is called a *pseudo-chain*. One can also view a pseudo-chain as a complete ordering of all the variables in a DCOP, which is used by SyncBB and in our descriptions later on. Finally, unless otherwise specified, we assume that each agent controls exactly one decision variable and thus use the terms "agent" and "variable" interchangeably.

Synchronous Branch-and-Bound (SyncBB): SyncBB [14] is a complete, synchronous, search-based algorithm that can be considered as a distributed version of a depth-first branch-and-bound algorithm. It uses a complete ordering of the agents to extend a *Current Partial Assignment* (CPA) via a synchronous communication process. The CPA holds the assignments of all the variables controlled by all the visited agents and, in addition, functions as a mechanism to propagate bound information. The algorithm prunes those parts of the search space whose solution quality is sub-optimal by exploiting the bounds that are updated at each step of the algorithm. In other words, an agent backtracks when the cost of its CPA is no smaller than the cost of the best complete solution found so far. The algorithm terminates when the root backtracks (i.e., the algorithm has explored or pruned the entire search space).

Anytime Local Search Algorithms: *Local search* DCOP algorithms (e.g., MGM, DSA) [18,39] are synchronous iterative processes, where, in each step of the algorithm, each agent sends its value assignment to all its neighbors in the constraint graph and waits to receive the value assignments of all its neighbors before deciding whether to change its value. In local search algorithms, agents are only aware of the cost of their own assignments and their neighbors' assignments. Therefore, no agent knows when a globally good solution is found. The *Anytime Local Search* (ALS) framework [40] enhances the local search algorithms by allowing them to detect when a globally better solution is found and return

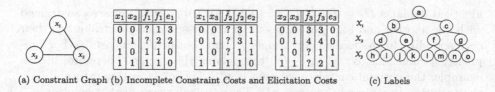

(a) Constraint Graph   (b) Incomplete Constraint Costs and Elicitation Costs   (c) Labels

Fig. 1. Incomplete DCOP with eicitation costs and search tree nodes

that solution upon termination (i.e., the anytime property). It uses a *Breadth-First Search spanning tree* (BFS-tree) of the constraint graph to aggregate costs up the tree to the root agent such that it is able to detect when a better solution is found. When such a solution is found, the root agent propagates the step number in which that solution is found down to its descendants. Therefore, upon termination, all the agents have a consistent view on when the best solution is found and take on their corresponding values.

3 Incomplete DCOPs

An *Incomplete DCOP* (I-DCOP) extends a DCOP by allowing some constraints to be partially specified. It is defined by a tuple $\langle \mathcal{A}, \mathcal{X}, \mathcal{D}, \mathcal{F}, \tilde{\mathcal{F}}, \mathcal{E}, \alpha \rangle$, where $\mathcal{A}$, $\mathcal{X}, \mathcal{D}, \mathcal{F}$, and α are exactly the same as in a DCOP. There are two key differences:

- The set of constraints $\mathcal{F}$ are not known to agents in an I-DCOP. Instead, only the set of partially-specified constraints $\tilde{\mathcal{F}} = \{\tilde{f}_i\}_{i=1}^m$ are known. Each partially-specified constraint is a function $\tilde{f}_i : \prod_{x \in \mathbf{x}^{f_i}} D_x \to \mathbb{R} \cup \{\infty, ?\}$, where ? is a special element denoting that the cost for a given combination of value assignment is not specified. The costs $\mathbb{R} \cup \{\infty\}$ that are specified are exactly the costs of the corresponding constraints $f_i \in \mathcal{F}$.
- $\mathcal{E} = \{e_i\}_{i=1}^m$ is the set of elicitation costs, where each elicitation cost $e_i : \prod_{x \in \mathbf{x}^{f_i}} D_x \to \mathbb{R}$ specifies the cost of eliciting the constraint cost of a particular ? in $\tilde{f}_i$.

An *explored solution space* $\tilde{\mathbf{x}}$ is the union of all solutions explored so far by a particular algorithm. The *cumulative elicitation cost* $\mathcal{E}(\tilde{\mathbf{x}}) = \sum_{e \in \mathcal{E}} e(\tilde{\mathbf{x}})$ is the sum of the costs of all elicitations conducted while exploring $\tilde{\mathbf{x}}$.

The *total cost* $\mathcal{F}(\mathbf{x}, \tilde{\mathbf{x}}) = \alpha_f \cdot \mathcal{F}(\mathbf{x}) + \alpha_e \cdot \mathcal{E}(\tilde{\mathbf{x}})$ is the weighted sum of both the cumulative constraint cost $\mathcal{F}(\mathbf{x})$ of solution $\mathbf{x}$ and the cumulative elicitation cost $\mathcal{E}(\tilde{\mathbf{x}})$ of the explored solution space $\tilde{\mathbf{x}}$, where $\alpha_f \in (0, 1]$ and $\alpha_e \in [0, 1]$ such that $\alpha_f + \alpha_e = 1$. The weights represent the tradeoffs between the importance of solution quality and the cumulative elicitation cost.

The goal is to find an optimal complete solution $\mathbf{x}^*$ while eliciting only a cost-minimal set of preferences from a solution space $\tilde{\mathbf{x}}^*$. More formally, the goal is to find $(\mathbf{x}^*, \tilde{\mathbf{x}}^*) = \operatorname{argmin}_{(\mathbf{x}, \tilde{\mathbf{x}})} \mathcal{F}(\mathbf{x}, \tilde{\mathbf{x}})$.

Figure 1(a) shows the constraint graph of an example I-DCOP that we will use as a running example in this paper. It has three variables x_1, x_2, and x_3 with identical domains $D_1 = D_2 = D_3 = \{0,1\}$. All three variables are constrained with one another and Fig. 1(b) shows the partially-specified constraints $\tilde{f}_i$, their corresponding fully-specified constraints f_i, and the elicitation costs e_i. For simplicity, assume that $\alpha_f = \alpha_e = 0.5$ throughout this paper. Therefore, in this example, the optimal complete solution is $\mathbf{x}^* = \langle x_1 = 1, x_2 = 1, x_3 = 0 \rangle$ and only that solution is explored (i.e., $\tilde{\mathbf{x}} = \mathbf{x}^*$). The constraint cost of that solution is 3 ($= f_1(\langle x_1 = 1, x_2 = 1 \rangle) + f_2(\langle x_1 = 1, x_3 = 0 \rangle) + f_3(\langle x_2 = 1, x_3 = 0 \rangle)$). The cumulative elicitation cost is 2 ($= e_2(\langle x_1 = 1, x_3 = 0 \rangle) + e_3(\langle x_2 = 1, x_3 = 0 \rangle)$). Thus, the total cost is $\alpha_f \cdot 3 + \alpha_e \cdot 2 = 0.5 \cdot 3 + 0.5 \cdot 2 = 2.5$.

4 Solving I-DCOPs

To solve I-DCOPs, one can easily adapt existing DCOP algorithms by allowing them to elicit unknown costs whenever those costs are needed by the algorithm. We describe below how to adapt SyncBB, a complete search algorithm, and ALS-MGM, a variant of the MGM [18] local search algorithm using the ALS framework [40], to solve I-DCOPs. We will also employ SyncBB and ALS-MGM as the underlying algorithms that use our proposed heuristics later.

4.1 SyncBB

The operations of SyncBB can be visualized with search trees. Figure 1(c) shows the search tree for our example I-DCOP shown in Figs. 1(a) and 1(b), where levels 1, 2, and 3 correspond to variable x_1, x_2, and x_3, respectively. Left branches correspond to the variables being assigned the value 0 and right branches correspond to the variables being assigned the value 1. Each non-leaf node thus corresponds to a partial solution and each leaf node corresponds to a complete solution. These nodes also correspond to unique CPAs of agents when they run SyncBB. We label each node of the search tree with an identifier so that we can refer to them easily below. When SyncBB evaluates a node n after exploring search space $\tilde{\mathbf{x}}$, it considers only the cumulative elicitation cost so far $\mathcal{E}(\tilde{\mathbf{x}})$ and the constraint costs of the CPA at node n, which we will refer to as g-values, denoted by $g(n)$.[1] We refer to the weighted sum of these values as f-values, denoted by $f(n, \tilde{\mathbf{x}}) = \alpha_f \cdot g(n) + \alpha_e \cdot \mathcal{E}(\tilde{\mathbf{x}})$.

Assume that all the agents know that there is a lower bound $\mathcal{L}$ on all the constraint costs. Before calculating $f(n, \tilde{\mathbf{x}})$ at node n, the algorithm *estimates* the total cost (i.e., constraint cost + elicitation cost) by replacing unknown constraint costs with $\mathcal{L}$ and summing them up with the elicitation cost thus far. If the estimated total cost is no smaller than the cost of the best solution found so far, SyncBB prunes node n. Otherwise, it elicits the unknown costs of node n and calculates its true total cost. By estimating the total costs, SyncBB only elicits unknown constraints when their costs are needed.

---

[1] We use A* notations [13] here.

4.2   ALS-MGM

Like for regular DCOPs, ALS-MGM for I-DCOPs also uses a *Breadth-First Search spanning tree* (BFS-tree) of the constraint graph to aggregate costs up the tree to the root agent such that it is able to detect when a better solution is found. When such a solution is found, the root agent propagates the step number in which that solution is found down to its descendants. Therefore, upon termination, all the agents have a consistent view on when the best solution is found and take on their corresponding values.

5   SyncBB Cost-Estimate Heuristic

To speed up SyncBB, one can use *cost-estimate heuristics* $h(n)$ to estimate the sum of the constraint and elicitation costs needed to complete the CPA at a particular node n. And if those heuristics are underestimates of the true cost, then they can be used to better prune the search space, that is, when $f(n, \tilde{\mathbf{x}}) = \alpha_f \cdot g(n) + h(n) + \alpha_e \cdot \mathcal{E}(\tilde{\mathbf{x}}) \geq \mathcal{F}(\mathbf{x}, \tilde{\mathbf{x}})$, where $\mathbf{x}$ is the best complete solution found so far and $\tilde{\mathbf{x}}$ is the current explored solution space.

We now describe below a cost-estimate heuristic that can be used in conjunction with SyncBB to solve I-DCOPs. This heuristic makes use of an estimated lower bound $\mathcal{L}$ on the cost of all constraints $f \in \mathcal{F}$. Such a lower bound can usually be estimated through domain expertise. In the worst case, since all costs are non-negative, for our running example we set the lower bound $(\mathcal{L})$ to 1. The more informed the lower bound, the more effective the heuristics will be in pruning the search space.

Additionally, this heuristic is parameterized by two parameters – a relative weight $w \geq 1$ and an additive weight $\epsilon \geq 0$. When using these parameters, SyncBB will prune a node n if:

$$w \cdot f(n, \tilde{\mathbf{x}}) + \epsilon \geq \mathcal{F}(\mathbf{x}, \tilde{\mathbf{x}}) \tag{1}$$

where $\mathbf{x}$ is the best complete solution found so far and $\tilde{\mathbf{x}}$ is the current explored solution space. Users can increase the weights w and ϵ to prune a larger portion of the search space and, consequently, reduce the computation time as well as the number of preferences elicited. However, the downside is that it will also likely degrade the quality of solutions found. Further, in I-DCOPs where elicitations are free (i.e., the elicitation costs are all zero), we theoretically show that the cost of solutions found are guaranteed to be at most $w \cdot OPT + \epsilon$, where OPT is the optimal solution cost.

Child's Ancestors' Constraints (CAC) Heuristic: This heuristic is defined recursively from the leaf of the pseudo-chain (i.e., last agent in the variable ordering) used by SyncBB up to the root of the pseudo-chain (i.e., first agent in the ordering). Agent x_i in the ordering computes a heuristic value $h(x_i = d_i)$ for each of its values $d_i \in D_i$ as follows: $h(x_i = d_i) = 0$ if x_i is the leaf of the pseudo-chain. Otherwise, $h(x_i = d_i)$ is:

$$\min_{d_c \in D_c} \Big[\alpha_f \cdot \hat{f}(x_i = d_i, x_c = d_c) + \alpha_e \cdot e(x_i = d_i, x_c = d_c) + h(x_c = d_c) \Big]$$

$$+ \sum_{x_k \in Anc(x_c) \backslash \{x_i\}} \min_{d_k \in D_k} \Big[\alpha_f \cdot \hat{f}(x_c = d_c, x_k = d_k) + \alpha_e \cdot e(x_c = d_c, x_k = d_k) \Big] \quad (2)$$

where x_c is the next agent in the ordering (i.e., child of x_i in the pseudo-chain), $Anc(x_c)$ is the set of variables higher up in the ordering that x_c is constrained with, and each estimated cost function $\hat{f}$ corresponds exactly to a partially-specified function $\tilde{f}$, except that all the unknown costs ? are replaced with the lower bound $\mathcal{L}$. Therefore, the estimated cost $\hat{f}(\mathbf{x})$ is guaranteed to be no larger than the true cost $f(\mathbf{x})$ for any solution $\mathbf{x}$.

For a parent x_p of a leaf agent x_l, the heuristic value $h(x_p = d_p)$ is then the minimal constraint and elicitation cost between the two agents, under the assumption that the parent takes on value d_p, and the sum of the minimal constraint cost of the leaf agent with its ancestors. As the heuristic of a child agent is included in the heuristic of the parent agent, this summation of costs is recursively aggregated up the pseudo-chain.

It is fairly straightforward to see that this heuristic can be computed in a distributed manner – the leaf agent x_l initializes its heuristic values $h(x_l = d_l) = 0$ for all its values $d_l \in D_l$ and computes the latter term in Eq. (2):

$$\sum_{x_k \in Anc(x_l)} \min_{d_k \in D_k} \Big[\alpha_f \cdot \hat{f}(x_l = d_l, x_k = d_k) + \alpha_f \cdot e(x_l = d_l, x_k = d_k) \Big] \quad (3)$$

for each of its values $d_l \in D_l$. It then sends these heuristic values and costs to its parent. Upon receiving this message, the parent agent x_p uses the information in the message to compute its own heuristic values $h(x_p = d_p)$ using Eq. (2), computes the latter term similar to Eq. (3) above, and sends these heuristic values and costs to its parent. This process continues until the root agent computes its own heuristic values, at which point it starts the SyncBB algorithm.

6   ALS-MGM Cost-Estimate Heuristic

Instead of having each agent in ALS-MGM choose its initial value randomly from its domains, one can also use cost-estimate heuristics to estimate costs for each value and have the agent choose the value that minimizes the estimated costs. Using cost-estimate heuristics helps ALS-MGM to find solutions with smaller costs faster since it starts with a better initial solution, which is more pronounced when there is not enough time to let the algorithm run until convergence.

Neighbors' Constraints (NHC) Heuristic: Each agent x_i computes a heuristic value $h(x_i = d_i)$ for each of its values $d_i \in D_i$ as follows:

$$\sum_{x_c \in Nh(x_i)} \min_{d_c \in D_c} \Big[\alpha_f \cdot \hat{f}(x_i = d_i, x_c = d_c) + \alpha_e \cdot e(x_i = d_i, x_c = d_c) \Big] \quad (4)$$

where x_c is a neighboring variable, $Nh(x_i)$ is the set of neighboring variables that x_i is constrained with, and each estimated cost function $\hat{f}$ corresponds exactly to a partially-specified function $\tilde{f}$, except that all the unknown costs ? are replaced with the lower bound $\mathcal{L}$. Therefore, the estimated cost $\hat{f}(\mathbf{x})$ is guaranteed to be no larger than the true cost $f(\mathbf{x})$ for any solution $\mathbf{x}$.

7 Theoretical Results

Theorem 1. *The computation of the CAC heuristic requires $O(|\mathcal{A}|)$ number of messages. The computation of the NHC heuristic requires no messages.*

Proof. The CAC heuristics is recursively computed starting from the leaf to the root and will take exactly $|\mathcal{A}| - 1$ number of messages. The NHC heuristic is not computed recursively and does not send any messages to compute its heuristic cost. ∎

Lemma 1. *When all elicitation costs are zero, the CAC and NHC heuristics are admissible.*

Proof. We only prove the admissibility of the CAC heuristic since the same proof applies to NHC. We prove that $h(n) \leq \mathcal{F}(\mathbf{x}_n) - \alpha_f \cdot g(n)$, where $\mathbf{x}_n$ is the best complete solution in the subtree rooted at node n, for all nodes n in the search tree. We prove this by induction from the leaf agent up the pseudo-chain:

- **Leaf Agent:** For a leaf agent x_i, $h(x_i = d_i) = 0$ for each of its values $d_i \in D_i$. Therefore, the inequality $h(n) = 0 \leq \mathcal{F}(\mathbf{x}_n) - \alpha_f \cdot g(n)$ trivially applies for all nodes n corresponding the agent x_i taking on its values $d_i \in D_i$.
- **Induction Assumption:** Assume that the lemma holds for all agents up to the $(k-1)$-th agent up the pseudo-chain.
- **The k-th Agent:** For the k-th agent x_k from the leaf:

$$
\begin{aligned}
&h(x_k = d_k) \\
&= \min_{d_c \in D_c} \left[\alpha_f \cdot \hat{f}(x_k = d_k, x_c = d_c) + \alpha_e \cdot e(x_k = d_k, x_c = d_c) + h(x_c = d_c) \right] \\
&\quad + \sum_{x_m \in Anc(x_c) \setminus \{x_k\}} \min_{d_m \in D_m} \left[\alpha_f \cdot \hat{f}(x_c = d_c, x_m = d_m) + \alpha_e \cdot e(x_c = d_c, x_m = d_m) \right]
\end{aligned}
\tag{5}
$$

where x_c is the next agent in the ordering (i.e., the $(k-1)$-th agent), $Anc(x_c)$ is the set of variables higher up in the ordering that x_c is constrained with. Based on our induction assumption the lemma holds for all agents up to the $(k-1)$-th agent up the pseudo-chain, hence, we have:

$$
h(n) \leq \mathcal{F}(\mathbf{x}_n) - \alpha_f \cdot g(n),
\tag{6}
$$

where node n corresponds to the $(k-1)$-th agent in the pseudo-chain, which we denote it as agent x_c. For each estimated cost function $\hat{f}$ in the CAC heuristic, it is easy to see:

$$\hat{f}(x_c{=}d_c, x_m{=}d_m) \leq f(x_c{=}d_c, x_m{=}d_m), \tag{7}$$

for any pair of agents x_c and x_m with any of their value combinations since all unknown costs ? are replaced with the lower bound $\mathcal{L}$ on all constraint costs. Thus, combined with the premise that elicitation costs are all zero and the induction assumption, we get:

$$
\begin{aligned}
&h(x_k{=}d_k) \\
&= \min_{d_c \in D_c} \left[\alpha_f \cdot \hat{f}(x_k{=}d_k, x_c{=}d_c) + \alpha_e \cdot e(x_k{=}d_k, x_c{=}d_c) + h(x_c{=}d_c) \right] \\
&\quad + \sum_{x_m \in Anc(x_c) \backslash \{x_k\}} \min_{d_m \in D_m} \left[\alpha_f \cdot \hat{f}(x_c{=}d_c, x_m{=}d_m) + \alpha_e \cdot e(x_c{=}d_c, x_m{=}d_m) \right]
\end{aligned}
\tag{8}
$$

$$
\begin{aligned}
&\leq \min_{d_c \in D_c} \left[\alpha_f \cdot f(x_k{=}d_k, x_c{=}d_c) + h(x_c{=}d_c) \right] \\
&\quad + \sum_{x_m \in Anc(x_c) \backslash \{x_k\}} \min_{d_m \in D_m} \alpha_f \cdot f(x_c{=}d_c, x_m{=}d_m) \leq \mathcal{F}(\mathbf{x}_n) - \alpha_f \cdot g(n) \tag{9}
\end{aligned}
$$

where node n corresponds to the agent x_k taking on its value $d_k \in D_k$. ∎

Theorem 2. *When all elicitation costs are zero, SyncBB with the CAC heuristic parameterized by a user-defined relative weight $w \geq 1$ and a user-defined additive weight $\epsilon \geq 0$ will return an I-DCOP solution whose cost is bounded from above by $w \cdot OPT + \epsilon$, where OPT is the optimal solution cost.*

Proof. The proof is similar to the proofs of similar properties [37] for other DCOP search algorithms that also use heuristics. The key assumption in the proofs is that the heuristics employed are admissible heuristics – and the CAC heuristic is admissible according to Lemma 1. ∎

8   Related Work

As our work lies in the intersection of constraint-based models, preference elicitation, and heuristic search, we will first focus on related work in this intersection before covering the three broader areas. Aside from the work proposed by Tabakhi *et al.* [29] discussed in Sect. 1, the body of work that is most related to ours is the work on *Incomplete Weighted CSPs* (IWCSPs) [9,10,32]. IWCSPs can be seen as centralized versions of I-DCOPs. Researchers have proposed a family of algorithms based on depth-first branch-and-bound and local search to solve IWCSPs including heuristics that can be parameterized like ours. Aside from IWCSPs, similar centralized constraint-based models include *Incomplete Fuzzy CSPs* and *Incomplete Soft Constraint Satisfaction Problems*.

In the context of the broader constraint-based models where constraints may not be fully specified, there are a number of such models, including *Uncertain CSPs* [38], where the outcomes of constraints are parameterized; *Open CSPs* [4], where the domains of variables and constraints are incrementally discovered; *Dynamic CSPs* [3], where the CSP can change over time; as well as distributed variants of these models [17,23].

Table 1. Varying number of agents $|\mathcal{A}|$

(a) SyncBB Without Heuristic

$|\mathcal{A}|$	# unk. costs	Without Elicitation Costs				With Elicitation Costs					
		#elic.	runtime	const. cost	#exp. nodes	#elic.	runtime	total cost	const. cost	elic. cost	#exp. nodes
10	43	39.92	7.55E-01	51.88	1.65E+03	18.08	5.60E-02	189.28	60.20	129.08	6.70E+01
12	62	58.80	2.59E+00	75.48	6.76E+03	25.52	9.28E-02	264.88	87.16	177.72	1.25E+02
14	86	81.88	9.18E+00	107.40	2.40E+04	35.16	7.68E-02	363.80	123.36	240.44	9.24E+01
16	115	110.92	3.58E+01	145.40	9.51E+04	48.56	1.13E-01	485.40	163.80	321.60	1.60E+02
18	146	139.80	1.24E+02	184.44	3.54E+05	60.76	1.29E-01	621.04	206.68	414.36	3.08E+02
20	182	175.56	5.52E+02	231.64	1.36E+06	72.84	1.63E-01	741.76	252.88	488.88	2.79E+02

(b) SyncBB with CAC Heuristic

$|\mathcal{A}|$	# unk. costs	#elic.	runtime	const. cost	#exp. nodes	#elic.	runtime	total cost	const. cost	elic. cost	#exp. nodes
10	43	37.96	3.63E-01	51.88	7.73E+02	12.96	2.15E-02	173.88	61.64	112.24	2.21E+01
12	62	57.32	1.22E+00	75.48	2.97E+03	18.32	1.82E-02	242.12	88.72	153.40	2.99E+01
14	86	80.32	3.58E+00	107.40	9.50E+03	26.08	3.69E-02	350.48	125.48	225.00	3.90E+01
16	115	110.80	1.61E+01	145.40	4.26E+04	35.56	3.03E-02	464.16	165.24	298.92	4.83E+01
18	146	139.12	4.16E+01	184.44	1.21E+05	44.84	4.75E-02	590.20	206.64	383.56	6.11E+01
20	182	164.76	3.67E+02	231.64	4.09E+05	54.52	6.29E-02	722.68	258.28	464.40	5.06E+01

In the context of the broader preference elicitation area, there is a very large body of work [11], and we focus on techniques that are most closely related to our approach. They include techniques that ask users a number of preset questions [28,29,33] as well as send alerts and notification messages to interact with users [2], techniques that ask users to rank alternative options or user-provided option improvements to learn a (possibly approximately) user preference function [1], and techniques that associate costs to eliciting preferences and takes these costs into account when identifying which preference to elicit as well as when to stop eliciting preferences [16,34]. The key difference between all these approaches and ours is that they identify preferences to elicit a priori before the search while we embed the preference elicitation in the underlying DCOP search algorithm.

Finally, in the context of the broader heuristic search area, starting with Weighted A* [26], researchers have long used weighted heuristics to speed up the search process in general search problems. Researchers have also investigated the use of dynamically-changing weights [27]; using weighted heuristic with other heuristic search algorithms like DFBnB [8], RBFS [15], and AND/OR search [20]; as well as extending them to provide anytime characteristics [12].

9   Empirical Evaluations

We evaluate SyncBB using the CAC heuristic and ALS-MGM using the NHC heuristic against their baselines without heuristics on I-DCOPs with and without elicitation costs. We evaluate them on random graphs where we measure the various costs of the solutions found – the cumulative constraint costs (i.e., const. cost), cumulative elicitation costs (i.e., elic. cost), and their aggregated total costs (i.e., total cost) – the number of unknown costs elicited (i.e., # elic.), the number of nodes expanded after SyncBB terminates (i.e., # exp. nodes), and the runtimes of the algorithms (in sec). In all experiments we set $\alpha_f = \alpha_e = 0.5$. Data points are averaged over 25 instances. We generate 25 random (binary) graphs, where we vary the number of agents/variables $|\mathcal{A}|$ from 10 to 180; the user-defined relative weight w from 1 to 10; and the user-defined additive weight ϵ from 0 to 50. The constraint density p_1 is set to 0.4, the tightness p_2 is set to 0; the fraction of unknown costs in the problem is set to 0.6. In our experiments below, we only vary one parameter at a time, setting the rest at their default values: $|\mathcal{A}| = 10$, $|D_i| = 2$, $w = 1$, and $\epsilon = 0$. All constraint costs are randomly sampled from $[2, 5]$ and all elicitation costs are randomly sampled from $[0, 20]$. As mentioned earlier, in the ALS-MGM algorithm, the number of steps that the algorithm needs to run before termination is equal to $m + H$, where m is the number of steps that a regular MGM algorithm would run and H is the height of the BFS tree. Since the ALS framework requires that $m \geq H$, we vary m from H to $H + 240$.

Table 1 tabulates our empirical results, where we vary the number of agents $|\mathcal{A}|$. Figure 2(b) plots the convergence rate of ALS-MGM when elicitation is free. We make the following observations:

- As expected, the runtimes and number of unknown costs elicited by all algorithms increase with increasing the number of agents $|\mathcal{A}|$. The reason is that the size of the problem, in terms of the number of constraints in the problem, increases with increasing $|\mathcal{A}|$.
- On problems without elicitation costs, SyncBB with CAC is faster than without CAC. The reason is the following: The CAC heuristic value includes estimates of not only all constraints between its descendant agents, but also constraints between any of its descendant agents with any of its ancestor agents. The CAC heuristic is thus likely to be more informed and provide better estimates.
- On problems with elicitation costs, SyncBB with CAC is still faster than without CAC. The reason is that the number of nodes expanded is significantly smaller with CAC than without CAC.
- Overall, the use of heuristics in conjunction with SyncBB reduces the number of unknown costs elicited by up to 22% and the runtime by up to 57% when elicitation is not free. Therefore, these results highlight the strengths of using our proposed heuristics for solving I-DCOPs.

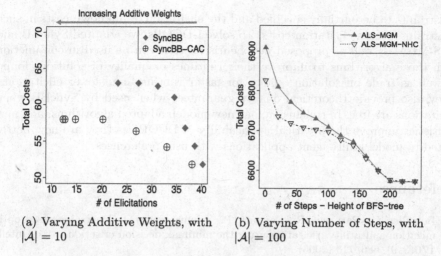

(a) Varying Additive Weights, with $|\mathcal{A}| = 10$

(b) Varying Number of Steps, with $|\mathcal{A}| = 100$

Fig. 2. Evaluation of ALS-MGM and SyncBB on random graphs

Figure 2(a) plots our empirical results, where we vary the user-defined additive bound (weight) ϵ for the problems when elicitation is free (i.e., all elicitation costs are zero). Additive weights increase from right to left on the top axis of the Figure. Each data point in the figures thus shows the result for one of the algorithms with one of the values of ϵ. Data points for smaller values of ϵ are in the bottom right of the figures and data points for larger values are in the top left of the figures. We plot the tradeoffs between total cost (=cumulative constraint and elicitation costs) and number of elicited costs. As expected, as the additive bound ϵ increases, the number of elicitations decreases. However, this comes at the cost of larger total costs. Between the two algorithms, SyncBB with CAC is the best. We omit plots of results where we vary the relative weight w as their trends are similar to those shown here, and we also omit plots of results with elicitation costs as their trends are similar to those without elicitation costs for both additive and relative weights.

Figure 2(b) clearly shows that the difference in the quality of solutions is largest at the start of the algorithm and decreases as the algorithm runs more steps. Therefore, the heuristic is ideally suited for time-sensitive applications with short deadlines, where there is not enough time to let ALS-MGM run for a long time until convergence.

10   Conclusions

Distributed Constraint Optimization Problems (DCOPs) have been used to model a variety of cooperative multi-agent problems. However, they assume that all constraints are fully specified, which may not hold in applications where constraints encode preferences of human users. To overcome this limitation, we proposed *Incomplete DCOPs* (I-DCOPs), which extends DCOPs by allowing some constraints to be partially specified and the elicitation of unknown costs in such constraints incurs elicitation costs. To solve I-DCOPs, we adapted SyncBB and ALS-MGM as well as proposed new heuristics that can be used in conjunction with those algorithms to improve their runtimes or quality of solutions found as well as trade off solution quality for faster runtimes and fewer elicitations. They also provide theoretical quality guarantees when used by SyncBB when elicitations are free. In conclusion, our new model, adapted algorithms, and new heuristics improve the practical applicability of DCOPs as they are now better suited to model multi-agent applications with user preferences.

References

1. Boutilier, C., Patrascu, R., Poupart, P., Schuurmans, D.: Constraint-based optimization and utility elicitation using the minimax decision criterion. Artif. Intell. **170**(8–9), 686–713 (2006)
2. Costanza, E., Fischer, J.E., Colley, J.A., Rodden, T., Ramchurn, S.D., Jennings, N.R.: Doing the laundry with agents: a field trial of a future smart energy system in the home. In: CHI, pp. 813–822 (2014)
3. Dechter, R., Dechter, A.: Belief maintenance in dynamic constraint networks. In: AAAI, pp. 37–42 (1988)
4. Faltings, B., Macho-Gonzalez, S.: Open constraint programming. Artif. Intell. **161**(1–2), 181–208 (2005)
5. Farinelli, A., Rogers, A., Petcu, A., Jennings, N.: Decentralised coordination of low-power embedded devices using the Max-Sum algorithm. In: AAMAS, pp. 639–646 (2008)
6. Fioretto, F., Pontelli, E., Yeoh, W.: Distributed constraint optimization problems and applications: a survey. J. Artif. Intell. Res. **61**, 623–698 (2018)
7. Fioretto, F., Yeoh, W., Pontelli, E.: A multiagent system approach to scheduling devices in smart homes. In: AAMAS, pp. 981–989 (2017)
8. Flerova, N., Marinescu, R., Dechter, R.: Weighted heuristic anytime search: new schemes for optimization over graphical models. Ann. Math. Artif. Intell. **79**, 77–128 (2017)
9. Gelain, M., Pini, M.S., Rossi, F., Venable, K.B., Walsh, T.: Elicitation strategies for soft constraint problems with missing preferences: properties, algorithms and experimental studies. Artif. Intell. **174**(3–4), 270–294 (2010)
10. Gelain, M., Pini, M.S., Rossi, F., Venable, K.B., Walsh, T.: A local search approach to solve incomplete fuzzy csps. In: ICAART, pp. 582–585 (2011)
11. Goldsmith, J., Junker, U.: Preference handling for artificial intelligence. AI Mag. **29**(4), 9–12 (2008)
12. Hansen, E.A., Zhou, R.: Anytime heuristic search. J. Artif. Intell. Res. **28**, 267–297 (2007)
13. Hart, P., Nilsson, N., Raphael, B.: A formal basis for the heuristic determination of minimum cost paths. IEEE Trans. Syst. Sci. Cybernet. SSC **4**(2), 100–107 (1968)
14. Hirayama, K., Yokoo, M.: Distributed partial constraint satisfaction problem. In: CP, pp. 222–236 (1997)
15. Korf, R.: Linear-space best-first search. Artifi. Intell. **62**(1), 41–78 (1993)
16. Le, T., Tabakhi, A.M., Tran-Thanh, L., Yeoh, W., Son, T.C.: Preference elicitation with interdependency and user bother cost. In: AAMAS, pp. 1459–1467 (2018)

17. Léauté, T., Faltings, B.: Distributed constraint optimization under stochastic uncertainty. In: AAAI, pp. 68–73 (2011)
18. Maheswaran, R., Pearce, J., Tambe, M.: Distributed algorithms for DCOP: a graphical game-based approach. In: PDCS, pp. 432–439 (2004)
19. Maheswaran, R., Tambe, M., Bowring, E., Pearce, J., Varakantham, P.: Taking DCOP to the real world: efficient complete solutions for distributed event scheduling. In: AAMAS, pp. 310–317 (2004)
20. Marinescu, R., Dechter, R.: AND/OR branch-and-bound search for combinatorial optimization in graphical models. Artif. Intell. **173**(16–17), 1457–1491 (2009)
21. Miller, S., Ramchurn, S., Rogers, A.: Optimal decentralized dispatch of embedded generation in the smart grid. In: AAMAS, pp. 281–288 (2012)
22. Modi, P., Shen, W.M., Tambe, M., Yokoo, M.: ADOPT: asynchronous distributed constraint optimization with quality guarantees. Artif. Intell. **161**(1–2), 149–180 (2005)
23. Nguyen, D.T., Yeoh, W., Lau, H.C.: Stochastic dominance in stochastic DCOPs for risk-sensitive applications. In: AAMAS, pp. 257–264 (2012)
24. Nguyen, D.T., Yeoh, W., Lau, H.C., Zivan, R.: Distributed Gibbs: a linear-space sampling-based DCOP algorithm. J. Artif. Intell. Res. **64**, 705–748 (2019)
25. Petcu, A., Faltings, B.: A scalable method for multiagent constraint optimization. In: IJCAI, pp. 1413–1420 (2005)
26. Pohl, I.: Heuristic search viewed as path finding in a graph. Artif. Intell. **1**(3–4), 193–204 (1970)
27. Sun, X., Druzdzel, M., Yuan, C.: Dynamic weighting A* search-based MAP algorithm for Bayesian networks. In: IJCAI, pp. 2385–2390 (2007)
28. Tabakhi, A.M.: Preference elicitation in DCOPs for scheduling devices in smart buildings. In: AAAI, pp. 4989–4990 (2017)
29. Tabakhi, A.M., Le, T., Fioretto, F., Yeoh, W.: Preference elicitation for DCOPs. In: CP, pp. 278–296 (2017)
30. Tabakhi, A.M., Xiao, Y., Yeoh, W., Zivan, R.: Branch-and-bound heuristics for incomplete DCOPs. In: AAMAS, pp. 1677–1679 (2021)
31. Tabakhi, A.M., Yeoh, W., Fioretto, F.: The smart appliance scheduling problem: A Bayesian optimization approach. In: (PRIMA), pp. 100–115 (2020)
32. Tabakhi, A.M., Yeoh, W., Yokoo, M.: Parameterized heuristics for Incomplete weighted CSPs with elicitation costs. In: AAMAS, pp. 476–484 (2019)
33. Trabelsi, W., Brown, K.N., O'Sullivan, B.: Preference elicitation and reasoning while smart shifting of home appliances. Energy Procedia **83**, 389–398 (2015)
34. Truong, N.C., Baarslag, T., Ramchurn, S.D., Tran-Thanh, L.: Interactive scheduling of appliance usage in the home. In: IJCAI, pp. 869–877 (2016)
35. Vinyals, M., Rodríguez-Aguilar, J., Cerquides, J.: Constructing a unifying theory of dynamic programming DCOP algorithms via the generalized distributive law. J. Autonom. Agents Multi-agent Syst. **22**(3), 439–464 (2011)
36. Xiao, Y., Tabakhi, A.M., Yeoh, W.: Embedding preference elicitation within the search for DCOP solutions. In: AAMAS, pp. 2044–2046 (2020)
37. Yeoh, W., Felner, A., Koenig, S.: BnB-ADOPT: an asynchronous branch-and-bound DCOP algorithm. J. Artif. Intell. Res. **38**, 85–133 (2010)
38. Yorke-Smith, N., Gervet, C.: Certainty closure: a framework for reliable constraint reasoning with uncertainty. In: CP, pp. 769–783 (2003)
39. Zhang, W., Wang, G., Xing, Z., Wittenburg, L.: Distributed stochastic search and distributed breakout: properties, comparison and applications to constraint optimization problems in sensor networks. Artif. Intell. **161**(1–2), 55–87 (2005)

40. Zivan, R., Okamoto, S., Peled, H.: Explorative anytime local search for distributed constraint optimization. Artif. Intell. **212**, 1–26 (2014)
41. Zivan, R., Yedidsion, H., Okamoto, S., Glinton, R., Sycara, K.: Distributed constraint optimization for teams of mobile sensing agents. Auto. Agents Multi-agent Syst. **29**(3), 495–536 (2014). https://doi.org/10.1007/s10458-014-9255-3

Securities Based Decision Markets

Wenlong Wang[✉] and Thomas Pfeiffer

Massey University, Auckland, New Zealand
{W.Wang1,T.Pfeiffer}@massey.ac.nz

Abstract. Decision markets are mechanisms for selecting one among a set of actions based on forecasts about their consequences. Decision markets that are based on scoring rules have been proven to offer incentive compatibility analogous to properly incentivised prediction markets. However, in contrast to prediction markets, it is unclear how to implement decision markets such that forecasting is done through the trading of securities. We here describe such a securities based implementation, and show that it offers the same expected payoff as the corresponding scoring rules based decision market. The distribution of realised payoffs however might differ, which allows more flexibility in shaping worst-case losses. Our analysis expands the knowledge on forecasting based decision making and provides novel insights for intuitive and easy-to-use decision market implementations.

Keywords: Decision markets · Prediction markets · Mechanism design

1 Introduction

Prediction markets [1,2,5,11,15–17,20,22,23,25] are popular tools for aggregating distributed information into often highly accurate forecasts. Participants in prediction markets trade contracts with payoffs tied to the outcome of future events. The pricing of these contracts reflects aggregated information about the probabilities associated with the possible outcomes. A frequently used contract type is Arrow-Debreu securities that pay $1 when a particular outcome is realised, and otherwise pay $0. If such a security is traded at $0.30, this can be interpreted as forecast for that outcome to occur at 30% chance. Potential caveats with the interpretation of prices in prediction markets as probabilities have been discussed in the literature [17,25], but are not seen as critical for typical applications [15,17,25]. Prediction markets have been extensively investigated in lab based experiments and real world settings [2,11,17,20–23,25].

In many practical prediction markets applications, such as recreational markets on political events, participants trade directly with each other, and one participant's gain is the other participant's loss. Prediction markets can, however, also be designed to offer net benefits to the participants. Such incentivised prediction markets can be used by a market creator who is willing to compensate the market participants for the information obtained from the market [5,15,16]. Incentivised prediction markets rely on market maker algorithms to trade with

© Springer Nature Switzerland AG 2022
J. Chen et al. (Eds.): DAI 2021, LNAI 13170, pp. 79–92, 2022.
https://doi.org/10.1007/978-3-030-94662-3_6

the participants, and on cost functions to update prices based on past trans-actions. These functions are closely related to proper scoring rules such as the Brier (or quadratic) scoring rule and the logarithmic scoring rule [3,13], which measure the accuracy of forecasts and allow rewarding a single expert based on forecast and actual outcome. The market maker in an incentivised prediction market subsidises the entire market rather than single experts; its worst-case loss is finite and its expected loss depends on how much the participants 'improve' on the information entailed by the initial market maker pricing [15].

Accurate forecasts, as obtained from prediction markets, can be of tremen-dous value for decision makers. Commercial companies, for instance, can benefit substantially from accurate forecasts regarding the future demand for their prod-ucts. However, many decision-making problems require conditional forecasts [8]. To decide, for instance, between alternative marketing campaigns, a company needs to understand how each of the alternatives will affect sales. In other words, it needs to predict, and choose between, 'alternative futures'. To implement such forecasting based decision making, Hanson [14] proposed so called decision mar-kets. While it is non-trivial to properly incentivise participants to provide their information in decision markets, it has been shown that this can be achieved [4,6–8,18,19].

Properly incentivised decision markets work in a stepwise process to select one among a number of mutually exclusive actions. First, forecasts about the expected future consequences of each action are elicited in a step analogous to incentivised prediction markets. Second, a decision rule is used to select an action based on the forecasted consequences. Once an action has been selected, and its consequences are revealed, payoffs are provided for the forecasts as elicited in the first step. Importantly, the decision rule in properly incentivised decision markets is stochastic, with each action being picked with a strictly positive probability [7,8]. Payoffs are scaled up to ensure that the participants' expected payoffs in decision markets remain analogous to those made in properly incen-tivised prediction markets [7,8], and that game-theoretical results on strategic interactions between participants in prediction markets [5] carry over.

The literature on decision markets has so far focused on implementations based on scoring rules. For prediction markets it is well established how to imple-ment properly incentivised forecasting such that forecasts are made through the trading of securities. A similar securities based decision market implementa-tion has however not yet been described. Such an implementation is important because participants in decision markets are likely familiar with ordinary asset trading, and it is thus convenient for them to report their forecasts through a securities trading interface. Furthermore, securities based decision markets sim-plify managing liabilities, because the payment for the purchased securities cov-ers the traders' worst-case loss. We here propose such a securities based decision market setting, and compare it to existing, scoring rule based decision markets.

The remaining manuscript is organised as follows: In Sect. 2 we briefly intro-duce scoring rules, sequentially shared scoring rules, prediction markets and scoring rule based decision markets. In Sect. 3, we describe a securities based market design and compare it with the existing scoring rule based decision markets. In Sect. 4 we investigate the difference between our design and the scoring rule based decision markets mechanism in terms of worst-case losses. Finally, in Sect. 5, we conclude and discuss our future work. An extended version of this paper can be found in [24], and elaborates further on the difference between the proposed design and scoring rule based decision markets.

2   Related Work and Notation

2.1   Scoring Rules

Let us define Ω as a finite set of mutually exclusive and exhaustive outcomes $\{\omega_1, \omega_2, ..., \omega_n\}$. A probabilistic prediction for those outcomes is denoted by $\vec{r} = (r_1, r_2, ..., r_n)$ with $\sum_{x=1}^{n} r_x = 1$ and $r_i \in [0, 1]$. A scoring function $s_i(\vec{r})$ allows to quantify the accuracy of prediction $\vec{r}$ once the outcome ω_i materialises [12].

Scoring rules allow to incentivise forecasters for predictions. Denoting the reported distribution as $\vec{r} = (r_1, r_2, \ldots, r_n)$ and the forecasters' belief as $\vec{p} = (p_1, p_2, \ldots, p_n)$, the expected payoff for a forecaster is $G(\vec{p}, \vec{r}) = \sum_{k=1}^{n} p_k s_k(\vec{r})$. A scoring rule is defined as proper if a forecaster maximises his/her expected payoff by truthfully reporting what he/she believes, i.e. $G(\vec{p}, \vec{p}) \geq G(\vec{p}, \vec{r})$. Furthermore, a scoring rule is strictly proper if $G(\vec{p}, \vec{p}) > G(\vec{p}, \vec{r})$ for all $\vec{r} \neq \vec{p}$.

2.2   Sequentially Shared Scoring Rules

Because information is often distributed across multiple agents, it is of interest to expand proper scoring to elicit forecasts from groups of forecasters. In his work on incentivised prediction markets, Hanson proposed a mechanism to sequentially elicit information from forecasters [15,16]. The mechanism keeps a current report $\vec{r}$ and offers a contract for a new report $^*\vec{r}$ to be scored as $s_i(^*\vec{r}) - s_i(\vec{r})$ if the outcome ω_i is observed. Note that $s_i(^*\vec{r}) - s_i(\vec{r})$ is a proper scoring rule if s_i is a proper scoring rule. Once a forecaster accepts the offer, the decision maker will update the current report from $\vec{r}$ to $^*\vec{r}$ and allow a next forecaster to further modify the new current report. Because the amount $s_i(^*\vec{r}) - s_i(\vec{r})$ can be negative, the offer requires the forecaster to demonstrate the capability to afford the worst-case loss, i.e., by depositing money in escrow. We refer to the necessity of such a demonstration as liabilities. Under such a sequentially shared scoring rule, forecasters are scored for how much they improve or worsen the current report. Such a mechanism uses incentives efficiently in that it avoids paying for the same information twice.

2.3   Securities Based Prediction Markets

The mechanism described in Sect. 2.2 involves a two-sided liability. That is, the decision maker is liable to pay each forecaster who improves a forecast, and forecasters are liable to pay the decision maker if they worsen a forecast.

It is often considered convenient to implement sequentially shared scoring rules through the trading of Arrow-Debreu securities [15,16]. In such an implementation, forecasters purchase securities from the market maker and their payments cover their liabilities. Another reason to use securities based trading is that the majority of existing real-world prediction markets, such as recreational markets on sports or political events, are trading securities in a double auction process. Traders who are familiar with these prediction markets will prefer an interface to be expressed in terms of trading with securities.

To incentivise traders, securities are bought and sold by a market creator. The market creator uses a market maker algorithm which keeps track of past trades and sets security prices derived from a cost function. The total amount spent on purchasing a particular quantity of securities can be calculated from this cost function. We denote the quantity of outstanding securities as $\vec{q} = (q_1, q_2, \ldots, q_n)$ for a market on n mutually exclusive and exhaustive outcomes Ω. Element q_i represents the number of securities sold by the market creator that pay if outcome ω_i is observed. The instantaneous prices of the securities with outstanding quantities $\vec{q}$ are denoted as $\vec{r}(\vec{q})$ and play the same role as reports in scoring rule based markets.

Assume a trader wants to change the outstanding securities distribution from $\vec{q}$ to $^*\vec{q}$ by buying securities to change the price from $\vec{r}$ to $^*\vec{r}$. The cost for the trader to purchase the amount of securities $^*\vec{q} - \vec{q}$ can be calculated from $C(^*\vec{q}) - C(\vec{q})$, where $C(\vec{q})$ denotes a cost function. Once the final event, i.e., ω_i is observed, the market maker will resolve the market by paying \$1 for each winning security. If the trader holds $^*q_i - q_i$ securities when the market is resolved, his/her payout will be \$$(^*q_i - q_i)$. Overall the realised payoff for the trader will be $(^*q_i - q_i) - (C(^*\vec{q}) - C(\vec{q}))$. Chen and Pennock generalised the relationship between cost functions, price functions and scoring rules, and proposed three equations that establish their equivalence [9]:

$$\begin{cases} s_i(\vec{r}) = q_i - C(\vec{q}) & \forall i \\ \sum_i r_i(\vec{q}) = 1 \\ r_i(\vec{q}) = \frac{\partial C(\vec{q})}{\partial q_i} \end{cases} \tag{1}$$

Furthermore, Chen and Vaughan proved that there exist a one-to-one mapping between any strictly proper scoring rule and cost function in securities based prediction markets and such a securities based market is incentive compatible [10]. Elicitation through scoring rule based and securities based prediction markets offers the same payoffs for participants when the markets start with the same initial forecasts and end with the same final forecasts. Cost functions $C(\vec{q})$ and price functions $r_k(\vec{q})$ have the following properties:

$$\begin{aligned} C(\vec{q} + \beta\vec{1}) &= \beta + C(\vec{q}) \\ r_k(\vec{q} + \beta\vec{1}) &= r_k(\vec{q}) \end{aligned} \tag{2}$$

where β is a real constant [9]. These properties imply that the same report can be made through different trades. If a trade $^*\vec{q} - \vec{q}$ changes market prices from $\vec{r}$ to $^*\vec{r}$, so does a trade $^*\vec{q} + \beta\vec{1} - \vec{q}$. This permits the trader to make any report by buying contracts from the market creator. Short selling is not required. The overall payoff will not be affected by the choice of β, because both costs of purchasing the contracts, and the payout from the contracts at resolution increase by the same amount.

2.4   Decision Markets

The design of decision markets expands prediction markets to use conditional forecasts for decision making. Decision markets consist of two components. The first component is a set of conditional prediction markets, each of which elicits the forecasts for one of the actions. The second component is the decision rule that defines—based on conditional prediction markets forecasts—how the final decision will be made. An example is the MAX decision rule [19] which is to always select the action that has the highest predicted probability for a desired outcome to occur. Decision markets with deterministic rules such as the MAX decision rule do not always properly incentivise a forecaster to truthfully report irrespective of the scoring rule it uses [7,19]. An intuitive example to illustrate how a trader can benefit from misreporting is given in [19]. Chen et al. described that a stochastic decision rule can myopically incentivise forecasters to truthfully report [8]. This approach is rephrased in the following in the notation used throughout this paper to allow for straight forward comparison with the securities-based implementation.

Definition 1. *In a decision market, the market creator has a finite set of m actions $\mathcal{A} = \{\alpha_1, \alpha_2, ..., \alpha_m\}$ to choose from. For each action α_j, there is a set of possible outcomes $\Omega_j = \{\omega_1^j, \omega_2^j, \ldots, \omega_{n_j}^j\}$, which n_j is the number of possible outcomes for action α_j. Both action set $\mathcal{A}$ and outcome sets Ω_j are collectively exhaustive and mutually exclusive. A stochastic decision rule $\vec{\phi}$ assigns a probability ϕ_k to each action α_k with $\phi_k > 0$ and $\sum_{k=1}^{m} \phi_k = 1$.*

Note that in our notation the outcome ω_i^j for action α_j can be unrelated to ω_i^k for action α_k. In other words, outcomes can be specific to the actions. The sets for the outcome of two actions can be completely disjoint. The decision rule can take the final report into account, i.e., $\vec{\phi} = \vec{\phi}(\vec{r}_1, \vec{r}_2, \ldots, \vec{r}_m)$ where $\vec{r}_1, \vec{r}_2, \ldots, \vec{r}_m$ are the final reports over the m different actions. It can, for instance, approximate the MAX decision rule by assigning high probabilities to actions with desirable outcomes.

Similar to scoring rules in prediction markets, decision scoring rules can be defined to map forecasts, decisions and outcomes to a real number. For simplicity, we will denote this score as $S_i^j(\vec{r}_j)$ that the selected action is α_j and the observed event is ω_i^j. Assume $s_i^j(\vec{r}_j)$ is a strictly proper scoring rule for conditional market j. A decision score for changing the current report from $\vec{r}_j$ to $^*\vec{r}_j$ is given by:

$$S_i^j(^*\vec{r}_j) - S_i^j(\vec{r}_j) = \frac{1}{\phi_j}\left(s_i^j(^*\vec{r}_j) - s_i^j(\vec{r}_j)\right) \tag{3}$$

The expected payoff G of a forecaster in a scoring rule based decision market as defined in Definition 1 is given by:

$$G = \sum_{j=1}^{m} \phi_j \sum_{i=1}^{n_j} p_i^j \frac{1}{\phi_j} \left(s_i^j({}^*\vec{r}_j) - s_i^j(\vec{r}_j) \right) = \sum_{j=1}^{m} \sum_{i=1}^{n_j} p_i^j \left(s_i^j({}^*\vec{r}_j) - s_i^j(\vec{r}_j) \right) \qquad (4)$$

where p_i^j denotes the belief of the forecaster which will be identical to ${}^*r_j$ if the scoring rule is strictly proper. The forecaster has the same expected payoff as if he/she participated in m independent and strictly proper prediction markets. Moreover, findings on strategic interaction between traders and incentives for instantaneous revelation of information from [5] apply as well.

Note that ϕ_j in Eqs. (3) and (4) is the probability for the selected action in the decision rule after the final report. Equation (4) shows that the value of ϕ_j does not affect the expected payoff that risk-neutral forecasters seek to maximise. This is important because for scoring rules that depend on the final report, no participant except for the final forecaster knows the value of ϕ_j. To provide truthful forecasts forecasters do not need the decision rule ϕ and its dependence on the final report as long as they can trust that the rule has full support.

3   Strictly Proper Securities Based Decision Markets

In Sect. 2.3, we discuss the advantages of implementing forecasting through the trading of securities. We here formulate a cost function for securities based decision markets that offers the same expected payoff for participants as a scoring rule based decision market. In a prediction market, the cost function and price function can be calculated by solving the Eqs. (1) [9]. However, because only decision markets with a stochastic decision rule are myopically incentive compatible, the stochastic decision rule needs to be accounted for.

3.1   Design

We adopt the cost function approach for prediction markets as described in Sect. 2.3. To account for the stochastic decision rule, the securities traded in this market have payoffs that depend on the selected action.

Definition 2. *In addition of the notation in Definition 1, we denote outstanding securities as $\vec{q}_j = (q_1^j, q_2^j, \ldots, q_{n_j}^j)$ for the conditional market for action α_j. Element q_i^j represents the number of securities sold by the market creator that pay if action α_j is selected and outcome ω_i^j is observed. The payout per security is denoted by v_j, and can depend on the selected action, but is the same for all traders and does not depend on the observed outcome. The payout for all other securities is zero. Cost function, price function and corresponding scoring rule for the conditional market on action α_j and outcome ω_i^j, are denoted by $C_j(\vec{q}_j)$, $r_i^j(\vec{q}_j)$ and $s_i^j(\vec{r}_j)$, respectively, and together fulfil Eq. 4.*

Theorem 1. *Let a trader in a securities based decision market as defined in Definition 2 make a trade $^*\vec{q}_j - \vec{q}_j$ to move prices from $\vec{r}_j(\vec{q}_j)$ to $\vec{r}_j(^*\vec{q}_j)$. Then the trader will have the same expected payoff as a forecaster who makes the same forecast in a scoring rule based decision market as described in Eq. (4) if and only if we set $v_j = 1/\phi_j$ for all action α_j.*

Proof. Let a forecaster in the a scoring rule based decision market change the reports from $\vec{r}_k$ to $^*\vec{r}_k$ for any action α_k. The expected payoff of the forecaster is denoted as G and is given in the Eq. (4). Let a trader in our securities based decision market change the outstanding securities distribution from $\vec{q}_k$ to $^*\vec{q}_k$ for each action α_k such that prices change from $\vec{r}_k$ to $^*\vec{r}_k$. Then the realised payoff the trader gains from such a trade is given by:

$$v_j\left(^*q_i^j - q_i^j\right) - \sum_{k=1}^m \left(C_k(^*\vec{q}_k) - C_k(\vec{q}_k)\right)$$

where the selected action is α_j and the observed outcome is ω_i^j. The expected payoffs of the trader is denoted as $\hat{G}$ and we obtain:

$$
\begin{aligned}
\hat{G} &= \sum_{j=1}^m \phi_j \sum_{i=1}^{n_j} p_i^j \left(v_j \left(^*q_i^j - q_i^j\right) - \sum_{k=1}^m \left(C_k(^*\vec{q}_k) - C_k(\vec{q}_k)\right) \right) \\
&= \sum_{j=1}^m \phi_j v_j \sum_{i=1}^{n_j} p_i^j \left(^*q_i^j - q_i^j\right) - \sum_{k=1}^m \left(C_k(^*\vec{q}_k) - C_k(\vec{q}_k)\right)
\end{aligned}
\tag{5}
$$

Substituting Eq. (1) into the Eq. (5), we obtain:

$$
\begin{aligned}
\hat{G} &= \sum_{j=1}^m \sum_{i=1}^{n_j} p_i^j \left(^*q_i^j - q_i^j\right) - \sum_{k=1}^m \left(C_k(^*\vec{q}_k) - C_k(\vec{q}_k)\right) + \sum_{j=1}^m (\phi_j v_j - 1) \sum_{i=1}^{n_j} p_i^j \left(^*q_i^j - q_i^j\right) \\
&= G + \underbrace{\sum_{j=1}^m (\phi_j v_j - 1) \sum_{i=1}^{n_j} p_i^j \left(^*q_i^j - q_i^j\right)}_{a}
\end{aligned}
\tag{6}
$$

The expected payoff $\hat{G}$ in a securities based market is equal to the expected payoff G in a scoring rule based market if and only if term a in Eq. (6) is zero. One way to achieve this for arbitrary trades is to set the payoffs of the contracts v_j to $1/\phi_j$. Thus $G = \hat{G}$ if $v_j = 1/\phi_j$

An alternative with $v_j \neq 1/\phi_j$ would be to choose v_j such that the vector (dot) product $\vec{a} \cdot \vec{b}$, with vector element a_j being defined as $\sum_{j=1}^m (\phi_j v_j - 1)$ and b_j being defined $\sum_{i=1}^{n_j} p_i^j \left(^*q_i^j - q_i^j\right)$, becomes zero. This however, would require to make v_j dependent on trade-specific quantities such as the $^*q_i^j$ and contradicts the properties of contract payoffs as defined in Definition 2.

3.2   Distribution of Realised Payoffs

Securities based decision markets and corresponding scoring rule based decision markets provide the identical expected payoff for participants under the same conditions. However, the actual distribution of payoffs for the participants are not necessarily the same. In this subsection, we will discuss the difference between securities based decision markets and the corresponding scoring rule based decision market in terms of realised payoffs for participants.

The realised payoffs for a forecaster who changes report $\vec{r_k}$ to $^*\vec{r_k}$ in a scoring rule based decision market is given by Eq. (3). In the securities based market, assume a trader makes a trade to change the price for any action α_k from $\vec{r_k}$ to $^*\vec{r_k}$. This trade changes the market creator inventory from $\vec{q_k}$ to $^*\vec{q_k}$ and has a cost given by $C_k(^*\vec{q_k}) - C_k(\vec{q_k})$. Let the market creator select the action α_j to execute and observe the outcome ω_i^j. Using Eq. (1) we obtain the realised payoffs for the trader in the securities based decision market:

$$
\frac{1}{\phi_j}\left(^*q_i^j - q_i^j\right) - \sum_{k=1}^{m}\left(C_k(^*\vec{q_k}) - C_k(\vec{q_k})\right)
$$
$$
= \frac{1}{\phi_j}\left(s_i^j(^*\vec{r_j}) - s_i^j(\vec{r_j})\right) + \frac{1}{\phi_j}\left(C_j(^*\vec{q_j}) - C_j(\vec{q_j})\right) - \underbrace{\sum_{k=1}^{m}\left(C_k(^*\vec{q_k}) - C_k(\vec{q_k})\right)}_{a}
$$

$$(7)$$

The term a in Eq. (7) shows that there is a difference in realised payoffs of participants between the securities based decision market and the scoring rule based decision market. This difference cancels out when computing the expected payoff of a participant. Although the sign of term a cannot be decided easily, term a will increase when the trader spends more in the selected conditional market and decrease when the trader spends more in the conditional markets that are not selected.

Equation (7) shows that the realised payoffs in our securities based decision market can be rewritten as a scoring rule with an additional 'lottery' that costs $I_k = C_k(^*\vec{q_k}) - C_k(\vec{q_k})$ and returns I_j/ϕ_j at probability ϕ_j. In the extended version of this paper we provide an example to illustrate the differences between the payoffs in securities based and scoring rule based decision markets [24]. In Sect. 4 we will show how the additional flexibility in the design of securities based decision markets can be used to reallocate worst-case losses between traders and the market creator.

4   Worst-Case Losses for Participants and Market Creator

An analysis of worst-case losses is crucial for practical implementation because it needs to be ensured that all liabilities can be properly resolved. A further purpose is to understand how liabilities can be distributed between market creator and participants.

4.1   Worst-Case Loss for Participants

Consider a forecaster in a scoring rule based decision market who reports $^*\vec{r_k}$ when the current prediction is $\vec{r_k}$ for each conditional market k. The worst-case loss for this report is $\min_{j,i}\left(S_i^j(^*\vec{r_j}) - S_i^j(\vec{r_j})\right) = \min_{j,i}\frac{1}{\phi_j}\left(s_i^j(^*\vec{r_j}) - s_i^j(\vec{r_j})\right)$. Thus we can tell that the worst-case loss for the forecaster depends on both the decision rule ϕ_j and the report $^*\vec{r_j}$ he/she made. The probability ϕ_j depends on the decision rule. Small probabilities in the decision rule, which may be in the interest of market creator to approximate deterministic scoring rules, increase the worst-case loss for the forecaster.

A trader in a security based decision market purchase securities from $\vec{q_k}$ to $^*\vec{q_k}$. Assuming again that forecasters cannot hold negative positions, the worst-case loss for the trader can be calculated as $\sum_{k=1}^{m}\left(C_k(\vec{q_k}) - C_k(^*\vec{q_k})\right)$ and it shows that the worst-case loss for a trader in the securities based decision market only depends on the cost the trader spent. In other words, the trader in the securities based market will not be exposed to any liabilities beyond the costs already paid for purchasing the assets. Therefore a securities based implementation has the advantage that it does not need to further track the liabilities on the side of the traders. Moreover, the worst-case loss of trader does not depend on the decision rule.

4.2   Worst-Case Loss for Market Creator

The loss of a market creator mirrors the profits gained by the participants. Apart from the distribution of realised payoffs for the participants, there is therefore a difference of the worst-case loss for the market creators between a scoring rule based decision market and the corresponding securities based decision market.

Carrying over the conditions from Eq. (3), the worst-case loss for a market creator of a scoring rule based decision market is $\min_{j,i}\left(S_i^j(\vec{r_j}) - S_i^j(^*\vec{r_j})\right) = \min_{j,i}\frac{1}{\phi_j}\left(s_i^j(\vec{r_j}) - s_i^j(^*\vec{r_j})\right)$. Thus, the worst-case loss for a market creator depends on three factors: initial report $\vec{r_j}$ and final report $^*\vec{r_j}$ for the selected conditional market and the decision rule ϕ_j. Among three factors, the market creator has control over the initial report $\vec{r_j}$ and the value that decision rule ϕ_j can take, but does not have control over which action is being picked. Even though the decision rule can be arbitrary as long as forecasters are convinced that it has full support, it is the interest of the market creator to take the final forecasts of the market into account. For instance, it does not fit the interest of the market creator to assign a small probability to the action that is predicted to most likely lead to a desirable outcome. The relationship between decision rule ϕ_j and the final score $s_i^j(^*\vec{r_j})$ can be complex, because it depends on how exactly the final forecast determines the decision rule $\vec{\phi}$. There is a suggestion about computing a minimal feasible decision rule for each action according to the budget of market creator [6].

Using the conditions for Eq. (7), the worst-case loss for a market creator of a securities based decision market is:

$$\sum_{k=1}^{m} \left(C_k(^*\vec{q_k}) - C_k(\vec{q_k}) \right) - \max_{i,j} \left(\frac{1}{\phi_j} \left(^*q_i^j - q_i^j \right) \right) \tag{8}$$

In Eq. (8), the term $\sum_{k=1}^{m}(C_k(^*\vec{q_k}) - C_k(\vec{q_k}))$ is the income from securities sales, which mirrors the cost spent by participants in order to move the inventory distribution from $\vec{q_k}$ to $^*\vec{q_k}$ in each conditional market k. The second term, $\max_{i,j}(\frac{1}{\phi_j}(^*q_i^j - q_i^j))$ in the equation is the maximal payout that can be won by participants. In order to compare the worst-case losses, we substitute the Eq. (1) into the Eq. (8) and obtain:

$$\sum_{k=1}^{m} \left(C_k(^*\vec{q_k}) - C_k(\vec{q_k}) \right) - \max_{i,j} \left(\frac{1}{\phi_j} \left(^*q_i^j - q_i^j \right) \right)$$

$$= \underbrace{\sum_{k=1,k\neq j}^{m} \left(C_k(^*\vec{q_k}) - C_k(\vec{q_k}) \right)}_{a} - \max_{i,j} \left(\underbrace{s_i^j(^*\vec{r_j}) - s_i^j(\vec{r_j})}_{b} - \underbrace{\frac{(1-\phi_j)}{\phi_j} \left(^*q_i^j - q_i^j \right)}_{c} \right)$$

$$\tag{9}$$

Term a of Eq. (9) is non-negative, and depends on the sales in all conditional markets except for the one representing the selected action. Term b is the scoring rule that corresponds to our cost function and is bounded. However, term c depends on final outstanding securities $^*q_i^j$ and $(1 - \phi_j)/\phi_j$. While the market creator has control over ϕ_j, the final outstanding securities is not known ex ante. Therefore no finite initial escrow can guarantee to cover the market creator's liabilities. This loss of a bound on the worst-case loss of a market maker differs from the loss of a bound from low probabilities in the decision rule as described in [6]. The final budget for a market maker in a decision market does not have an upper limit because traders can buy arbitrarily large numbers of shares q_i^j on the selected action, while buying fewer (or no) shares on the other actions. Note that it is in the interest for the market creator to assign a small probability ϕ_j to actions that are not preferred. The term $(1 - \phi_j)/\phi_j$ increases rapidly as ϕ_j approaches zero. In summary, the advantage of a worst-case loss for the participants that does not depend on the decision rule thus comes at the disadvantage that the worst-case loss for the market creator cannot be known ex ante. A numeric example, which further demonstrates the difference in worst-case losses between different market designs, can be found in the extended version [24].

4.3   Re-allocation of Worst-Case Losses

Compared to scoring rule based decision markets, securities based decision markets offer additional flexibility to shape the distribution of realised payoffs. This flexibility arises because in a securities based decision market, each report $^*r$ can be realised through infinitely many trades. As outlined in Eq. (2), in prediction markets, if a trader purchases β security for each outcome, the cost will be β. The prices, i.e., the current forecast for the probability distribution over the outcomes, remains the same. The payout from these additional securities will be exact β regardless of outcomes, and the net realised payoff for such a trade will be zero.

In a securities based decision market this is, however, not the case. Assume a trader in a securities based decision market purchases a number of β_k of each outcome in conditional market k, for any k. We refer to such a trade as purchasing a 'bundle' of securities. Let the market creator select action α_j. Regardless of which outcome is observed, the realised payoff of the trader from the 'bundle' of securities can be obtained as $\frac{1}{\phi_j}\beta_j - \sum_{k=1}^{m}\beta_k$. While the prices for each outcome in all conditional markets remain the same, the realised payoff for the trader in a decision market is affected by these trades and depends on which action is selected. This property allows a trader to adjust the distribution of realised payoffs through purchasing bundles of securities without changing the reported probability distributions. Purchasing the same number of β_k of each outcome in conditional market k can also be viewed as the trader purchasing a 'lottery' ticket that costs β_k and returns β_k/ϕ_k at a probability of ϕ_k. The realised payoff of a trader in a securities based decision market can be rewritten as:

$$\frac{1}{\phi_j}(^*q_i^j - q_i^j) - \sum_{k=1}^{m}(C_k(^*\vec{q}_k) - C_k(\vec{q}_k)) + \underbrace{\frac{1}{\phi_j}\beta_j - \sum_{k=1}^{m}\beta_k}_{\text{Zero expected payoff}} \qquad (10)$$

For each specific report there is a manifold of trading strategies that link to it, each of which leading to a different distribution of payoffs. This property can be used to re-allocate worst-case losses between traders and market creator, or even move it to specialised traders acting as insurers. The flexibility to shape the distribution of realised payoffs can be used to design trades such that the payoffs in securities based decision markets match exactly those in scoring rule based markets. However, this requires the traders to accept negative positions, i.e., to short sell securities, and re-introduces two-sided liabilities. Let β_k in Eq. (10) to be substituted by $-\left(C_k(^*\vec{q}_k) - C_k(\vec{q}_k)\right)$ for all k. The realised payoff for such a trader simplifies to $\left(^*s_i^j(\vec{r}_j) - s_i^j(\vec{r}_j)\right)/\phi_j$, i.e., the realised payoff in the corresponding scoring rule based decision market. With such a trade, a trader makes a report through longing securities that she/he believes are under-priced and shorting securities on over-priced outcomes to meet the cost. The net cost for such a trade is zero.

Moreover, we can design the market such that it allocates worst-case losses entirely to the traders' side. This is done by allowing the traders to take on only short positions. In such a market traders short the securities that they believe are over-priced with the market creator. The market creator essentially purchases securities from the traders and covers the worst-case loss by paying for the cost of the securities.

Lastly, specialised traders that only accept negative position can act like insurers. In this case, worst-case losses can be separated from both traders who only long securities and the market creator. Such insurers essentially offers a 'lottery'. For real-world applications it might not be realistic to assume that this is done without additional costs. However, paying a small additional fee to such insurers might be in the interest of the market creator, because it allows to establish a bound for the worst-case loss. With such a bound in place, the market creator will be able to better approximate a deterministic decision rule.

5   Conclusion and Discussion

We introduce a setting for securities based decision markets that can be conveniently deployed in practical applications. In such a setting, a trader will report a forecast through trading securities. For the securities that represent the selected action and observed outcome, a trader will receive $1/\phi_j$ payoff per share, where ϕ_j is the probability in the decision rule corresponding to the selected action. Other shares pay zero, including those purchased in the unselected conditional markets. We prove that under the same condition, specifically, the same action space and the same outcome space, a securities based decision market in our setting has the same expected payoff for participants as the corresponding scoring rule based decision market.

A comparison between our setting and the corresponding scoring rule based decision market demonstrates that the realised payoffs depend on how much the participants report or trade in the selected conditional market. We find that the forecaster in scoring rule based decision market will have no cost for reporting in unselected conditional markets while this is not the case in the securities based decision market. Furthermore, with an additional 'lottery', a forecaster in scoring rule based decision market can have the identical realised payoffs as a trader in the corresponding securities based decision market. Similarly, the realised payoffs of a trader in a securities based decision market can recover the realised payoffs of the corresponding scoring rule based market, but this requires the forecasters to 'short-sell' securities, i.e., to hold negative positions.

By being equivalent to a scoring rule based decision market with an additional zero-mean 'lottery', the securities based mechanism described here offers an additional set of parameters that allow to shape the distribution of payoffs beyond what can be achieved based on scoring rules alone. This allows to re-allocate worst-case losses between forecasters and the market creator. As illustrated in Sect. 4, in a market where forecasters only purchase positive positions (no short selling), their liabilities are covered when paying for the purchased securities. Moreover, in contrast to scoring rule based decision markets, their worst-case losses do not depend on the probabilities used in the decision rule. A securities based decision market design might thus be of advantage for a market creator who aims to attract forecasters who are concerned about limiting their worst-case losses. Further empirical studies will be of value in determining how to shape trading to obtain the most accurate forecasts.

References

1. Arrow, K.J., et al.: The promise of prediction markets. Science **320**(5878), 877–878 (2008). https://doi.org/10.1126/science.1157679
2. Berg, J.E., Nelson, F.D., Rietz, T.A.: Prediction market accuracy in the long run. Int. J. Forecast. **24**(2), 285–300 (2008). https://doi.org/10.1016/j.ijforecast.2008.03.007
3. Bickel, J.E.: Some comparisons among quadratic, spherical, and logarithmic scoring rules. Decis. Anal. **4**(2), 49–65 (2007). https://doi.org/10.1287/deca.1070.0089
4. Boutilier, C.E.: Eliciting forecasts from self-interested experts: scoring rules for decision makers. In: 11th International Conference on Autonomous Agents and Multiagent Systems 2012, AAMAS 2012: Innovative Applications Track, vol. 2, pp. 1008–1015. International Foundation for Autonomous Agents and Multiagent Systems (IFAAMAS), Department of Computer Science, University of Toronto, Toronto (2012)
5. Chen, Y., et al.: Gaming prediction markets: equilibrium strategies with a market maker. Algorithmica **58**(4), 930–969 (2010). https://doi.org/10.1007/s00453-009-9323-2
6. Chen, Y., Kash, I., Ruberry, M., Shnayder, V.: Decision markets with good incentives. In: Chen, N., Elkind, E., Koutsoupias, E. (eds.) WINE 2011. LNCS, vol. 7090, pp. 72–83. Springer, Heidelberg (2011). https://doi.org/10.1007/978-3-642-25510-6_7
7. Chen, Y., Kash, I.A.: Information elicitation for decision making. In: 10th International Conference on Autonomous Agents and Multiagent Systems, vol. 1, pp. 161–168. International Foundation for Autonomous Agents and Multiagent Systems (IFAAMAS) (2011)
8. Chen, Y., Kash, I.A., Ruberry, M., Shnayder, V.: Eliciting predictions and recommendations for decision making. ACM Trans. Econ. Comput. **2**(2), 1–27 (2014). https://doi.org/10.1145/2556271
9. Chen, Y., Pennock, D.M.: A utility framework for bounded-loss market makers. In: Proceedings of the 23rd Conference on Uncertainty in Artificial Intelligence. Conference on Uncertainty in Artificial Intelligence, UAI 2007, pp. 49–56 (2007)
10. Chen, Y., Vaughan, J.W.: A new understanding of prediction markets via no-regret learning. In: Proceedings of the ACM Conference on Electronic Commerce, pp. 189–198 (2010). https://doi.org/10.1145/1807342.1807372
11. Dreber, A., et al.: Using prediction markets to estimate the reproducibility of scientific research. Proc. Natl. Acad. Sci. **112**(50), 15343–15347 (2015). https://doi.org/10.1073/PNAS.1516179112
12. Garthwaite, P.H., Kadane, J.B., O'Hagan, A.: Statistical methods for eliciting probability distributions. J. Am. Stat. Assoc. **100**(470), 680–701 (2005). https://doi.org/10.1198/016214505000000105
13. Gneiting, T., Raftery, A.E.: Strictly proper scoring rules, prediction, and estimation. J. Am. Stat. Assoc. **102**(477), 359–378 (2007). https://doi.org/10.1198/016214506000001437
14. Hanson, R.D.: Decision markets. IEEE Intell. Syst. **14**(3), 16–19 (1999)
15. Hanson, R.D.: Combinatorial information market design. Inf. Syst. Front. **5**(1), 107–119 (2003). https://doi.org/10.1023/A:1022058209073
16. Hanson, R.D.: Logarithmic market scoring rules for modular combinatorial information aggregation. J. Predict. Markets **1**(1), 3–15 (2007). https://doi.org/10.5750/jpm.v1i1.417

17. Manski, C.F.: Interpreting the predictions of prediction markets. Econ. Lett. **91**(3), 425–429 (2006). https://doi.org/10.1016/j.econlet.2006.01.004
18. Oesterheld, C., Conitzer, V.: Decision scoring rules. In: 16th International Conference on Web and Internet Economics (2020)
19. Othman, A., Sandholm, T.: Decision rules and decision markets. In: Proceedings of the International Joint Conference on Autonomous Agents and Multiagent Systems, vol. 1, pp. 625–632 (2010)
20. Plott, C.R.: Markets as information gathering tools. South. Econ. J. **67**(1), 2–15 (2000). https://doi.org/10.2307/1061610
21. Plott, C.R., Chen, K.Y.: Information aggregation mechanisms: concept, design and implementation for a sales forecasting problem (2002)
22. Plott, C.R., Wit, J., Yang, W.C.: Parimutuel betting markets as information aggregation devices: experimental results. Econ. Theor. **22**(2), 311–351 (2003). https://doi.org/10.1007/s00199-002-0306-7
23. Tziralis, G., Tatsiopoulos, I.: Prediction markets: an extended literature review. J. Predict. Markets **1**(1), 75–91 (2007). https://doi.org/10.5750/jpm.v1i1.421
24. Wang, W., Pfeiffer, T.: Securities based decision markets. arXiv e-prints arXiv:2103.10011 (2021)
25. Wolfers, J., Zitzewitz, E.: Interpreting prediction market prices as probabilities. Technical report, Working paper 12200, National Bureau of Economic Research, May 2006. https://doi.org/10.3386/w12200

MARL for Traffic Signal Control in Scenarios with Different Intersection Importance

Liguang Luan[1], Yu Tian[1], Wanqing Fang[1], Chengwei Zhang[1(✉)], Wanli Xue[2], Rong Chen[1], and Chen Sang[1]

[1] School of Information Science and Technology, Dalian Maritime University, Dalian, China
{llg2020,ty97,fangwanqing,chenvy,rchen,lzdlmu}@dlmu.edu.cn
[2] School of Computer Science and Engineering, Tianjin University of Technology, Tianjin, China
xuewanli@email.tjut.edu.cn

Abstract. Recent efforts that applied Multi-Agent Reinforcement Learning (MARL) to the adaptive traffic signal control (ATSC) problem have shown remarkable progress. However, those methods assume that all agents in the cooperative games are isomorphic, which ignores the situation that different agents can play heterogeneous roles in the ATSC scenario. The tolerance of vehicles at different intersections in the same area is different, *e.g.*, traffic congestion near hospitals or schools will affect the timely treatment of patients or the safety of children and definitely need to be paid more attention than ordinary congestions. Motivated by the human wisdom in cooperative behaviours (e.g. team members will execute the action according to the strategy implemented by the team leader), we present a leader-follower paradigm based Markov game model which taking into account both the overall and special intersections. Specifically, the leader-follower paradigm control intersections in a traffic scenario by two kinds of agents, *i.e.*, leader agent controlling intersections that need special attention, and follower agents controlling ordinary intersections. Then a multi-agent reinforcement learning framework, named Breadth First Sort Hysteretic DQN (BFS-HDQN) is proposed to train the optimal control policy of the proposed ATSC model. BFS-HDQN consists of two parts, an independent MARL algorithm (here we use Hysteretic DQN as the base algorithm) to train different kinds of agents, and a communication mechanism based on Breadth First Sort (BFS) to generate observation information of each agent. We evaluate our methods empirically in two synthetic and one real-world traffic scenarios. Experimental results show that, compared with the state-of-the-art methods, BFS-HDQN can not only ensure the optimal overall performance, but also obtain better performance at special intersections, in almost all metrics commonly used in ATSC.

The work is supported by the National Natural Science Foundation of China (Grant Nos.: 61906027, 61906135), China Postdoctoral Science Foundation Funded Project (Grant No.: 2019M661080).

© Springer Nature Switzerland AG 2022
J. Chen et al. (Eds.): DAI 2021, LNAI 13170, pp. 93–106, 2022.
https://doi.org/10.1007/978-3-030-94662-3_7

Keywords: Adaptive Traffic Signal Control · Multiagent reinforcement learning · Leader-follower paradigm

1   Introduction

Traffic congestion is one of the main factors affecting people's daily life. How to control the switching of a traffic light to improve transportation efficiency, *i.e.*, the Adaptive Traffic Signal Control (ATSC) problem, is becoming more and more crucial, especially when there are multiple intersections connected.

Specifically, the objective of ATSC is to minimize traffic congestion by adaptively adjust signal phases. Early traffic signal control methods [4,5,7] mostly depended on pre-defined rules by expert strategy, and cannot be adaptively adjusted to the real traffic situation. With the development of artificial intelligence technology, deep reinforcement learning (DRL) has got a lot of attention in adaptive traffic signal control problems. Many efforts have been performed to design strategies both in isolated traffic signal control and cooperative multi-agent traffic control. This work concentrates on the cooperation of multiple intersections, which has been categorized by Literature [17] into two classes: joint action learners and independent learners. Joint action learners use a single agent to control multiple intersections by takes the whole state as input directly and learns the joint policy of all intersections [11,14,19]. These methods can result in the curse of dimensionality because the action space is exponential growth with the increase of intersection scale. Different from common MARL scenarios like games and robot control, there may be thousands of intersections in a real-world ATSC scenario that coordinate together to optimize urban transportation. The state transition of the traffic environment and rewards received by intersections depends on the joint action of all intersections. It is more natural to formulate traffic signal control as a cooperative Multi-Agent game, where each intersection is controlled by a single agent with local observations than well-known works of MARL literature that use the centralized training distributed execution (CTDE) mechanism [6,12,13]. Not surprisingly, to date most existing works on the multi-agent perspective for ATSC concentrate on independent optimization-based methods using local observations and messages from other coordinated agents [1–3,8,15,16].

In these algorithms, intersections are isomorphic, and all intersections are treated equally. However, the importance of each intersection in real world scenario is different. Many places have special traffic control measures near hospitals and schools, such as speed limit or detour. Another example is the ongoing outbreak of novel coronavirus, in which the demand for vehicles in the epidemic area is much higher than that in ordinary areas. Under these circumstances, it is obviously unreasonable to treat all intersections equally. While optimizing the overall goal of the road network, it is also needed to take into account the local interests of some individual intersections.

Motivated by the human wisdom in cooperative behaviours (e.g. the core player in a football team have a higher decision-making power as the captain, while other players cooperate with the captain to achieve the goal), in this paper we propose a new model named Leader Follower Markov Game (LF-MG) to define goals of different intersections in ATSC. Specifically, the leader-follower paradigm based model intersections in a traffic scenario into two categories, *i.e.*, leader agents and follower agents, and adopts different optimization objectives for the two kinds of agents. Leader agents represent intersections that need special attention which only consider optimizing the traffic conditions near them, while follower agents need to consider the congestion of themselves and their connected neighbors at the same time. On this basis, we designed an independent MARL framework to learn the multi-objective optimization strategy mentioned above, which we name it Breadth First Sort Hysteretic DQN (BFS-HDQN). BFS-HDQN models the ATSC problem as a Neighbor-Aware Markov Game where each agent in the framework decentralized control of an intersection based on its local observations and information from its connected neighbors. The MARL framework consists of two parts, an independent MARL algorithm (here we use Hysteretic DQN [10] as the base algorithm, which is an independent MARL algorithm designed to learn optimal joint strategy in cooperative multiagent games) to train different kinds of agents, and a communication mechanism based on graph Breadth First Sort (BFS) is to generate interactive information of each agent. The communication mechanism uses a leader-following action selection paradigm that enables a successor agent to select the individual action after obtaining precursor's strategy, which determines the order and content of information transmission between agents. We evaluate our methods with a real world scenarios and two synthetic scenarios. Experimental results show that, compared with the state-of-the-art methods, our algorithm can not only ensure the optimal overall performance, but also obtain better performance at special intersections, in the three most commonly used criteria, *i.e.*, travel time, throughput, and queue length.

The rest of this paper is organized as follows. Section 2 introduces necessary notations in RL and ATSC scenario. The proposed LF-MG is introduced in Sect. 3. The MARL framework BFS-HDQN is proposed in Sect. 4. We introduce the simulation studies in Sect. 5, and finally, we conclude this paper in Sect. 6.

2 Basic Notation

2.1 Adaptive Traffic Signal Control

Adaptive traffic signal control network includes multiple interactions and roads between connected intersections. Naturally, the relationship between interactions can be defined as a graph $G(\mathcal{N}, \mathcal{E})$, in which $\mathcal{N}$ is the set of interactions and $ij \in \mathcal{E}$ indicates that the two intersections $i, j \in \mathcal{N}$ are directly connected. For convenience, we denote $\mathcal{N}_i$ as the set of i's neighbors, $\mathcal{N}_i = \{j \in \mathcal{N} | ij \in \mathcal{E}\}$.

Lanes that vehicles enter the intersection are incoming lanes of the intersection. An intersection has 12 incoming lanes (marked by colored boxes) categorized into four incoming directions: West, East, North, and South ('W', 'E', 'N', 'S' for short). Formally, the incoming lanes of i is defined by $L_i = \bigcup_{j \in \mathcal{N}_i} L_{j \to i}$, where $L_{j \to i}$ is the set of incoming lanes coming from one of i's neighbor j. In this work, we use the number of vehicles in each incoming line of an intersection as well as its connected neighbors as observation inputs of each RL agent.

The action set of each intersection is defined by signal phases, *i.e.*, the combination of traffic signals in all incoming lanes. Commonly, traffic signals indicate three traffic movements: Left, Through, and Right ('L', 'T', and 'R' for short), which can constitute eight non-conflicting signal phases. Considering that the right-turn vehicle passes regardless of the signal, and go straight and turn left are usually controlled with the same signal, here we consider four signal phases only, *i.e.*, NT-ST, WT-ET, SL-NL, and WL-EL as the action space of RL agents.

2.2 Network Markov Game

The Markov game (MG) is the standard multi-agent RL setting, commonly be defined as a 7-tuple $\langle \mathcal{N}, \mathcal{S}, \mathcal{O}, \mathcal{A}, P, R, \gamma \rangle$. Here $\mathcal{N}$ is the agent set where $N = |\mathcal{N}|$ is the set size. $\mathcal{S}$ is the state space. $\mathcal{O} = \langle \mathcal{O}_1, ..., \mathcal{O}_N \rangle$ is the observation space in which $\mathcal{O}_i$ is the local observation that agent i can observe. Similarly, $\mathcal{A} = \langle \mathcal{A}_1, ..., \mathcal{A}_N \rangle$ and $\mathcal{A}_i$ are the joint action space of all agents and the local action space of agent i respectively. $P : S \times \mathcal{A} \times \mathcal{S} \to [0, 1]$ represents the state transition function. $R = \langle r_1, ..., r_N \rangle$ is the joint reward function. γ is the discount factor. Especially, a Markov game is a fully cooperative game if every agents uses the same reward, *i.e.*, $\forall i, r_i = r$. The Networked Markov Game (NMG) takes the relationship between agents into account, which replace $\mathcal{N}$ by a graph $\mathcal{G}(\mathcal{N}, \mathcal{E})$ in MG, where $ij \in \mathcal{E}$ is the communication link between agent i and j. Besides, the observation $\mathcal{O}_i$ of agent i in NMG is defined by local information of agent i and messages $\{M_{ij}\}_{ij \in \mathcal{E}}$ received from its neighbors.

The policy $\pi_i : O_i \to \Delta(A_i)$ of agent i maps the observation space to a probability distribution of action space. Similarly, we denote $\pi = \langle \pi_i, ..., \pi_N \rangle$ as the joint policy of all agents. The expected return (or expected sum of future rewards) with joint policy π for agent i on state s is defined by the value function or the state-value function, also known as Q value function: $Q_{i,\pi}(s, a) = \mathbb{E}_\pi[\sum_{k=0}^{\infty} \gamma^k r_i^{(t+k+1)} | s_t = s, a_t = u]$, where $r_i^{(t)}$ is the reward received by agent i at time t.

2.3 Deep Q-Learning and HDQN

Deep reinforcement learning (DRL), *e.g.* DQN [9], combines deep neural networks with traditional RL by representing the value function with deep neural networks. To train neural network, DRL use a *replay memory buffer* to store the experience $\langle s, a, r, s' \rangle$, where the r and s' are reward and next state received after taking action a in state s respectively. Denote θ as parameters of a neural network used in a DRL agent. For convenience, we also use θ directly to refer to a neural network. In DQN [9], θ are updated by minimising the TD error δ_i:

$$L_{DQN}(\theta) = \sum_{i=1}^{b} \delta_i^2,$$
$$\delta_i = r + \gamma \max_{a'} Q(s', a'; \theta^-) - Q(s, a; \theta) \tag{1}$$

where b is the batch size, θ^- is the target network periodically copied from θ and keeping constant during each copy.

Hysteretic DQN (HDQN) [10] is an independent learning algorithm designed for cooperative multi-agent environments. Two learning rates are used, α and $\alpha' = \alpha h$, with $0 < h < 1$. Whenever an update would reduce a Q-value, HDQN uses the smaller learning rate αh, otherwise, use the big one. This results in an optimistic update by pays more attention to positive experiences, which has shown to be useful in many fully cooperative multi-agent tasks. Formally,

$$L_{HDQN} = \sum_{i=1}^{b} \bar{\delta}_i^2, \bar{\delta}_i = \begin{cases} \delta_i & \delta_i > 0 \\ h\delta_i & \delta_i \leq 0 \end{cases} \tag{2}$$

In this work, we use HDQN as the basic RL framework to learn the joint optimal policy of multiple intersections. We will show in the experiment section that with a reasonable game design, using this simple RL algorithm can also achieve excellent performance in dealing with ATSC problems. The major notations used in the paper are listed in Table 1.

Table 1. Summary of notations

Notation	Definition
$\mathcal{N}$	Set of intersections
i and j	Intersection (agent) i and j
$\mathcal{N}_i$	Set of intersection i's neighbors
$\mathcal{N}_{i-}$	Precursors of i's neighbors
L_i	Set of incoming lanes of intersection i
$wave\,[l]$	Waiting vehicles numbers alone incoming line l
$phase_i$	Phase of intersection i
r_i	Individual reward of intersection i
$\boldsymbol{r}_i$	Individual rewards of intersection i and i's neighbors
o_i	Observation of agent i
$\bar{o}_i$	Neighbor-aware observation of agent i

3   Leader Follower Markov Game

In this section, we detail the Leader-Follower Markov Game (LF-MG) that distinguish intersections in an ATSC scenario into special intersections and ordinary intersections. The LF-MG is built based on a special Networked Markov game, in which an intersection can observe the situation of vehicles along each incoming lane of the ego intersection and its connected neighbors. This assumption is supported in many related works [2,8] where information of neighbor intersections can be understood as messages received from adjacency agents defined in NMG. In our previous work [18], we have verified experimentally that neighborhood cooperation can strengthen the cooperation between intersections.

As mentioned in previous sections, the goal of this work is to learn a joint policy that maximize both returns of the overall network and some special intersections. LF-MG defines two kind of agents, *i.e.*, leader agents and follower agents, where the former only considers its own interests, while the latter comprehensively considers the interests of itself and its neighbors. Here we define reward of the two kinds of agent as following,

$$r_i^{(t)} = \begin{cases} \tilde{r}_i^{(t)} & i \in \mathcal{N}_L \\ \frac{1}{1+|\mathcal{N}_i|}(\tilde{r}_i^{(t)} + \sum_{j \in \mathcal{N}_i} w\tilde{r}_j^{(t)}) & i \in \mathcal{N}_F \end{cases} \tag{3}$$

where $\tilde{r}_i^{(t)}$ is the current congestion near an intersection i at time t as the negative of the waiting vehicles number, *i.e.*, $r_i^{(t)} = -\sum_{l \in L_i} wave[l]^{(t+1)}$, which can be obtained directly from the observation of agent i. $\mathcal{N}_L$ and $\mathcal{N}_F = \mathcal{N} - \mathcal{N}_L$ are sets of leader intersections and followers intersections respectively. $w \in (0,1)$ is a discount factor to weigh the importance between a follower agent and its neighbors. Obviously, the greater w is, the more social aware the agent is, and the goals of all agents tend to be the same. On the contrary, the smaller w is, the more selfish the agent is. When $w = 0$, the agent is equivalent to a leader agent. In this work, we fixed it to 1 if the neighbor is a leader otherwise fixed it to 0.5. The definition of follower reward mainly takes into account that this kinds of agent aims at group interests and the communication limitation between agents under the networked scenarios. The goal of each agent in LF-MG is to maximize its cumulative rewards (return), *i.e.*, $\max_\pi \sum_{0 \le t \le T} r_i^{(t)}$, by finding a joint policy π.

For the observation and action definitions of different agents, we do not distinguish them. Formally, the observation $o_i \in \mathcal{O}_i$ of an agent i is defined by information of intersection i and its neighbors,

$$o_i^{(t)} = \{phase_i^{(t)}, \{wave[l]^{(t)}\}_{l \in L_i \cup L_{\mathcal{N}_i}}\} \tag{4}$$

where $phase_i^{(t)}$ is its phase at time t, L_i and $L_{\mathcal{N}_i}$ are sets of incoming lines of intersection i and its neighbors respectively, and $wave[l]^{(t)}$ is the number of waiting vehicles alone incoming line l at time t. Action $a_i \in \mathcal{A}_i$ of agent i is the signal phase defined in Sect. 2.

Algorithm 1. Multiple Start BFS

Input: Graph $\mathcal{G}(\mathcal{N}, \mathcal{E})$, set of leaders $\mathcal{N}_L$.
Output: Sorted sequence sq
 1: Initialize a empty sequence sq.
 2: Initialize a empty queue q and enqueue all $i \in \mathcal{N}_L$ to q.
 3: **while** q is not empty **do**
 4:     Dequeue front element u from q.
 5:     Enqueue all unvisited neighbors of u.
 6:     Add u to sq.
 7: **end while**

4 Breadth First Sort Hysteretic DQN

In this section, we introduce the Breadth First Sort Hysteretic DQN framework (BFS-HDQN) that enables two kinds of agents to learn a joint policy that maximizes both returns of the overall network and some special intersections. Realistic tasks sometimes require agents considering the information of other agents in the action selection phase in order to appropriately cooperate. For instance, in human cooperative tasks, there is usually a team leader who has stronger ability and priority than other team members. And in the process of cooperation, the leader first tells other players what he wants to do, and other team members make corresponding cooperation according to the leader's decision. This will give priority to protecting the interests of the leader. At the same time, other members as collaborator, will consider how to maximize the interests of the team after knowing the action to be performed by the leader. As a result, the final strategy takes into account the common interests of the leader and the team.

BFS-HDQN is motivated by the above organization behavior. Considering the graph characteristics in ATSC problem, the influence of actions between intersections is gradually transmitted along the edges between each pair of connected neighbors. BFS-HDQN takes leader intersections as starting nodes and sorts intersections in an ATSC scenario by Breadth First Sort method, a classical graph sorting algorithm, to determine the order of action selection at each intersection. Considering that there may be more than one leader in an ATSC scenario, we made a slight change to the original BFS algorithm (Algorithm 1).

Based on the sorted sequence of intersections, we can determine which agents have announced the actions they want to perform before any agent i. Here we denote $\mathcal{N}_{i-} = \{j \in \mathcal{N} | j \succ i\}$ as precursors of i's neighbors. Then, for agents of ordinary intersections, when they choose actions at each time, they have obtained local observations of themselves and their neighbors, as well as the actions of their precursor neighbors. For leader agents, we define their precursor set as empty. Based on this, a BFS-HDQN agent i can select according to a conditional estimation when interacting with the environment or train their policy, $i.e.$,

$$Q_i(o_i^{(t)}, a; \theta_i, a_{j \in \mathcal{N}_{i-}}^{(t)}) \tag{5}$$

Algorithm 2. BFS-HDQN

1: Initialize Q network θ_i and target Q network $\tilde{\theta}_i$ for each agent;
   Initialize replay memory $D_i = \emptyset$ and a empty trajectory τ_i .
2: Sort all intersections by Alg.1
3: **while** episode$\leq M$ **do**
4:    **while** $t \leq T$ **do**
5:       **for** intersection i from sq_1 to sq_N **do**
6:          Observe $o_i^{(t)}$ and joint action $a_{j \in \mathcal{N}_{i-}}^{(t)}$ of its precursors;
7:          With probability ε select a random action $a_i^{(t)}$, otherwise select $a_i^{(t)} = \arg\max_a Q_i(o_i^{(t)}, a; \theta_i, a_{j \in \mathcal{N}_{i-}}^{(t)})$ (Eq.5);
8:          Broadcast $a_i^{(t)}$ to i's successors.
9:       **end for**
10:       All intersections interact with environment simultaneously.
11:       **for** intersection $i \in \mathcal{N}$ **do**
12:          Observe reward $r_i^{(t)}$ after all agents execute their actions;
13:          Add $(< o_i^{(t)}, a_{j \in \mathcal{N}_{i-}}^{(t)} >, a_i^{(t)}, r_i^{(t)})$ in τ_i;
14:       **end for**
15:    **end while**
16:    **for** intersection $i \in \mathcal{N}$ **do**
17:       Generate transitions $(< o_i^{(t)}, a_{j \in \mathcal{N}_{i-}}^{(t)} >, a_i^{(t)}, r_i^{(t)}, < o_i^{(t+1)}, a_{j \in \mathcal{N}_{i-}}^{(t+1)} >)$ from τ_i
       and store it to D_i.
18:       Update DQN network θ_i by Eq.2;
19:       Every N_θ steps reset $\tilde{\theta}_i \leftarrow \theta_i$
20:    **end for**
21:    reset τ_i to empty.
22: **end while**

---

where θ_i is the network parameter of agent i, $a_{j \in \mathcal{N}_{i-}}^{(t)}$ is the joint actions of i's processor neighbors. To estimate the joint action of the ego agent and its precursors when modeling the DQN network of each agent, DQN network of a BFS-HDQN agent i's takes actions of precursors neighbors as a part of its state input, and output Q values of all actions of its. Considering that to train DQN network, it is also needed to calculate the greedy value for the next state (Eq. 1), definition above cannot get the experience directly after each step as traditional DQN. Here, we use a temporary memory τ_i to temporarily store the trajectory information observed by the agent under each episode. Experiences are generated through this trajectory at the end of the episode. In the training process, all agents use their experiences to train their networks independently.

Algorithm 2 shows details of BFS-HDQN. For each agent i, BFS-HDQN has two randomly initialized neural networks, *i.e.*, a DQN evaluation network θ_i and a DQN target network $\tilde{\theta}_i$. Agents choose their actions by their neighbor-aware observations and actions from their precursors using the ε-greedy strategy one by one in order (line 7), where ε is gradually decreased from 1 to 0 with the increase of interaction times. Then they broadcast the action to their successors. After all agents execute their actions, each of them obtains a new observation and a reward (line 12). At the end of each episode, each agent generates experiences from the trajectory τ_i it observed at this episode (line 17). Finally, all networks are trained as normal DRL algorithms (Line 18–19).

5   Experiment

We implement our methods using in a real-world traffic networks imported from Jinan and two synthetic traffic grids. We first tested BFS-HDQN and a state-of-the-art MARL methods MA2C [3] and Colight [16] to show the effectiveness of our methods. We also do ablation experiments with HDQN and a BFS-HDQN without weight setting in 3, noted by BFS-HDQN (non), to show that both the leader-follower setting and the BFS communication mechanism in BFS-HDQN are useful in improving the performance of ATSC.

5.1   Scenarios Setting

We use three public available traffic networks, *i.e.*, a real-world traffic networks, and two 4×4 synthetic traffic grids based on Cityflow[1], a commonly used traffic simulator, to evaluate our methods. As shown in Fig. 1, we set up different numbers of special intersections in the three ATSC scenarios for completeness. Intersections in all the three scenarios are homogeneous and have the same action space as defined in Sect. 2. An episode in all the three scenarios simulates peak-hour traffic. Considering the reality that it is not allowed to change traffic lights too often, each intersection controls traffic signals synchronously at an interval of 10 s. One control step corresponds to 10 s in the traffic environment, which results in the episode length of each scenario is fixed to 360.

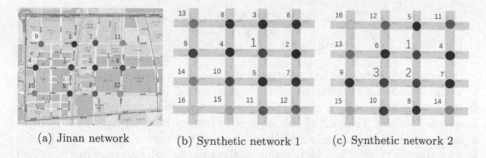

(a) Jinan network   (b) Synthetic network 1   (c) Synthetic network 2

Fig. 1. The three traffic network settings. Red dots and blue dots in each networks indicate special intersections (leaders) and ordinary intersections (followers) respectively. Road networks (a) and (b) have only one special intersection and (c) has three. The special intersection (leader) and ordinary intersection (follower) are numbered according to the order (colored from dark to light) calculated by Algorithm 1. (Color figure online)

---

[1] https://cityflow-project.github.io/.

Vehicle flow of the three traffic scenarios contains information of vehicles, including the start and target locations, arrival time, and driving route. Vehicle data of the two real-world networks, *i.e.*, vehicle IDs and their driving routes, was created using cameral data captured at intersections of the real traffic environment. In the two synthetic scenario, the vehicle date was created randomly. Specifically, vehicle number is generated by Gaussian distribution. The driving routes for each vehicle start from one of the edge nodes of the network and randomly turn directions whenever through an intersection. The turning ratio is set as 60% straight ahead, 10% left turn and 30% right turn at each turn. We summary detailed vehicle and intersection statistics in Table 2.

Table 2. Data statistics of the three scenarios

Scenarios	Intersections	Leaders	Arrival num (vehicles/300 s)			
			Mean	Std	Max	Min
Jinan	12	1	524.58	98.53	672	256
Synthetic 1	16	1	935.92	17.47	960	896
Synthetic 2	16	3	935.92	17.47	960	896

Following the most existing researches, we choose three metrics: (1) travel time m_t: the average travel time of all vehicles, the most common metric used to evaluate different methods; (2) queue length m_q: the average length of vehicles in each lane of all intersections, and (3) throughput m_{th}: total number of vehicles that reach their destinations. Specifically,

$$
\begin{aligned}
m_t &= \frac{1}{|\mathcal{V}_{in}|} \sum_{v \in \mathcal{V}_{in}} (t_v^{out} - t_v^{in}) \\
m_q &= \frac{1}{|\mathcal{N}|T} \sum_{t \leq T} \sum_{i \in \mathcal{N}} \sum_{l \in L_i} \frac{wait^{(t)}[l]}{|L_i|} \\
m_{th} &= |\mathcal{V}|
\end{aligned}
\tag{6}
$$

t_v^{in} and t_v^{out} indicate the time that vehicle v enters and leaves, respectively. $\mathcal{V}$ is the vehicle set that arrives at their destination in an hour. $\mathcal{V}_{in}$ is vehicles that entering the traffic network. $|\mathcal{N}|$ is the intersection numbers. The episode length is $T = 360$. The goal of this work is to learn a joint strategy that to learn a joint policy that maximizes both returns of the overall network and some special intersections. Thus we test our method in both the average metrics of all intersections and special intersections (leaders).

5.2   Training Setting

We use DQN architecture [9] as a basis of algorithms. The Q networks of HDQN and BFS-HDQN consist of two MLP layers with each layer contains 100 neurons and an output neuron for each action. All the three algorithms use the same hyper-parameters. For fairness, we ran MA2C and Colight[2] with source codes released by the authors. Hyper-parameter values are summarized in Table 3.

Table 3. Parameter setting

Component	Hyper-parameter	Setting
DQN-optimization	Learning rate α	0.001
	Discount γ	0.99
	ERM size	200000
	Network optimizer	Adam
	Activation function	Relu
ϵ-greedy exploration	Initial and final ϵ	1.0 and 0.001
	Decay rate d_ϵ	1.0/20000
HDQN	Hysteretic	0.5

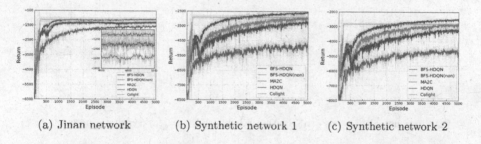

(a) Jinan network   (b) Synthetic network 1   (c) Synthetic network 2

Fig. 2. Average return dynamics of all interactions in the three traffic scenarios.

5.3 Performance Comparison

Figure 2 and 3 refer to the learning curves of MARL methods mentioned above in the three traffic scenarios of average overall returns and leader returns respectively. Lines and shadows around them are average returns and error ranges of those methods during the learning episodes. From Fig. 2, we can see that BFS-HDQN has stable and almost highest returns in all environments, followed by Colight, BFS-HDQN (non) and HDQN. MA2C performed worst. The results show that our algorithm is effective in improving the overall benefit and all innovations of the algorithm are meaningful. From Fig. 3, the comparison results of BFS-HDQN with MA2C and HDQN are obvious, while BFS-HDQN (non)

---

[2] MA2C, Colight: https://traffic-signal-control.github.io/.

shows almost the same performance with BFS-HDQN (even slightly better than BFS-HDQN in the two synthetic scenarios). The reason is intuitive, for the weight in Eq. 3 is added to the return of the follower, not the leader. However, in Fig. 2, The overall performance of BFS-HDQN is significantly higher than that of BFS-HDQN (non), which shows the effectiveness of our reward mechanism. Besides, in Fig. 2 and Fig. 3, we can see BFS-HDQN (non) is outperformed than HDQN in almost all game settings, which indicates that the BFS communication mechanism in BFS-HDQN (non) is helpful to improve the performance of ATSC problems. The reason is that in the BFS communication mechanism, a higher priority is been given to special intersections, while ordinary intersections act as a collaborator by obtain the actions of special intersections. The BFS communication mechanism strengthens the communication between agents, where agents can cooperate better to optimize traffic strategies. The BSF communication mechanism emphasized the priority of special intersections, resulting better performance in returns of those intersections, as shown in Fig. 3. To compare execution performance, we also showed the final statistical results of trained MARL policies in Table 4 and 5, which shows that our algorithm achieves the best results in all environments.

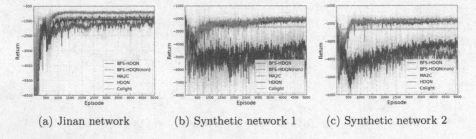

(a) Jinan network   (b) Synthetic network 1   (c) Synthetic network 2

Fig. 3. Average return dynamics of leader interactions in the three traffic scenarios.

Table 4. Execution comparison over trained policies of overall performances. Top two values are in red and blue respectively.

Metrics		MA2C	HDQN	BFS-HDQN (non)	Colight	BFS-HDQN
Jinan	m_t	290.2 ± 3.4	311.0 ± 2.7	287.75 ± 2.4	277.6 ± 4.2	272.3 ± 1.1
	m_q	1.2 ± 0.0	1.4 ± 0.1	1.21 ± 0.02	1.19 ± 0.1	1.03 ± 0.30
	m_{th}	5736.3 ± 3	5684 ± 24	5747 ± 9	5772 ± 54	5798 ± 8
Synthetic 1	m_t	321.9 ± 10.5	239.2 ± 12.1	297.2 ± 2.9	294 ± 37	272.7 ± 2.3
	m_q	3.1 ± 0.1	4.5 ± 0.2	2.74 ± 0.01	2.71 ± 0.1	2.38 ± 0.04
	m_{th}	9724 ± 41	8298 ± 60	9850 ± 9	9960 ± 156	10145 ± 38
Synthetic 2	m_t	319.3 ± 2.5	450.6 ± 12.0	291.7 ± 6.5	287 ± 32	269.3 ± 1.3
	m_q	3.0 ± 0.1	4.6 ± 0.2	2.68 ± 0.03	2.66 ± 0.74	2.32 ± 0.02
	m_{th}	9784 ± 9	8118 ± 36	9888 ± 72	9993 ± 163	10188 ± 19

Table 5. Execution comparison over trained policies of leader performances on metric queue length. Top two values are in red and blue respectively.

Metrics		MA2C	HDQN	Colight	BFS-HDQN (non)	BFS-HDQN
Jinan	m_q	1.2 ± 0.0	1.5 ± 0.3	1.16 ± 0.1	0.86 ± 0.01	0.87 ± 0.07
Synthetic 1	m_q	3.7 ± 0.5	4.5 ± 1.3	2.89 ± 0.2	1.89 ± 0.04	2.05 ± 0.18
Synthetic 2	m_q	2.9 ± 0.1	3.9 ± 0.3	2.44 ± 0.2	1.77 ± 0.17	1.82 ± 0.10

6 Conclusion

To solve the ATSC problem, this paper modeled the problem as LF-MG, a Leader Follower paradigm Markov game model considering both the performance of overall and special intersections. Based on LF-MG, we proposed a decentralized learning multi-agent cooperative methods extended by HDQN, named BFS-HDQN, to learn the cooperative control policy for optimizing the overall return of multiple intersections and intersections requiring special attention. We evaluate our method in a real-world and two synthetic traffic scenarios comparing with SOTA methods. The effectiveness of our proposed model and algorithms are fully demonstrated by experimental results.

As a future work, we will design algorithms that can establish heterogeneous traffic environments to more truly reflect the real environment. Another meaningful direction is to design hierarchical algorithms to adaptively split a very large traffic network into regions. Considering the cost and timeliness of vehicle data in a real environment, designing MARL algorithms that can train effective strategies with few data is also a meaningful direction.

References

1. Chen, C., et al.: Toward a thousand lights: decentralized deep reinforcement learning for large-scale traffic signal control. In: Proceedings of the AAAI Conference on Artificial Intelligence, vol. 34, pp. 3414–3421 (2020)
2. Chu, T., Chinchali, S., Katti, S.: Multi-agent reinforcement learning for networked system control. In: International Conference on Learning Representations (2019)
3. Chu, T., Wang, J., Codeca, L., Li, Z.: Multi-agent deep reinforcement learning for large-scale traffic signal control. IEEE Trans. Intell. Transp. Syst. **21**(3), 1086–1095 (2020)
4. Cools, S., Gershenson, C., Dhooghe, B.: Self-organizing traffic lights: a realistic simulation. arXiv: Adaptation and Self-Organizing Systems (2006)
5. Gartner, N.H., Assmann, S.F., Lasaga, F., Hous, D.L.: Multiband-a variable-bandwidth arterial progression scheme. Transportation Research Record (1287) (1990)
6. Lowe, R., Wu, Y., Tamar, A., Harb, J., Abbeel, P., Mordatch, I.: Multi-agent actor-critic for mixed cooperative-competitive environments. In: Proceedings of the 31st International Conference on Neural Information Processing Systems, NIPS 2017, pp. 6382–6393 (2017)

7. Luk, J.Y.: Two traffic-responsive area traffic control methods: SCAT and SCOOT. Traffic Eng. Control **25**(1), 14–22 (1984)

8. Ma, J., Wu, F.: Feudal multi-agent deep reinforcement learning for traffic signal control. In: Proceedings of the 19th International Conference on Autonomous Agents and Multiagent Systems (AAMAS) (2020)

9. Mnih, V., et al.: Human-level control through deep reinforcement learning. Nature **518**(7540), 529 (2015)

10. Omidshafiei, S., Pazis, J., Amato, C., How, J.P., Vian, J.: Deep decentralized multi-task multi-agent reinforcement learning under partial observability. In: Proceedings of the 34th International Conference on Machine Learning, vol. 70, pp. 2681–2690. JMLR. org (2017)

11. Van der Pol, E., Oliehoek, F.A.: Coordinated deep reinforcement learners for traffic light control. In: Advances in Neural Information Processing Systems (2016)

12. Son, K., Kim, D., Kang, W.J., Hostallero, D.E., Yi, Y.: QTRAN: learning to factorize with transformation for cooperative multi-agent reinforcement learning. In: Proceedings of the 36th International Conference on Machine Learning, vol. 97, pp. 5887–5896 (2019)

13. Sunehag, P., et al.: Value-decomposition networks for cooperative multi-agent learning. arXiv: Artificial Intelligence (2017)

14. Tan, T., Bao, F., Deng, Y., Jin, A., Dai, Q., Wang, J.: Cooperative deep reinforcement learning for large-scale traffic grid signal control. IEEE Trans. Cybernet. **50**(6), 2687–2700 (2019)

15. Wei, H., et al.: PressLight: learning max pressure control to coordinate traffic signals in arterial network. In: Proceedings of the 25th ACM SIGKDD International Conference on Knowledge Discovery & Data Mining, pp. 1290–1298 (2019)

16. Wei, H., et al.: CoLight: learning network-level cooperation for traffic signal control. In: Proceedings of the 28th ACM International Conference on Information and Knowledge Management, pp. 1913–1922 (2019)

17. Wei, H., Zheng, G., Gayah, V., Li, Z.: Recent advances in reinforcement learning for traffic signal control: a survey of models and evaluation. SIGKDD Explor. **22**(2), 12–18 (2021)

18. Zhang, C., Jin, S., Xue, W., Xie, X., Chen, S., Chen, R.: Independent reinforcement learning for weakly cooperative multiagent traffic control problem. IEEE Trans. Veh. Technol., 1 (2021). https://doi.org/10.1109/TVT.2021.3090796

19. Zhu, F., Aziz, H.M.A., Qian, X., Ukkusuri, S.V.: A junction-tree based learning algorithm to optimize network wide traffic control: a coordinated multi-agent framework. Transp. Res. Part C Emerg. Technol. **58**, 487–501 (2015)

Safe Distributional Reinforcement Learning

Jianyi Zhang[1] and Paul Weng[1,2][✉]

[1] University of Michigan-Shanghai Jiao Tong University Joint Institute,
Shanghai Jiao Tong University, Shanghai, China
{zhangjy97,paul.weng}@sjtu.edu.cn
[2] Department of Automation, Shanghai Jiao Tong University, Shanghai, China

Abstract. Safety in reinforcement learning (RL) is a key property in both training and execution in many domains such as autonomous driving or finance. In this paper, we formalize it with a constrained RL formulation in the distributional RL setting. Our general model accepts various definitions of safety (e.g., bounds on expected performance, CVaR). To ensure safety during learning, we extend a safe policy optimization method to solve our problem. The distributional RL perspective leads to a more efficient algorithm while additionally catering for natural safe constraints. We empirically validate our propositions against appropriate state-of-the-art safe RL algorithms.

Keywords: Reinforcement learning · Safe exploration · Risk constraint

1 Introduction

Reinforcement learning (RL) has shown great promise in various applications [28,34]. As such techniques start to be deployed in real applications, safety in RL [19] starts to be recognized as a key consideration not only during learning, but also during execution after training. Indeed, in many domains from medical applications [22] to autonomous driving [21] to finance [7], the actions chosen by an RL agent can have disastrous consequences and therefore the corresponding risks need to be controlled both during training, but also during execution.

While traditional RL does not take safety into account, recent work has started to study it more actively. Safety takes various definitions in the literature. In its simplest sense, it means avoiding bad states [20], but it can take more general meaning such as decision-theoretic risk aversion [9], or risk constraints [30], satisfaction of logic specifications [2], but also simple bounds on expected cumulated costs [44].

For a given definition of safety, one may want to learn a policy that satisfies it, without constraining the training process. Such approach would provide a safe policy for deployment after training. In contrast, recent work in safe RL aims at enforcing safety during learning as well, which is a difficult task as the RL agent needs to explore.

© Springer Nature Switzerland AG 2022
J. Chen et al. (Eds.): DAI 2021, LNAI 13170, pp. 107–128, 2022.
https://doi.org/10.1007/978-3-030-94662-3_8

This paper follows this latter trend and safety is formulated as the satisfaction of a set of general constraints on distributions of costs or rewards. Thus, a safe policy is defined as a policy that respects some constraints in expectation or in probability. Our goal is to learn among safe policies and find one that optimizes the usual expected discounted sum of rewards. Furthermore, we also require safe learning, i.e., the safety constraints shall be satisfied during training as well.

To that aim, we first propose a general framework that accepts various safety formulations from bounds on Conditional Value at Risk (CVaR) to variance, to probability of reaching bad states. This general framework is made tractable by formulating the problem in the distributional RL setting, where distributions of returns are learned in contrast to their expectations. Based on this general distributional formulation, we extend an existing safe RL algorithm, Interior-Point Policy Optimization (IPO) [25], to the distributional setting, for which we formulate a performance bound.

Contributions. Our contributions are threefold: (1) We propose a general framework for safe RL where safety is expressed as the satisfaction of risk constraints, which is enforced during and after training. A risk constraint can be expressed as any (sub)differentiable function of a random variable representing a cumulative reward or cost. (2) In order to obtain a practical algorithm, we formulate our problem and solution method in the distributional RL setting. (3) Our proposition, called Safe Distributional Policy Optimization (SDPO), is empirically validated on multiple domains with various risk constraints against relevant state-of-the-art algorithms.

2 Related Work

Safe RL is becoming an important research direction in RL [19]. In this paper, we distinguish three main non-exclusive aspects for safe RL: policy safety, algorithmic safety, and exploration safety during training.

Policy safety corresponds to the goal of learning a safe policy such that its execution would avoid/limit the occurrence of bad outcomes (e.g., probability of reaching bad states or bound on performance). In that sense, safe RL is related to risk-sensitive RL [9,13,15] where the goal is to learn a policy that optimizes a risk-sensitive objective function. Safety can also be modeled as additional constraints or penalization, thus, it is also related to constrained RL [1,25,26,38] where the goal is to learn a policy that satisfies some constraints, and risk-constrained RL [8,11,14,20,30,41], which in some sense combines the previous

Table 1. Summary of related algorithms: which constraints are accepted, whether safety is guaranteed during learning/execution. PD (primal-dual) actually corresponds to several algorithms.

Algorithm	PD	CPO	IPO	PCPO	SDPO
Expectation	✓	✓	✓	✓	✓
Variance	✓	✗	✗	✗	✓
CVaR	✓	✗	✗	✗	✓
(Sub)differentiable fun	✗	✗	✗	✗	✓
Safe learning	✗	✓	✓	✓	✓
Safe execution	✓	✓	✓	✓	✓

settings. The works in those three areas, with a few exceptions (Constrained Policy Optimization (CPO) [1], IPO [25], Projection-Based Constrained Policy Optimization (PCPO) [43]), do not provide any safety guarantee during learning. They are based on a primal-dual approach (PD). For the exceptions, they can only accept simple constraints on expected discounted total costs. Notably, our algorithm SDPO builds on IPO [25] and extends it to the distributional RL setting, which then allows the formulation of sophisticated constraints. See Table 1 for a summary.

Algorithmic safety corresponds to the idea that running a safe RL algorithm should also guarantee some safety property, e.g., continuous policy improvement [29], convergence to stationary point [44], satisfaction of logic specifications [2], satisfaction of constraints [1, 43] during learning. However, none of those propositions can take into account sophisticated safety constraints (e.g., risk measure).

Exploration safety focuses on an important aspect of safe RL: the exploration problem during learning in order to limit/avoid selecting dangerous actions. In this context, safety is generally modeled as avoiding bad states. One main line of work [6, 12, 39, 40] tries to prevent the choice of a bad action by learning a model. Other directions have been explored, for instance, by using a verification method [18] or by correcting a chosen action [17]. However, this type of approaches requires the assumption that the environment is deterministic.

Although research has been active in safe RL, to the best of our knowledge, no efficient algorithm has been proposed for the general framework that we propose. In particular, our proposition can learn a risk-constrained policy while ensuring the satisfaction of the risk constraint during learning. Our proposition is based on distributional RL [5], which has demonstrated that estimating distributions of returns instead of their expectations can ensure better overall performance of RL algorithms. Most work [16, 42] in this area focuses on value-based methods, extending mostly the DQN algorithm [27]. However, there are also works [24, 35] in risk-sensitive RL that combine distributional RL to optimize CVaR. One recent work [4] has also extended policy optimization to distributional RL, but not in the risk setting. Our work is based on the IQN algorithm [16] instead of more recent propositions (e.g., [42]) because of its simplicity and because it perfectly fits our purposes. Note that in IQN, the authors consider optimizing a risk-sensitive objective function, but they do not consider constraints, as we do.

3   Background

In this section, we present the notations, recall the definition of a Markov Decision Process (MDP) as well as its extension to Constrained Markov Decision Process (CMDP), and review the notions (e.g., CVaR) and the related deep RL algorithms, which we use to formulate our method.

Notations. For any set X, $\Delta(X)$ denotes the set of probability distributions (or densities if X is continuous) over X. For any function $f : Y \to \Delta(X)$ and any $(x, y) \in X \times Y$, $f(x \mid y)$ denotes the probability (or density value if X

is continuous) of obtaining x according to $f(y)$. For any $n \in \mathbb{N}$, $[n]$ denotes $\{1, 2, \ldots, n\}$. Vectors (resp. matrix) will be denoted in bold lowercase (resp. uppercase) with their components in normal font face with indices. For instance, $\boldsymbol{v} = (v_1, \ldots, v_n) \in \mathbb{R}^n$ or $\boldsymbol{M} = (m_{ij})_{i \in [n], j \in [m]} \in \mathbb{R}^{n \times m}$.

MDP and CMDP Model. A *Markov Decision Process* (MDP) [36] is defined as a tuple $(\mathcal{S}, \mathcal{A}, P, r, \boldsymbol{\mu}, \gamma)$, where $\mathcal{S}$ is a set of states, $\mathcal{A}$ is a set of actions, $P : \mathcal{S} \times \mathcal{A} \to \Delta(\mathcal{S})$ is a transition function, $r : \mathcal{S} \times \mathcal{A} \to \mathbb{R}$ is a reward function, $\boldsymbol{\mu} \in \Delta(\mathcal{S})$ is a distribution over initial states, and $\gamma \in [0, 1)$ is a discount factor. In this model, a policy $\pi : \mathcal{S} \to \Delta(\mathcal{A})$ is defined as a mapping from states to distributions over actions. We also use notation π_θ to emphasize that the policy is parameterized by $\boldsymbol{\theta}$ (e.g., parameters of neural network). In the remaining, we identify π_θ to its parameter $\boldsymbol{\theta}$ for ease of notation. The usual goal in an MDP is to search for a policy that maximizes the expected discounted total reward:

$$J(\boldsymbol{\theta}) = \mathbb{E}_{\mu, P, \pi_\theta}\left[\sum_{t=0}^{\infty} \gamma^t r(s_t, a_t)\right] \tag{1}$$

where $\mathbb{E}_{\mu, P, \pi_\theta}$ is the expectation with respect to the distribution $\boldsymbol{\mu}$, the transition function P, and π_θ. We define the *(state) value function* of a policy π_θ for state s as:

$$V^\theta(s) = \mathbb{E}_{P, \pi_\theta}\left[\sum_{t=0}^{\infty} \gamma^t r(s_t, a_t) | s_0 = s\right] \tag{2}$$

where $\mathbb{E}_{P, \pi_\theta}$ is the expectation with respect to the transition function P and π_θ. The *(action) value function* is defined as follows:

$$Q^\theta(s, a) = \mathbb{E}_{P, \pi_\theta}\left[\sum_{t=0}^{\infty} \gamma^t r(s_t, a_t) | s_0 = s, a_0 = a\right] \tag{3}$$

and the *advantage function* is defined as: $A^\theta(s, a) = Q^\theta(s, a) - V^\theta(s)$. As there is no risk of ambiguity, to avoid clutter we drop $\boldsymbol{\mu}$ and P in the notation of the expectation from now on.

Reinforcement learning (RL) is based on MDP, but in RL, the transition and reward functions are not assumed to be known. Thus, in (online) RL, an optimal policy needs to be learned by trial and error.

The MDP model can be extended to the *Constrained MDP* (CMDP) setting [3] in order to handle constraints. In a CMDP, m cost functions $c_i : \mathcal{S} \times \mathcal{A} \to \mathbb{R}$ for $i \in [m]$ are introduced in addition to the original rewards. For each cost function c_i, the corresponding value functions can be defined. They are denoted with a subscript, e.g., J_{c_i}, V_{c_i}, or Q_{c_i}. For a CMDP, the goal is to find a policy that maximizes the expected discounted total reward while satisfying constraints on the expected costs $J_{c_i}(\boldsymbol{\theta})$:

$$\max_{\boldsymbol{\theta}} J(\boldsymbol{\theta}) \text{ s.t. } J_{c_i}(\boldsymbol{\theta}) \leq d_i \quad \forall i \in [m], \tag{4}$$

where each $d_i \in \mathbb{R}, \forall i \in [m]$ is a fixed constraint bound.

Proximal Policy Optimization. The family of policy gradient methods constitutes the standard approach for tackling an RL problem when considering parameterized policies. Such a method iteratively updates a policy parameter in the direction of a gradient given by [36]: $\nabla_\theta J(\theta) = \mathbb{E}_{(s,a) \sim d^{\pi_\theta}}[A^\theta(s,a)\nabla_\theta \log \pi_\theta(a \mid s)]$ where the expectation is taken with the respect to the state-action visitation distribution d^{π_θ} of π_θ. One issue in applying a policy gradient method is the difficulty of estimating A^θ online. This issue motivates the use of an actor-critic scheme where an actor (π_θ) and a critic (e.g., A^θ or V^θ depending on the specific algorithm) are simultaneously learned. Learning the value function can help the policy update, such as reducing the gradient variance.

Proximal Policy Optimization (PPO) [33] is a state-of-the-art actor-critic algorithm that optimizes instead a clipped surrogate objective function $J_{PPO}(\theta)$:

$$\sum_{t=0}^\infty \min(\omega_t(\theta)A^{\bar\theta}(s_t, a_t), \text{clip}(\omega_t(\theta), \epsilon)A^{\bar\theta}(s_t, a_t)), \tag{5}$$

where $\bar\theta$ is the current policy parameter, $\omega_t(\theta) = \frac{\pi_\theta(a_t|s_t)}{\pi_{\bar\theta}(a_t|s_t)}$, and $\text{clip}(\cdot, \epsilon)$ is the function to clip between $[1 - \epsilon, 1 + \epsilon]$. This surrogate function approximates the one used in TRPO [32], which was introduced to ensure monotonic improvement after a policy parameter update. Although PPO is more heuristic than TRPO, its advantages are its simplicity and lower sample complexity.

Distributional Reinforcement Learning. The key idea in distributional RL [5] is to learn a random variable to represent the discounted return $Z^\theta(s,a) = \sum_{t=0}^\infty \gamma^t r_t$ where r_t is the random variable representing the immediate reward received at time step t when applying action a in state s and policy π_θ thereafter. In contrast, standard RL algorithms directly estimate the expectation of $Z^\theta(s,a)$, since $Q^\theta(s,a) = \mathbb{E}_{Z^\theta}[Z^\theta(s,a)]$ where the expectation is with respect to the distribution of $Z^\theta(s,a)$.

Recall that any real random variable Z can be represented by its cumulative distribution denoted $F_Z(z) = \mathbb{P}(Z \le z) \in [0,1]$, or equivalently by its quantile function (inverse cumulative distribution) denoted $F_Z^{-1}(p) = \inf\{z \in \mathbb{R} \mid p \le F_Z(z)\}$ for any $p \in [0,1]$. For ease of notation, Z_p denotes the p-*quantile* $F_Z^{-1}(p)$. In the *Implicit Quantile Network* (IQN) algorithm, [16] proposed to approximate the quantile function of $Z(s,a)$ with a neural network and to learn it using quantile regression [23].

Concretely, the quantile function of $Z(s,a)$ can be learned as follows. Denote $\hat Z(s,a)$ as the approximated random variable whose quantile function is given by a neural network $\Psi(s,\tau)$, which takes as input a state s and a probability $\tau \in [0,1]$ and returns the corresponding τ-quantile $\hat Z_\tau(s,a)$ for each action a. After observing a transition (s,a,r,s'), Ψ can be trained by sampling $2N$ values $\tau = (\tau_1, \ldots, \tau_N)$ and $\tau' = (\tau_1', \ldots, \tau_N')$ with the uniform distribution on $[0,1]$. By inverse transform sampling, sampling τ amount is equivalent to sampling N values from $\hat Z(s,a)$ corresponding to $(\hat Z_{\tau_1}(s,a), \ldots, \hat Z_{\tau_N}(s,a))$, and similarly for τ' and sampling from $\hat Z(s', \pi(s'))$ where π is the current policy. Those samples define N^2 TD errors in the distributional setting: $\delta_{ij} = r + \gamma \hat Z_{\tau_j'}(s', \pi(s')) - \hat Z_{\tau_i}(s,a)$. The *distributional Bellman operator* $T^\pi Z(s,a) = r(s,a) + \gamma Z(S', \pi(S'))$ (with $S' \sim P(s,a)$) has been proved [5] to be a γ-contraction under the p-Wasserstein distance $W_p(U, U') = (\int_0^1 |F_U^{-1}(w) - F_{U'}^{-1}(w)|^p dw)^{1/p}$. The quantile function $Z(s,a)$ can then be estimated by minimizing the 1-Wasserstein distance to the distributional Bellman target following quantile regression [23]. Thus, the loss function for training the neural network Ψ in (s, a, r, s') is given by

$$L_{IQN} = \frac{1}{N} \sum_{i \in [N]} \sum_{j \in [N]} \xi^{\kappa}_{\tau_i}(\delta_{ij}) \tag{6}$$

where for any $\tau \in (0, 1]$, $\xi^{\kappa}_{\tau}(\delta_{ij}) = |\tau - \mathbb{I}(\delta_{ij} < 0)| \frac{L_{\kappa}(\delta_{ij})}{\kappa}$ is the quantile Huber loss with threshold κ with $L_{\kappa}(\delta) = \frac{1}{2}\delta^2$ for $|\delta| \leq \kappa$ or $\kappa(|\delta| - \frac{1}{2}\kappa)$ otherwise.

Interior-Point Policy Optimization. In the CMDP setting, Interior-point Policy Optimization (IPO) [25] is a recent RL algorithm to maximize an expected discounted total rewards while satisfying constraints on some expected discounted total costs. To deal with a constraint, IPO augments PPO's objective function with a logarithmic barrier function applied to it, which provides a smooth approximation of the indicator function.

The constrained problem then becomes an unconstrained one with an augmented objective function:

$$\max_{\theta} J_{IPO}(\theta) = J_{PPO}(\theta) + \sum_{i \in [m]} \frac{\ln(d_i - J_{c_i}(\theta))}{\eta}, \tag{7}$$

where η is a hyper-parameter. The objective J_{IPO} is differentiable, therefore, we can apply a gradient-based optimization method to update the policy.

4   Problem Formulation

Let $\Delta(\mathbb{R})$ denote the set of real random variables. Therefore, $\boldsymbol{Z} \in \Delta(\mathbb{R})^S$ denotes a function from states to random variables. Given an (unknown) CMDP, the problem tackled in this paper can be expressed as a constrained optimization problem formulated in the distributional RL setting:

$$\max_{\theta} \quad \mathbb{E}_{s_0 \sim \mu, \boldsymbol{Z}^{\theta}}[\boldsymbol{Z}^{\theta}(s_0)] \tag{8}$$

$$\text{s.t.} \quad \rho_i(\boldsymbol{Y}^{\theta}_i) \leq d_i \quad \forall i \in [m], \tag{9}$$

where $\boldsymbol{Z}^{\theta}(s)$ corresponds to the return distribution generated by policy π_{θ} from the reward function, for all $i \in [m]$, $\boldsymbol{Y}^{\theta}_i(s)$ corresponds to the cumulated cost distribution from cost function c_i, and $\rho_i : \Delta(\mathbb{R})^S \to \mathbb{R}$ is a (sub)differentiable function. Note that this formulation is strictly more general than problem (4) thanks to the possibly non-linear functions ρ_i's.

We recall a few common cases for ρ_i in Table 2. The expectation is a simple example. For episodic MDPs with absorbing bad states, another simple example is the probability of bad states, which is defined like the expectation, but applied to an undiscounted cost equal to 1 for a bad state and 0 otherwise. CVaR is a widely-used risk measure in finance. In this context, the α-CVaR of a portfolio is intuitively its expected return in the worst $\alpha \times 100\%$ cases. Here, we adapted the definition to rewards instead of costs, thus the integration is on the left tail. Naturally, a CVaR of an additional cost would also be possible. In contrast to previous methods, our framework can accept any (sub)differentiable definitions for ρ_i (e.g., coherent risk measures).

Note that we chose to take the mean (over initial states) of the CVaRs instead of the CVaR of the mean. The latter would have been possible as well, but because CVaR is a convex risk measure, our definition is an upperbound of the CVaR of the

Table 2. Common examples for ρ_i.

ρ_i	Definition
Expectation	$\mathbb{E}_{s_0 \sim \mu_Y}[Y(s_0)]$
Prob. of bad states	$\mathbb{E}_{s_0 \sim \mu_Y}[Y(s_0)]$
α-CVaR of rewards	$\mathbb{E}_{s_0 \sim \mu}[\frac{1}{\alpha} \int_0^\alpha Y_\zeta(s_0)d\zeta]$
Variance	$\mathbb{E}_{s_0 \sim \mu}[\mathbb{E}_Y[Y(s_0)^2] - \mathbb{E}_Y[Y(s_0)]^2]$

mean, which means that our formulation is more conservative and in that sense, safer. The same trick applies if ρ_i were defined based on any other coherent risk measure, of which CVaR is only one instance. Similarly, for the variance, we use the mean (over initial states) of variances instead of the other way around. Since the initial states are sampled in an independent way, the $Y(s_0)$'s are independent. This means that our definition upperbounds the variance of the mean of the $Y(s_0)$'s, leading to a more cautious formulation, which is more desirable for safe RL.

In this paper, we define a *safe policy* as a policy satisfying constraints (9). Our goal is to learn a policy maximizing an expected discounted total rewards (8) among all safe policies (i.e., safe execution). Besides, we require that any policy used during learning should be safe (i.e., safe learning).

The formulation of (8)–(9) in the distributional RL setting serves two purposes. First, as observed in distributional RL, estimating the distributions of the cumulated rewards improves the overall performance. Second, many safety constraints (9), such as CVaR, become natural and simple to express in the distributional setting.

5   Proposed Method

To solve problem (8)–(9) in the safe RL setting, we extend IPO to the distributional RL setting and combine it with an adaptation of IQN. Next, we explain the general principle of our approach, and then discuss some techniques to obtain a concrete efficient implementation.

5.1   General Principle

To adapt IPO, we rewrite the surrogate objective function used in PPO in the distributional setting:

$$J_{PPO}(\theta) = \sum_{t=0}^{\infty} \min(\omega_t(\theta)\mathbb{E}_\theta[Z^{\bar\theta}(s_t, u_t) - Z^{\bar\theta}(s_t)], \mathrm{clip}(\omega_t(\theta), \epsilon)\mathbb{E}_\theta[Z^{\bar\theta}(s_t, a_t) - Z^{\bar\theta}(s_t)]).$$

(10)

Problem (8)–(9) can then be tackled by iteratively solving the following problem with this surrogate function:

$$\max_\theta \quad J_{PPO}(\theta) \quad \text{s.t.} \quad \rho_i(Y_i^\theta) \le d_i \quad \forall i \in [m].$$

(11)

Now, following IPO, using the log barrier function, we reformulate problem (11) as an unconstrained problem:

$$\max_{\boldsymbol{\theta}} \quad J_{PPO}(\boldsymbol{\theta}) + \sum_{i \in [m]} \frac{\ln(d_i - \rho_i(\boldsymbol{Y}_i^{\theta}))}{\eta_i}. \tag{12}$$

In contrast to convex optimization [10], we introduce a constraint-dependent hyperparameter η_i to better control the satisfaction of each constraint separately.

Finally, we propose to solve problem (12) with an actor-critic architecture where both the actor and the critic are approximated with neural networks. For the critic, we adapt the approach proposed for IQN [16] to learn random returns $\boldsymbol{Z}$ and random cumulated costs $\boldsymbol{Y}_i$'s. For the actor, parameter $\boldsymbol{\theta}$ of policy $\pi_{\boldsymbol{\theta}}(a|s)$ is updated in the direction of the gradient of the objective function in (12):

$$\nabla_{\boldsymbol{\theta}} J_{PPO}(\boldsymbol{\theta}) - \sum_{i \in [m]} \frac{1}{\eta_i} \frac{\nabla_{\boldsymbol{\theta}} \rho_i(\boldsymbol{Y}_i^{\theta})}{d_i - \rho_i(\boldsymbol{Y}_i^{\theta})}. \tag{13}$$

This gradient raises one difficulty with respect to the computation of $\nabla_{\boldsymbol{\theta}} \rho_i(\boldsymbol{Y}_i^{\theta})$, which corresponds to the gradient of a critic with respect to the parameters of the actor. When ρ_i is linear (i.e., for expectation constraints), the policy gradient theorem [37] applies and specifies how to compute $\nabla_{\boldsymbol{\theta}} \rho_i(\boldsymbol{Y}_i^{\theta})$. However, when ρ_i is non-linear (i.e., for more sophisticated risk constraints), the gradient in (13) cannot be obtained easily. To solve this issue, we propose a simple and generic solution, which consists in

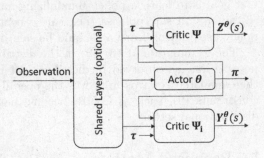

Fig. 1. Architecture of SDPO where critic Ψ corresponds to the objective function and critic Ψ_i corresponds to constraint i. Both critics outputs a distribution.

connecting the actor network to any critic network with a non-linear ρ_i (see Fig. 1 for an illustration where only one critic corresponding to non-linear ρ_i is displayed). Using this construct, the exact gradient of $\rho_i(\boldsymbol{Y}_i^{\theta})$ can be computed by automatic differentiation if ρ_i is (sub)differentiable and $\boldsymbol{Y}_i^{\theta}$ is approximated with a neural network, as we assume. Note that in previous work, [16] who proposed to optimize a risk measure in IQN did not face this gradient issue because their algorithm is based on DQN [27] and therefore does not have an actor network. As a side note, this construct could be used to deal with a more general problem than (8)–(9) where a non-linear transformation is also applied on the objective function. For instance, one may want to optimize the CVaR of some rewards subject to some other risk constraints, which is as far as we know a completely novel problem. We leave this investigation to future work.

5.2   Techniques for Efficient Implementation

In this section, to simplify notations, we do not write the superscript θ for the random variables Z and Y_i's.

To make our final algorithm more efficient, we propose to learn $Z(s)$ only, instead of $Z(s, a)$ as it is the usual practice in distributional RL. This serves two purposes: (1) a state-dependent distribution is easier to learn, and (2) the advantage function can be estimated from a state value function alone. Note that for the constraints only $Y_i(s)$ is needed for any $i \in [m]$. Recall that the two random variables $Z(s)$ and $Z(s, a)$ are related by the following equation:

$$Z(s, a) = R(s, a) + \gamma \mathbb{E}_{s' \sim P(\cdot | s, a)}[Z(s')]. \tag{14}$$

Following IQN, random variable $Z(s)$ is approximated by a random variable $\hat{Z}$, which is represented by a neural network. The expectation of $Z(s)$ can then be approximated by that of $\hat{Z}(s)$ with τ randomly uniformly sampled in $[0, 1]$:

$$\mathbb{E}_\theta[Z(s)] \approx \sum_{i=1}^N (\tau_i - \tau_{i-1}) \hat{Z}_{\tau_i}(s). \tag{15}$$

setting $\tau_0 = 0$ by convention and assuming $0 < \tau_1 < \tau_2 < \ldots < \tau_N < 1$.

The exact handling of the constraints depend on the definition of ρ_i. As illustrative examples, we explain how they can be computed for some concrete cases. If ρ_i is simply defined as an expectation, it can be dealt with like the objective function. For CVaR, it can be estimated as follows for a random variable $Y(s_0)$:

$$c_\alpha(Y) \approx c_\alpha(\hat{Y}) = \frac{1}{\alpha} \sum_{i | \tau_i \leq \alpha} (\tau_i - \tau_{i-1}) \hat{Y}_{\tau_i}(s_0). \tag{16}$$

Here, in contrast to the standard expectation (e.g., (15)), an implementation trick consists in sampling τ in $[0, \alpha]$ such as $\tau_1 < \tau_2 < \ldots < \tau_N = \alpha$ since (16) corresponds to the expectation conditioned on event "$\hat{Y} \leq \hat{Y}_\alpha$". For the variance, $\rho_i(Y)$ can be estimated by:

$$\sum_{i=1}^N (\tau_i - \tau_{i-1}) \hat{Y}_{\tau_i}(s_0)^2 - \left(\sum_{i=1}^N (\tau_i - \tau_{i-1}) \hat{Y}_{\tau_i}(s_0) \right)^2. \tag{17}$$

The pseudo code of our method is shown in Algorithm 1. A theoretical performance bound is discussed in Appendix A.

6   Experimental Results

The experiments are carried out in two different domains: safety gym and financial investment. More experiments are shown in Appendix B due to the limited space. In Appendix B.1, we consider two cases (a constraint over the variance and a constraint over CVaR) and compare with primal-dual methods. In Appendix B.2, we show more experimental results in safety gym. All the settings and hyper-parameters can be found in Appendix C.

Algorithm 1. SDPO

Require: Constraint bound d, Initial policy network π_{θ_0}, Initial IQN network Ψ_0,
  Hyperparameters ϵ for PPO clip rate and η_i for each logarithmic barrier function.
1: **for** $k = 0, 1, \ldots$ **do**
2:     $\mathcal{B} \leftarrow$ run policy π_θ for N trajectories
3:     Sample $\tau_1 < \ldots < \tau_N$ from $\mathcal{U}[0, 1]$
4:     **for** $i, j \in [N]$ **do**
5:         $\delta_{ij} = r + \gamma \hat{Z}_{\tau'_j}(s', \pi(s')) - \hat{Z}_{\tau_i}(s, a)$
6:     **end for**
7:     Update Ψ_{k+1} with ∇L_{IQN} (see (6)) using $\mathcal{B}$
8:     Update θ_{k+1} with $\nabla J(\theta_k)$ defined in (13) using $\mathcal{B}$
9: **end for**

6.1 Safety Gym

Safety gym [31] includes a set of environments designed for evaluating safe exploration in RL. They all correspond to navigation problems where an agent (i.e., Point, Car, Doggo) moves to some random goal positions to perform some tasks (i.e., Goal, Button, Push) while avoiding entering dangerous hazards or bumping into fragile vases. The three tasks, Goal, Button and Push with two difficulty levels (i.e., 1 or 2) are tested. Task Goal is to move the agent to a series of goal positions, while task button is to press a series of goal buttons, and task push is to move a box to a series of goal positions. For detailed explanation of these tasks, please refer to [31]. With a constraint on expected total cost $\rho_1(Y_1) = \mathbb{E}_{s_0 \sim \mu, Y_1}[Y_1(s_0)] \leq d_1$, we are able to compare SDPO with other safe RL algorithms like CPO, PCPO and IPO. CPO takes the loss of TRPO to calculate the direction of the policy update, and enforce the constraints every update by adjusting the step size. PCPO is a two-step approach enlightened by CPO. In the first step, the policy is updated in the direction to improve the objective function in the trust region. In the second step, PCPO projects the potentially infeasible policy back to the constraint set.

Among all these tasks, we only show the results for Point-Goal1 in Figs. 2(a) and 2(b). For other tasks, please refer to Appendix B.2. According to Figs. 2(a) and 2(b), SDPO, PCPO and IPO can explore safely, while CPO cannot satisfy the constraint well. This latter observation regarding CPO may be surprising since CPO was designed to solve CMDPs, but similar results were also reported in previous work [31]. Among these three latter algorithms, SDPO and IPO performs the best. In all the Safety-gym environments (see Appendix B.2), SDPO dominates IPO in terms of either returns or convergence rates (and sometimes both), which confirms the positive contribution of the distributional critics.

To demonstrate that SDPO can satisfy multiple constraints, the safety gym environment is used again, but with a variation. We modify the hazard area to be end states where an agent receives a cost $c_2 = 1$, and the episode is terminated. In addition to the previous constraint, another one is enforced: $\rho_2(Y_2) = \mathbb{E}_{s_0 \sim \mu, Y_2}[Y_2(s_0)] \leq d_2$, where Y_2 is the undiscounted cumulative cost distribution from cost function c_2. Here, we set the bounds: $d_1 = 10$ and $d_2 = 0.1$.

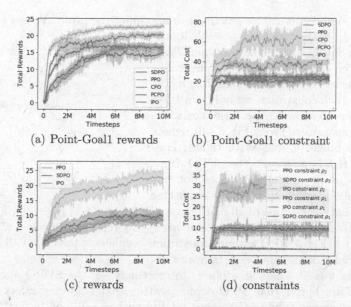

(a) Point-Goal1 rewards   (b) Point-Goal1 constraint

(c) rewards   (d) constraints

Fig. 2. Figures 2(a) and 2(b): Average performance over 10 runs of PPO, SDPO, CPO, PCPO and IPO under Point-Goal1. They are bounded by the dashed line $d_1 = 25$ in Fig. 2(b). Figure 2(c): Average performance over 5 runs of PPO, SDPO and IPO under Point-Goal2. Figure 2(d): Average costs of PPO, SDPO and IPO under Point-Goal2.

Note that among all the baselines, only SDPO and IPO can accept multiple constraints, but IPO can only accept less sophisticated constraints. Thus, we choose c_1 and c_2 to conduct this experiment. From Figs. 2(c) and 2(d), PPO achieves much more goals, but at the cost of violating all the constraints. For constraint ρ_2, both SDPO and IPO agents can avoid entering into hazards during training. For constraint ρ_1, SDPO converges faster than IPO because of the adaption to distributional RL. For more sophisticated constraints, IPO will not work.

6.2   Stock Investment

The second domain is the financial stock market. The RL agent can observe the close prices of the stocks in one day, i.e., the observation $o_t = \boldsymbol{p}_t = (1, p_{1,t}, ..., p_{N,t})$ for N selected stocks where the first component corresponds to cash, which is always 1. The agent can hold the cash to avoid high risks. We further assume that all transactions are dealt at these prices. The action of the agent is defined by a portfolio vector, which corresponds to allocation weights over cash and stocks, i.e., $a_t = \boldsymbol{w}_{t+1} = (w_{0,t+1}, ..., w_{N,t+1})$, w_0 (resp. w_i for $i \in [N]$) is the weight for cash (resp. stock i) and $\sum_{i=0}^{N} w_{i,t} = 1$. Naturally, for each stock, we want to maximize the profit. Thus, with reward function $r_t = \ln \sum_{i=0}^{N} w_{i,t} \frac{p_{i,t}}{p_{i,t-1}}$, optimizing the undiscounted cumulative rewards can maximize the profit. We set the CVaR boundary $d_1 = 0$ to avoid any possible loss.

We run SDPO with a constraint on CVaR defined over rewards using different confidence levels α. As none of the baseline methods accept such constraint, we only run PPO as a baseline to show the performance without any constraints. From Fig. 3, all agents manage to make profits. With tighter constraint on risk (smaller α), the SDPO agent makes less profit. While PPO does not satisfy the constraint as expected, the curves for the constraint satisfaction of all SDPO agents are all similar. We therefore plot their average directly in Fig. 3. PPO without constraint cannot avoid risk and thus suffers from fluctuation and loss at some time point. Interestingly, all the SDPO agents eventually perform better than PPO demonstrating that enforcing safety does not necessarily prevent good performance. SDPO with $\alpha = 0.1$ performs best.

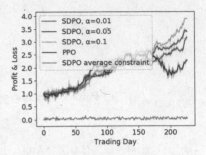

Fig. 3. Average performance over 10 runs of PPO and SDPO with confidence level $\alpha = 0.01, 0.05, 0.1$.

7 Conclusion

We presented a general framework for safe RL that encompasses many previous propositions. The novelty of our approach is the exploitation of a distributional RL formulation that allows us to deal with sophisticated risk constraints in a natural and efficient way for policy optimization. Our algorithm, SDPO, is shown to perform well in diverse environments and is competitive with previous algorithms in situations when they can be applied. However, SDPO can cover a larger range of safety formulations.

Acknowledgements. This work is supported in part by the program of National Natural Science Foundation of China (No. 61872238), the program of the Shanghai NSF (No. 19ZR1426700), and a Yahoo FREP grant.

A Performance Guarantee Bound

For fixed η, solving (12) instead of (11) may incur a performance loss, which can be bounded under natural conditions, which we discuss below. Since this result uses weak Lagrange duality, we first recall the definition of the Lagrangian of (11):

$$\mathcal{L}(\boldsymbol{\theta}, \boldsymbol{\lambda}) = J_{PPO}(\boldsymbol{\theta}) + \sum_{i \in [m]} \lambda_i (d_i - \rho_i(\boldsymbol{Y}_i^{\boldsymbol{\theta}}))$$

and its dual function: $g(\boldsymbol{\lambda}) = \max_{\boldsymbol{\theta}} \mathcal{L}(\boldsymbol{\theta}, \boldsymbol{\lambda})$. The following bound can be proven:

Theorem 1. *If $\boldsymbol{\theta}_1^*$ is an optimal solution of (11), $\boldsymbol{\theta}_2^*$ is the strictly feasible optimal solution of (12) and the unique stationary point of $\mathcal{L}(\cdot, \boldsymbol{\lambda}^*)$ with $\lambda_i^* = \frac{1}{\eta_i(d_i - \rho_i(\boldsymbol{Y}_i^{\theta_2^*}))}$ then:*

$$J_{PPO}(\boldsymbol{\theta}_1^*) - J_{PPO}(\boldsymbol{\theta}_2^*) \leq \sum_{i \in [d]} \frac{1}{\eta_i} \qquad (18)$$

Proof. This result generalizes Theorem 1 of [25], whose proof implicitly uses convexity (which does not hold in deep RL) and follows from the discussion in page 566 of [10].

We adapt the proof to our more general setting. We have:

$$J_{PPO}(\boldsymbol{\theta}_1^*) \leq g(\boldsymbol{\lambda}^*) \qquad (19)$$

$$= J_{PPO}(\boldsymbol{\theta}_2^*) + \sum_{i \in [m]} \lambda_i^*(d_i - \rho_i(\boldsymbol{Y}_i^{\theta_2^*})) \qquad (20)$$

$$= J_{PPO}(\boldsymbol{\theta}_2^*) + \sum_{i \in [m]} \frac{1}{\eta_i} \qquad (21)$$

Step (19) holds by weak duality because $\lambda_i^* \geq 0$ for all $i \in [m]$ (since $\boldsymbol{\theta}_2^*$ is strictly feasible). Step (20) holds because we have by definition of $\boldsymbol{\theta}_2^*$:

$$\nabla_\theta J_{PPO}(\boldsymbol{\theta}_2^*) - \sum_{i \in [m]} \frac{\nabla_\theta \rho_i(\boldsymbol{Y}_i^{\theta_2^*})}{\eta_i(d_i - \rho_i(\boldsymbol{Y}_i^{\theta_2^*}))} = 0 \qquad (22)$$

which implies that $\boldsymbol{\theta}_2^*$ maximizes $\mathcal{L}(\cdot, \boldsymbol{\lambda}^*)$ since $\boldsymbol{\theta}_2^*$ is assumed to be its unique stationary point. Step (21) holds by definition of $\boldsymbol{\lambda}^*$.

The conditions in this theorem are natural. In order to apply an interior point method, the constrained problem needs to be strictly feasible. The condition on the stationarity of $\boldsymbol{\theta}_2^*$ is reasonable and can be controlled by setting ϵ (used in the clipping function of J_{PPO}) small enough.

As a direct corollary, this result implies that if (12) could be solved exactly, the error made by algorithm SDPO is controllable via setting appropriate η_i's. Naturally, in the online RL setting, this assumption does not hold perfectly, but this result still provides some theoretical foundation to our proposition.

B   More Results

In all our experiments, all the agents are initialized so that they are in a feasible region at the beginning. In practice, an initial safe policy can be defined using domain knowledge or by an expert, e.g., in the safety-gym domain, the agent can be initialized to stay and doing nothing. For fairness, the PPO agent is also initialized with the same safe policy as all other agents. Two policy gradient algorithms with CVaR and variance constraints respectively, PD-CVaR [13] and PD-VAR, which is modified from Algorithm 2 in [30] are used as baselines in the first domain. SDPO is compared with CPO [1], PCPO [43], and IPO [25] in the safety-gym domain. PPO [33] is evaluated on all domains to serve as a non-safe RL method. Note that in contrast to our architecture SDPO, none of those algorithms can tackle the problem defined in (8)–(9) in its most general form.

B.1   Random CMDP

To compare SDPO with methods based on Lagrangian relaxation, we design this random CMDP environment. Random CMDPs are CMDPs with N states and M actions, where transition probabilities $P(s' \mid s, a)$ are randomly assigned with $\lceil \ln N \rceil$ positive values for each pair of state-action, and rewards are sampled from a uniform distribution, i.e., $r(s, a) \sim \mathcal{U}[0, 1]$. In the experiments, we set $N = 1000$ and $M = 10$. For each state, we randomly choose $\lceil \ln N \rceil = 7$ successor states. We consider two cases: a bound over the variance or a bound over the CVaR. For simplicity, the cost function is defined as the negative reward: $c = -r$. The CVaR constraint on the associated distribution $\boldsymbol{Y}$ is defined as $\rho(\boldsymbol{Y}) = \rho(-\boldsymbol{Z}) = -\mathbb{E}_{s_0 \sim \mu}[c_\alpha(\mathbb{E}_{\boldsymbol{Z}}[\boldsymbol{Z}(s_0)])] \leq d$. The variance is bounded by $\rho(\boldsymbol{Y}) = \mathbb{E}_{s_0 \sim \mu}[\mathbb{E}_{\boldsymbol{Z}}[\boldsymbol{Z}(s_0)^2] - \mathbb{E}_{\boldsymbol{Z}}[\boldsymbol{Z}(s_0)]^2] \leq d$.

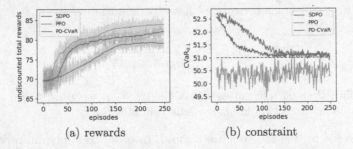

(a) rewards   (b) constraint

Fig. 4. Figure 4(a): Average performance over 10 runs of PPO, SDPO and PD-CVaR under the random CMDP for $N = 1000$. Figure 4(b): 0.1-CVaR bounded by 51. Both SDPO and PD-CVaR converge to the level indicated by the dashed line.

We perform some experiments on this domain with either a constraint on CVaR or a constraint on variance. Both of the constraints are based on the rewards. Note that the cost does not have to be independent of the reward. In this case, applying ρ to the reward r enables us to train a risk-averse policy to avoid worst cases of getting low rewards. Therefore, the first needs to be lower-bounded, while the second needs to be upper-bounded. The confidence level is fixed to $\alpha = 0.1$ and the bound for CVaR is set to 51 and that for the variance is set to 2. The bounds were chosen so that they are not too restrictive.

From the results in Fig. 4, as expected, PPO without constraint achieves the best total rewards and converges faster than the constrained ones. When the CVaR value is bounded, PD-CVaR and SDPO both converge to a slightly worse but safe policy, however SDPO converges faster. From the results in Fig. 5,

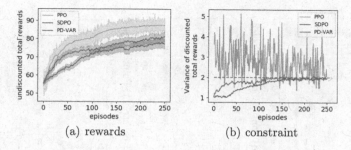

Fig. 5. Figure 5(a): Average performance over 10 runs of PPO, SDPO and PD-VAR under the random CMDP for $N = 1000$. Figure 5(b): Variance bounded by 2. Both SDPO and PD-VAR converge to the level indicated by the dashed line.

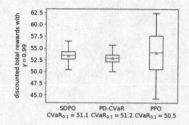

Fig. 6. Evaluation of Z^{π_θ} of trained policy over 1000 runs.

similar observations can be drawn for PPO, PD-VAR, and SDPO. With regards to safety, we can again conclude than SDPO is superior.

To evaluate the converged policies in Fig. 4(a) with CVaR constraint, we run them for 1000 episodes each. Figure 6 indicates that PPO without constraint can reach the highest result but suffers from the risk of getting the lowest reward. Both SDPO and PD-CVaR receives a lower mean reward but much higher CVaR values, which indicates lower risk.

B.2   Safety Gym

In safety gym, the agent is penalized by receiving a cost $c_1 = 1$ when touching a fragile vase. With a constraint on expected total cost $\rho_1(Y_1) = \mathbb{E}_{s_0 \sim \mu, Y_1}[Y_1(s_0)] \le d_1$, we are able to compare SDPO with other safe RL algorithms like CPO, PCPO and IPO. Here we show the results for all tasks in safety gym.

From the results in Figs. 7, 8 and 9, SDPO, IPO and PCPO can explore the environment safely, but CPO may failed to learn a safe policy as the tasks go harder (i.e., with the car agent). In experiments, when the CPO agents violate the constraints, Equation (14) in CPO [1] failed to purely decrease the constraint value, and thus learn an unsafe policy. For the other three agents, SDPO converges faster to a slightly better policy than IPO and PCPO. An interesting future work would be to extend PCPO to the distributional setting as well.

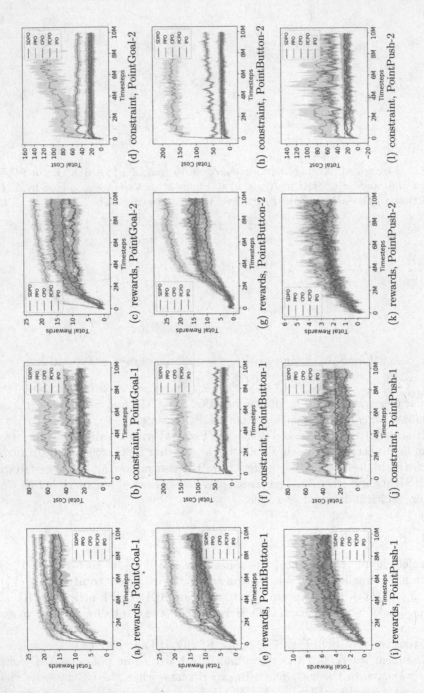

Fig. 7. Average performance of the point agent over 10 runs of PPO, SDPO, PCPO and IPO under Safety-Gym. Both SDPO, PCPO and IPO converge to the level indicated by the dashed line.

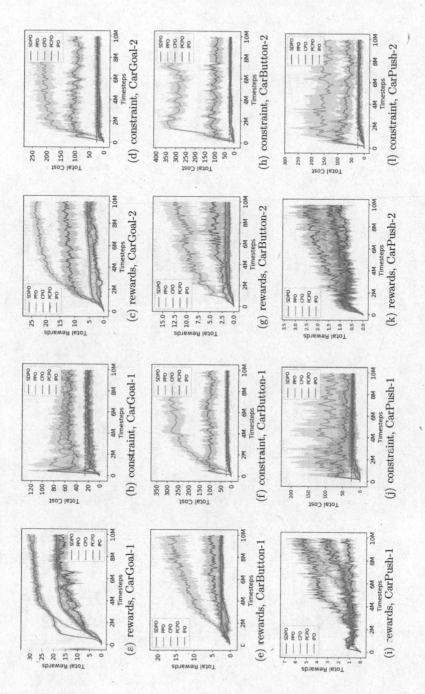

Fig. 8. Average performance of the car agent over 10 runs of PPO, SDPO, PCPO and IPO under Safety-Gym. Both SDPO, PCPO and IPO converge to the level indicated by the dashed line.

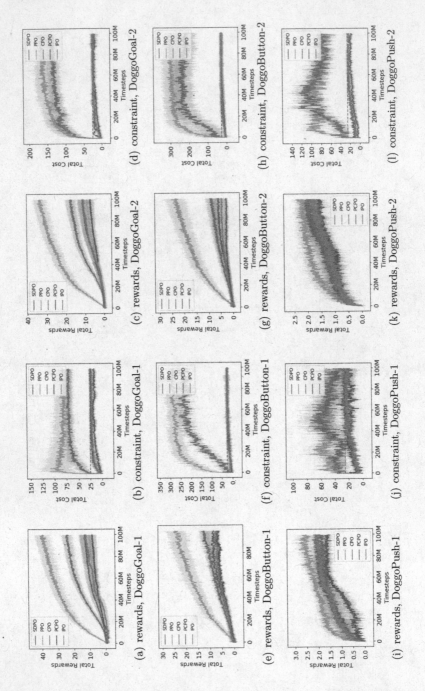

Fig. 9. Average performance of the doggo agent over 10 runs of PPO, SDPO, PCPO and IPO under Safety-Gym. Both SDPO, PCPO and IPO converge to the level indicated by the dashed line.

C   More Details on Experiments

C.1   Random CMDP

In the random CMDP domain, we constructed it with $N = 1000$ states and 10 actions. The number of randomly-chosen possible successor states is $\lceil \ln N \rceil = 7$. An episode in this CMDP is terminated after 100 time-steps. To achieve the results in Figs. 4 and 5, we trained the agents 5 times for both random seeds 5 and 10. The ADAM optimizer is used. The hyper-parameters are listed in Table 3.

Table 3. Hyperparameters for experiments on random CMDP.

	PPO	SDPO	PD-CVaR	PD-VAR
Discount factor γ	0.99	0.99	0.99	0.99
Batch size	1000	1000	1000	1000
Learning rate (actor)	1e–4	1e–4	1e–4	1e–4
Learning rate (critic)	1e–3	1e–3	/	/
Hidden sizes	(64, 64)	(64, 64)	(64, 64)	(64, 64)
GAE factor λ	0.9	0.9	/	/
Clip range ϵ	0.2	0.2	/	/
η	/	20	/	/
# quantile atoms N	/	128	/	/
Quantile dimension[a]	/	256	/	/

[a] Refer to Equation (4) in [16]

C.2   Stock Transaction

For the experiment on stock market, we use Quandl[1] in Python to load all market data. The trading agent is assumed to have zero market impact and zero transaction cost. When conducting this experiment, We choose 9 stocks in SP500 (AAPL, CSCO, DOW, GE, GS, JNJ, JPM, MMM, MSFT). The agents are initialized with a safe policy that always holding cash, and then trained in a rolling bias in year 2019 to evaluate the offline performance, i.e., at time step t, prices from $t - 15$ to t are used for training. The shared part of the actor and critic network is implemented as an LSTM network. The hyper-parameters are listed in Table 4.

C.3   Mujoco Simulator

For the parameters and other settings of the Point-Goal2 domain we used the default values set in the source code of Safety-Gym (see line 108 in safety-gym/ safety_gym/envs/suite.py). The hyper-parameters are listed in Table 5.

---

[1] https://quandl.com.

Table 4. Hyperparameters for experiments on stock transaction.

	PPO	SDPO
Discount factor γ	0.99	0.99
Batch size	1280	1280
Learning rate (actor)	1e–4	1e–4
Learning rate (critic)	1e–3	1e–3
Hidden sizes	$(64, 64)$	$(64, 64)$
GAE factor λ	0.9	0.9
Clip range ϵ	0.1	0.1
η	/	60
# quantile atoms N	/	128
Quantile dimension	/	256

Table 5. Hyperparameters for experiments on Point-Goal2.

	PPO	SDPO	IPO
Discount factor γ	0.99	0.99	0.99
Discount factor for constraints γ_1, γ_2	1	1	1
Batch size	30000	30000	30000
Learning rate (actor)	1e–4	1e–4	1e–4
Learning rate (critic)	1e–3	1e–3	1e–3
Hidden sizes (actor)	$(256, 256)$	$(256, 256)$	$(256, 256)$
Hidden sizes (critic)	$(256, 256)$	$(256, 256)$	$(256, 256)$
GAE factor λ	0.9	0.9	0.9
Clip range ϵ	0.1	0.1	0.1
η_1	/	40	60
η_2	/	60	60
# quantile atoms N	/	128	/
Quantile dimension	/	256	/

References

1. Achiam, J., Held, D., Tamar, A., Abbeel, P.: Constrained policy optimization. In: International Conference on Machine Learning (2017)
2. Alshiekh, M., Bloem, R., Ehlers, R., Könighofer, B., Niekum, S., Topcu, U.: Safe reinforcement learning via shielding. In: AAAI (2018)
3. Altman, E.: Constrained Markov Decision Processes. CRC Press, Boca Raton (1999)
4. Barth-Maron, G., et al.: Distributed distributional deterministic policy gradients. In: International Conference on Learning Representations (2018)

5. Bellemare, M.G., Dabney, W., Munos, R.: A distributional perspective on reinforcement learning, pp. 449–458 (2017)
6. Berkenkamp, F., Turchetta, M., Schoellig, A.P., Krause, A.: Safe model-based reinforcement learning with stability guarantees. In: NeurIPS (2017)
7. Bertoluzzo, F., Corazza, M.: Reinforcement learning for automated financial trading: basics and applications. In: Bassis, S., Esposito, A., Morabito, F.C. (eds.) Recent Advances of Neural Network Models and Applications. SIST, vol. 26, pp. 197–213. Springer, Cham (2014). https://doi.org/10.1007/978-3-319-04129-2_20
8. Borkar, V., Jain, R.: Risk-constrained Markov decision processes. IEEE Trans. Autom. Control 59(9), 2574–2579 (2014)
9. Borkar, V.S.: Learning algorithms for risk-sensitive control. In: International Symposium on Mathematical Theory of Networks and Systems (2010)
10. Boyd, S., Boyd, S.P., Vandenberghe, L.: Convex Optimization. Cambridge University Press, New York (2004)
11. Brázdil, T., Chatterjee, K., Novotný, P., Vahala, J.: Reinforcement learning of risk-constrained policies in Markov decision processes. In: AAAI (2020)
12. Cheng, R., Orosz, G., Murray, R.M., Burdick, J.W.: End-to-end safe reinforcement learning through barrier functions for safety-critical continuous control tasks. In: AAAI (2019)
13. Chow, Y., Ghavamzadeh, M.: Algorithms for CVaR optimization in MDPs. In: NeurIPS (2014)
14. Chow, Y., Ghavamzadeh, M., Janson, L., Pavone, M.: Risk-constrained reinforcement learning with percentile risk criteria. J. Mach. Learn. Res. 18(1), 1–51 (2017)
15. Chow, Y., Tamar, A., Mannor, S., Pavone, M.: Risk-sensitive and robust decision-making: a CVaR optimization approach. In: NeurIPS (2015)
16. Dabney, W., Ostrovski, G., Silver, D., Munos, R.: Implicit quantile networks for distributional reinforcement learning, pp. 1096–1105 (2018)
17. Dalal, G., Dvijotham, K., Vecerik, M., Hester, T., Paduraru, C., Tassa, Y.: Safe exploration in continuous action spaces. ArXiv (2018)
18. Fulton, N., Platzer, A.: Safe reinforcement learning via formal methods. In: AAAI Conference on Artificial Intelligence (2018)
19. García, J., Fernández, F.: A comprehensive survey on safe reinforcement learning. J. Mach. Learn. Res. 16, 1437–1480 (2015)
20. Geibel, P., Wysotzki, F.: Risk-sensitive reinforcement learning applied to control under constraints. J. Artif. Intell. Res. 24, 81–108 (2005)
21. Isele, D., Nakhaei, A., Fujimura, K.: Safe reinforcement learning on autonomous vehicles. In: IROS, pp. 1–6. IEEE (2018)
22. Jia, Y., Burden, J., Lawton, T., Habli, I.: Safe reinforcement learning for sepsis treatment. In: 2020 IEEE International Conference on Healthcare Informatics (ICHI), pp. 1–7. IEEE (2020)
23. Koenker, R., Hallock, K.F.: Quantile regression. J. Econ. Perspect. 15(4), 143–156 (2001)
24. Lim, S.H., Malik, I.: Distributional reinforcement learning for risk sensitive policies (2021). https://openreview.net/forum?id=19drPzGV691
25. Liu, Y., Ding, J., Liu, X.: IPO: interior-point policy optimization under constraints. In: AAAI Conference on Artificial Intelligence (2020)
26. Miryoosefi, S., Brantley, K., Daumé III, H., Dudík, M., Schapire, R.: Reinforcement learning with convex constraints. In: NeurIPS (2019)
27. Mnih, V., et al.: Human-level control through deep reinforcement learning. Nature 518(7540), 529–533 (2015)

28. Pilarski, P.M., Dawson, M.R., Degris, T., Fahimi, F., Carey, J.P., Sutton, R.S.: Online human training of a myoelectric prosthesis controller via actor-critic reinforcement learning. In: 2011 IEEE International Conference on Rehabilitation Robotics, pp. 1–7. IEEE (2011)

29. Pirotta, M., Restelli, M., Pecorino, A., Calandriello, D.: Safe policy iteration. In: International Conference on Machine Learning (2013)

30. Prashanth, L.A., Ghavamzadeh, M.: Variance-constrained actor-critic algorithms for discounted and average reward MDPs. Mach. Learn. **105**(3), 367–417 (2016). https://doi.org/10.1007/s10994-016-5569-5

31. Ray, A., Achiam, J., Amodei, D.: Benchmarking Safe Exploration in Deep Reinforcement Learning (2019)

32. Schulman, J., Levine, S., Abbeel, P., Jordan, M., Moritz, P.: Trust region policy optimization. In: ICML, pp. 1889–1897. PMLR (2015)

33. Schulman, J., Wolski, F., Dhariwal, P., Radford, A., Klimov, O.: Proximal policy optimization algorithms. CoRR (2017). http://arxiv.org/abs/1707.06347

34. Silver, D., et al.: Mastering the game of go without human knowledge. Nature **550**(7676), 354–359 (2017)

35. Singh, R., Zhang, Q., Chen, Y.: Improving robustness via risk averse distributional reinforcement learning. In: Learning for Dynamics and Control, pp. 958–968. PMLR (2020)

36. Sutton, R.S., Barto, A.G.: Reinforcement Learning: An Introduction. MIT Press, Cambridge (2018)

37. Sutton, R.S., McAllester, D.A., Singh, S.P., Mansour, Y., et al.: Policy gradient methods for reinforcement learning with function approximation. In: NeurIPS (2000)

38. Tessler, C., Mankowitz, D.J., Mannor, S.: Reward constrained policy optimization. In: International Conference on Learning Representations (2019)

39. Turchetta, M., Berkenkamp, F., Krause, A.: Safe exploration in finite Markov decision processes with gaussian processes. In: NeurIPS (2016)

40. Wachi, A., Sui, Y., Yue, Y., Ono, M.: Safe exploration and optimization of constrained MDPs using gaussian processes. In: AAAI (2018)

41. Xie, T., et al.: A block coordinate ascent algorithm for mean-variance optimization. In: NeurIPS (2018)

42. Yang, D., Zhao, L., Lin, Z., Qin, T., Bian, J., Liu, T.: Fully parameterized quantile function for distributional reinforcement learning. In: NeurIPS (2019)

43. Yang, T.Y., Rosca, J., Narasimhan, K., Ramadge, P.J.: Projection-based constrained policy optimization. In: ICLR (2020)

44. Yu, M., Yang, Z., Kolar, M., Wang, Z.: Convergent policy optimization for safe reinforcement learning. In: NeurIPS (2019)

The Positive Effect of User Faults over Agent Perception in Collaborative Settings and Its Use in Agent Design

Reut Asraf[✉] [ID], Chen Rozenshtein [ID], and David Sarne [ID]

Bar-Ilan University, Ramat Gan, Israel

Abstract. This paper studies the effect of user's own task-related faults over her satisfaction with a fault-prone agent in a human-agent collaborative setting. Through a series of extensive experiments we find that user faults make the user more tolerant to agent faults, and consequently more satisfied with the collaboration, in particular compared to the case where the user is performing faultlessly. This finding can be utilized for improving the design of collaborative agents. In particular, we present a proof-of-concept for such augmented design, where the agent, whenever in charge of allocating the tasks or can pick its own tasks, deliberately leave the user with a relatively difficult task for increasing the chance for a user fault, which in turn increases user satisfaction.

Keywords: Intelligent user interfaces · Human-agent collaboration · Human-agent interaction

1 Introduction

Computer systems (in their wider term that also includes remote services, Internet platforms and robots) are becoming more and more abundant. With the increase in their intelligence and uses (providing advice and information, collaborating in a joint task, etc.), they are now an integral part of almost every aspect of our daily modern life, substantially enriching us by their content and functionality [2]. However, these increases in intelligence and uses typically dictate architectures and designs that rely on complex machine learning algorithms and advanced hardware [21]. Furthermore, many of the environments these systems are now operating in lack structure and are highly unpredictable [3,30]. Consequently, these systems have become more prone to failures [11]. Such failures can take various forms. For example, a hardware failure can result in a robot malfunction, a software bug can lead to a system restart, and a failure to identify user's intentions can produce a wrong action. In many cases, users may even consider poor outcomes due to choices or recommendations made by the system

Preliminary results of this work (in particular, the results of experiments T2 and T3) were presented as a poster at HAI 2021 [1].

as "failure", even if these are fully attributable to external factors. For example, a GPS navigation system such as *Waze* and *Google Maps* may be considered to perform poorly if recommending a route which eventually turns to be very congested due to a car accident (that obviously could not be predicted at the time of making the original recommendation).

System failure has been recognized in prior work as strongly affecting user's satisfaction, adoption, and churn [11,14,25]. As such, much research has been devoted to studying agent failure in the context of human-agent interaction (see a detailed review in the following section). In many cases, however, especially when the interaction with the system is inherently collaborative, it is common that users themselves may cause a failure as well. Considering the GPS navigation system example, while the system is the one that is responsible for providing the navigation instructions, the ability of the user (i.e., the driver) to follow these instructions is crucial to navigating successfully. As a result, successful navigation can be viewed as a collaborative effort between the user and the system, where each has the potential to fail. For example, the system can suggest an erroneous route, report a wrong estimated time of arrival or use a road segment which is currently reserved for public transportation. Similarly, the user can get distracted and at times fail to follow the instructions received. Obviously, a user's local failure at some points along the joint mission will influence her perception of the collaborative experience. Still, to the best of our knowledge, the effect of users' own faults over their tolerance to and perception of agent's faults (and consequently their satisfaction with a fault-prone agent in a collaborative setting) has not been studied to date.

In this paper, we report the results of extensive experimentation aiming to reason about such effects. Using a web-based infrastructure that emulates a collaborative human-agent Captchas solving task, we apply four experimental treatments primarily differing in the way the faults influence task progress and attract players' focus. Subjects encountering the different settings are asked to rate their satisfaction with the agent and the collaborative task. The analysis of the results of 536 subjects participating in our experiments suggests that users' own faults play a significant role in the way they perceive agent's faults and consequently their satisfaction with the collaborative agent. Thus, users that made at least one fault are more tolerable to agent faults and assign a substantially greater satisfaction score when grading the agent and its competence.

The above finding is instrumental for agent design and its fault management logic. We illustrate the benefit of taking the phenomenon into account in agent design by evaluating the performance of an agent that instead of always signing for the more difficult tasks (as the optimal, task-duration-minimizing, strategy prescribes), allocates, at times, a rather difficult task to the user as well, hoping for a user fault to occur. We find that the proposed design results in a significantly improved user satisfaction with the agent, despite its counter-intuitiveness (as the additional user fault actually extends the task duration, which is supposed to negatively affect user satisfaction).

2   Related Work

System failure has been long recognized as strongly affecting user's perception of the system [14]. In particular, research has suggested (see Honig and Oron-Gilad [11]) that failures reduce a system's perceived sincerity [9], competence [19], reliability [24], trustworthiness [5,12], understandability [24], and intelligence. All these result in a decreased user satisfaction [6] that has a strong correlation with undesired behaviors such as user churn, ignoring the system (or preferring to work without it), limiting the system's capabilities, allowing it to act only upon a request, and ignoring its actions despite some of them possibly being useful [14]. These phenomena motivated the need for effective management of failure (and more broadly, the unexpected behavior of the system).

For a system failure to have some effect it needs to be apparent. In that sense, failure symptoms are numerous. For example, in human-robot interaction (HRI) these may include: the robot not completing a given task, running into obstacles, performing the wrong action, performing the right action incorrectly or incompletely, producing no action or speech (unresponsiveness), timing speech improperly, producing unexpected or erratic behavior, making knowledge-based mistakes and requiring additional information [11]. As such, much effort has been devoted to classifying failures according to severity [23], the ability to recover from them [22], their type (e.g., technical vs. social norm violation) [8], and their source (e.g., physical, interaction, and algorithmic failures) [28].

Honig and Oron-Gilad [11] proposed a human-robot failure taxonomy that distinguishes between technical and interaction failures, categorizing failure events according to their functional severity, social severity, relevance to the system, frequency of occurrence, context in which the failure occurred, and indicators used to identify the failure. Overall, despite the rich literature dealing with failure taxonomy, most studies considering system failures focused on technical ones, including a number of hardware errors and software bugs [5,11], as well as social norms violations [17,31], as opposed to interaction failures and those attributed to system decision making and intelligence.

Various studies have focused on the strategies that systems should apply after a failure, and found that the use of recovery strategies after a system failure can mitigate the effect of the failures [5,20]. These included setting expectations concerning the system's abilities [13], justifying the failure [5,20], apologizing [18], expressing regret [10,29], or responding non-verbally to a fault (e.g., looking away [27]). Some work has considered system recovery-from-failure methods, for cases where recovery is possible [16], and for intentionally communicating failures and their causes to users [4,26].

Despite the rich work cited above, the incorporation of failure-related considerations in agent design has not been considered in prior work. An exception can be found in our recent prior work [25], which studies how the tradeoffs between agent's performance (e.g., its contribution to the shared task) and its number of failures throughout affect user satisfaction. In that work, we show that agent failures directly affect user satisfaction, beyond the indirect effect through their impact on the completion of the task, and suggested a new paradigm for designing collaborative agents that takes this effect into account. Still, the described work does not consider the effect of users' own faults on their perception of the (fault-prone) agent. To the best of our knowledge, the current paper is the first to study this latter effect and incorporate it into agent design.

3 The Model

We consider a user-agent collaborative setting where both the (human) user and the agent need to separately execute tasks in order to achieve some specified shared goal. In this research we limit the tasks to be of a binary success nature, meaning that each task carried out by one of the players can either succeed or fail. A failure (or "fault" as used onwards) will hinder the overall mission (yet not completely fail it), e.g., delay its completion or reduce its performance score. Hence the agent may consider this possible outcome when reasoning about its desired course of action. The model assumes that both the agent and the user can observe the actions of one another, and identify in real time the occurrence of a fault (both own fault and faults of the other party). The goal of the agent is to maximize user satisfaction with the agent collaborator, which can be measured by various metrics [6,25,32].

The above model maps to various real-life settings, including the motivating example of the GPS navigation system discussed in the introduction section. Here, the user and agent share the goal of navigating successfully, minimizing journey time. This goal is achieved through the execution of separate tasks - the agent (i.e., the system) produces and maintains routing instructions to the destination, while the user follows them. Faults are possible on both sides. The agent can produce a futile instruction (for example, instructing the user to get off the main road, and then direct her to merge back to the main road at the exact interchange of departure), which will delay arrival time. The user, on the other hand, can take a wrong turn or miss a turn, resulting in a route re-plan that will also delay the arrival time.

4 Research Hypotheses

We hypothesize that user's own faults and their interaction with agent's faults have a substantial influence over user's satisfaction with the collaborative agent. In particular, we hypothesize that users that account for faults in task execution, will be more tolerant to agent's faults (compared to users that perform faultlessly) and more satisfied with the agent overall. Taking this phenomenon into account, many existing collaborative agent designs can be improved in a way that the agent's decision making incorporates the possible effect of user and agent faults on the resulting user satisfaction.

5 Experimental Framework

The experimental framework used in this research is based on a two-player turn-based game called "Solve the Captcha", which emulates a captcha-solving collaborative mission. One of the players in the game is human (termed "user" onward) while the other is a computer agent (termed "agent"), visually presented as a robot. Each player on her turn is provided with a captcha composed of randomly generated letters, symbols and digits, and needs to solve it, or at least attempt to solve it, within a reasonable time limit (for a screenshot of the game interface see Appendix A). Score (individual and joint) is correlated with the number of captchas solved. The task is considered collaborative in the sense that the (joint) ultimate goal is throughput-maximization. This can take the form of either maximizing the number of captchas successfully solved by the two players within a fixed time interval, maximizing joint score (which may take into account penalties for faults made) or minimizing the time it takes them to successfully solve some pre-specified number of captchas or reach some pre-specified score.

The framework is easily configurable, and in particular enables full control over the way faults are handled. For example, we can control the way we focus the user's attention to the fault through the use of pop-ups with textual mentions of the fault, or playing different sounds upon successfully solving a captcha or failing to solve it. Similarly, we can control the consequences of failing to solve a captcha correctly, e.g., switching the turn, applying a score penalty, introducing to the player a new captcha to solve, or some logical combination of the three. In terms of presenting the players' progress in the task, we can either present individual scores, joint score, or both.[1] Finally, we can dictate the way arriving captchas are assigned to the user and to the agent (e.g., according to arrival order, according to captcha's difficulty if some relevant signal can be obtained) and control the agent's ability to successfully solve captchas (e.g., increase the chance of solving wrong a captcha according to some properties).

6   Experimental Design

The "Solve the Captcha" task was developed as an interactive web-based game, using ASP.NET. Data was collected through the use of Amazon Mechanical Turk. We restricted participation to AMT workers who live in the US and have already completed more than 1000 tasks in the platform, and at least 98% of the tasks they have worked on were successfully approved. To prevent carryover effects, we employed a between-subjects design, assigning each participant to one treatment only.

Each participant in our experiments first received thorough instructions (and an illustration) regarding the task rules and her goal in the task. As part of the instructions, the agent was also introduced, referred to as a virtual player. Following the instructional phase, participants had to successfully solve a captcha and a quiz containing number of questions to ensure they understood the task instructions. Then, participants were directed to the actual task. Finally, participants

---

[1] Or, as an alternative to score, present the number of captchas successfully solved so far.

were asked to complete a questionnaire designed to evaluate their satisfaction with the agent and the way they perceive the interaction in general.

We hereby describe the experimental treatments used and the measures collected through the questionnaire.

6.1   Experimental Treatments

For studying user perception of agent faults given own faults we used three treatments (denoted T1, T2 and T3), primarily differing in the way the players' faults influence the task progress and attract players' focus (see Table 1 in Appendix B):

- T1 (Equal Effect) - here, the effect of a fault (either user's or agent's) is equal in the sense that the player will need to solve another captcha in order to proceed, delaying the joint task which is solving a total of ten captchas. Meaning that the turn is switched to the other player only upon solving a captcha. The chance of having the agent make a mistake in solving a captcha was set to 15% (per-captcha), yet if the agent reached the seventh round and has not made a fault so far then it was forced to fail solving the captcha, to guarantee at least one agent fault.
- T2 (Greater Severity, little focus) - here, an agent's fault has a greater (negative) effect over task completion (compared to the effect of a user's fault). This is reflected by having an agent's fault results in a decrease of three points in the joint score (i.e., three additional captchas need to be solved in order to reach the goal), whereas user's faults do not influence the joint score. The turn always switches to the other player upon submission of the captcha's solution, regardless of whether it was correct or not. The goal is to jointly reach a score of 24 solved captchas and the agent is pre-programmed to make two mistakes overall - the first when reaching (for the first time) a joint score of 8 or 9 solved captchas and the second when reaching (for the first time) a joint score of 17 or 18.
- T3 (Greater Severity, high focus) - same as T2, except that the agent is pre-programmed to make a mistake only once, the first time the joint score reaches 17, and turn transition takes place only upon solving a captcha (as otherwise the player has to continue trying to solve new captchas until succeeding). Seemingly, this latter change is minor, however it carries much weight in terms of the focus placed on the fault made. In T2 the players move on and the load of solving the three additional captchas until reaching again to the original score is primary on the non-faulty player (two out of the three captchas, as the turn switched right away). In T3, the player who made the mistake continues solving captchas until managing to succeed in solving, hence that player takes most of the load. Overall, these two factors (the number of agent faults and the turn-switching rule applied) have a similar effect, in the sense that in T3, they both influence positively on user's perception of the agent, and vice versa in T2, therefore there is no worry that conflicting influences will eliminate one another. We note that T2 and T3 were designed to cover

different aspects of agent characteristics so that findings can be generalized, rather than to draw conclusions based on a direct comparison of their results.

A list of state machines associated with each treatment flow can be found in Appendix C.

Unlike with treatments T2 and T3, where same type of captchas are presented to the user and to the agent to solve, in treatment T1 users received significantly easier captchas to solve, compared to the agent's captchas. This choice was made for two primary reasons. First, by having an unequal-captcha-difficulty treatment we would be possibly gaining further insight regarding the effect of this aspect over user satisfaction within the context of player's faults. Second, and more importantly, with this variation in captcha allocation, we were able to use T1 as a baseline for evaluating the performance of a design that takes user's faults (or lack of) into account when deciding on captcha allocation, as we explain in the following paragraphs.

For providing a proof-of-concept for the superiority of designs that take into account user's faults as a factor influencing her satisfaction with the collaborating agent, we used an additional treatment denoted T4. Treatment T4 is similar to T1 except for the following (see state machine in Fig. 6(d)): unlike with the agent used in T1, when the agent in T4 makes a mistake and finds that the user performed faultlessly so far, it changes its captchas allocating method for the following turn, switching from allocating the easier captchas to the user to allocating the "more challenging" captcha among those remaining.[2] The idea in assigning a challenging captcha to the user is to increase the chance she will make a fault, hence increase her tolerance to the agent's faults based on our hypothesis. We emphasize that while having the user make a mistake may have a positive effect over her tolerance to agent's mistakes, this also carries some disadvantages as it hinders the task performance. Meaning that overall there might also be a secondary (negative) effect over the user's satisfaction with the collaborative process as the task will take longer to execute. Still, we believe the first effect is more substantial, hence the proposed change in agent design. Since T4 differs from T1 only in the sense that the agent design incorporates the chance for user's fault in its captcha allocation reasoning, the comparison of the user satisfaction measure between the two can provide an effective evaluation for the proposed design.

6.2   Measures

At the end of the task, participants were asked to complete a short question-naire designed to evaluate their satisfaction with the agent they interacted with, and gain insights on how they perceived both the agent and the collaborative

---

[2] A challenging captcha is one that combines some characters that are difficult to distinguish. For example in "bXF0yrl" it is not clear if the middle character is the letter "o" or the number 0. Or, one where it is difficult to distinguish between lower case "L" and upper case "I". While these are difficult to humans to distinguish, a computer agent will easily learn their pattern and distinguishing pixels.

experience as a whole. In order to assess satisfaction from different perspectives, a numerical rating was requested for the following three measures, on a scale of 1 (lowest) to 10 (highest): Competence, Satisfaction and Recommendation (see details in Appendix D). Additionally, participants were asked to provide a textual explanation for their choice of the numeric satisfaction rating, as well as to detail their thoughts about the agent. Both questions were open-ended.

7   Results and Analysis

Overall, we had 536 participants taking part in our experiments according to the following breakdown: 136 in T1, 103 in T2, 161 in T3 and 136 in T4. All participants were from the US. The average age was 41.13 (ranging between 18–76), of which 50% were men and 50% were women.

We divide the results analysis into three parts. First, we show the effect of user's own faults over her satisfaction with the collaborative (though fault-prone) agent. Then we provide the results of user's satisfaction with the agent which design takes into account the former effect. Finally, we bring insights from the textual responses received. For space considerations, we provide results and graphs only based on the user satisfaction measure, which we consider to be the primary one. Similar graphs for all other measures are provided in Appendix E, depicting similar behaviors and phenomena, qualitatively.

While the results analysis relies primarily on average satisfaction, we will be providing also its complementary measure - the average dissatisfaction. Since satisfaction is bounded by a score 10, we can consider the distance between the average satisfaction and the maximum value 10 as the extent of user's dissatisfaction (and similarly with the other measures). Meaning that when comparing the results of two treatments we can measure both the relative increase in user satisfaction and the complementing relative decrease in user's dissatisfaction under different conditions.[3]

7.1   The Effect of User's Own Faults over Her Satisfaction with the Collaborative Agent

We begin with treatment T1, where user and agent faults have equal effect. Since the nature of treatments T1 and T4 is identical and the only difference is in pushing the user, at times, to make a mistake, we can augment the results of T1 with those of T4 in order to strengthen the statistical validity of findings. Figure 1a depicts the average user satisfaction for different combinations of user faults and agent faults as reflected in the two treatments.[4]

---

[3] The actual increase in satisfaction equals the decrease in dissatisfaction, and vice versa. However the relative increase and decrease are different, as the calculation takes a different baseline to begin with.

[4] Out of the 272 subjects of T1 and T4, only 14 experienced more than two agent faults, hence their results were added to the pool of 73 subjects who experienced two agent faults, forming the category "two or more agent faults". Only eight subjects made two and more faults hence their category is excluded from the graph.

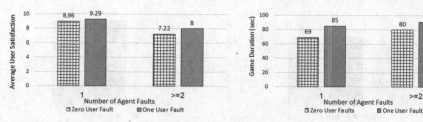

(a) Average user satisfaction as a function of the number of user and agent faults.

(b) Average task duration as a function of the number of user and agent faults.

Fig. 1. Average user satisfaction (left) and average task duration (right) in treatments T1 + T4.

From the graph we observe that for both levels of agent faults, user satisfaction with the agent increases whenever the user herself was faulty, compared to when the user performed faultlessly. The relative increase in average user satisfaction is 4% (considering the ratings given by 85 subjects exhibited no faults, and 95 subjects exhibited one fault) and 11% (32 subjects vs. 52 subjects, similarly) for one and two-and-above agent faults, respectively. Similarly, the relative decrease in average dissatisfaction is 32% and 28%. In all cases the difference is statistically significant based on t-test ($p < 0.05$). Non-surprisingly, for any number of user faults, the user satisfaction decreases as the number of agent faults increases. We emphasize that the agent faults directly influence the time it takes to complete the joint task. Therefore, the decrease in satisfaction can also be attributed to that factor. Still, from Fig. 1b, which depicts the average duration of the task (in seconds) for any combination of user faults and agent faults, we observe that the increase in task duration due to an increase in the number of faults made by the agent is 6–11 s. This increase is quite moderate, from the user's point of view, especially when taking into account also the time it takes to read the task instructions and successfully complete the quiz. Therefore, the decrease in user satisfaction is mostly attributed to the absolute number of faults made by the agent.

We emphasize that an increase in task's duration typically has a negative effect on user satisfaction overall. This phenomenon is probably more apparent in our setting where subjects are AMT users and place much importance on the time they spend. Therefore, the actual increase in the average user satisfaction in cases where a fault was made by the user, as reflected in Fig. 1a, is merely a lower bound, as it also combines the overall decrease in user satisfaction due to the increase in task duration.

Figures 2 and 3 depict the average user satisfaction for different numbers of user faults, in treatments T2 and T3, respectively. Here, as explained above, the agent accounts for a fixed number of faults in each treatment: two faults in T2 and one fault in T3. The graphs reflect the same pattern as in treatment T1, according to which user satisfaction increases whenever the user makes a fault, compared to when she performs faultlessly. Here, the increase in average

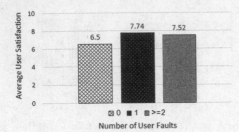

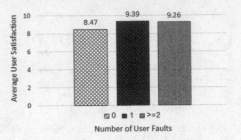

Fig. 2. Average user satisfaction as a function of the number of user faults in treatment T2.

Fig. 3. Average user satisfaction as a function of the number of user faults in treatment T3.

satisfaction and the corresponding decrease in average dissatisfaction are quite impressive (average satisfaction increase of 19% in T2 considering 44 subjects who had no faults vs. 38 subjects who had one fault, and 11% in T3 considering 32 subjects vs. 51 subjects, similarly, and dissatisfaction decrease of 35% and 60%, respectively, all statistically significant based on t-test, $p < 0.002$). Furthermore, it is apparent from the graphs that the average user satisfaction with the agent in T2 is lower than in T3 for all numbers of user faults. We suggest two possible explanations for this phenomenon. The first is that the number of agent faults during the game is higher in T2 compared to T3 (two faults vs. one fault, respectively). This strengthens the claim made based on the results of T1 and T4, according to which user satisfaction decreases as the number of agent faults increases. Our second suggestion relates to how the agent recovers from a fault, which varies among these treatments. In T3, whenever the agent fails it must solve a new captcha before the turn can be passed to the user; however, in T2 the turn is passed immediately after the agent fails. We believe that the mechanism applied in T3 can be viewed as a recovery strategy adopted by the agent that helps mitigate the negative effect of its faults over user satisfaction.

One interesting finding reflected in Figs. 2 and 3 is that further increase in the number of user faults (i.e., beyond the transition from zero to one user fault) does not necessarily result in further improvement in user satisfaction. In our settings, the comparison of the average satisfaction between two and more faults to one fault reveals a (non-statistically-significant) decrease of 3% and 1% in treatments T2 and T3, respectively. While further research is required to investigate the effect of different numbers of human faults, we speculate that the above phenomenon is the result of having the second fault reducing the uniqueness of faults in one's mind (e.g., if experiencing it twice then perhaps it is a common thing in this type of task). Hence it is primarily the event of first user fault that positively affect users' tolerance to agent's faults.

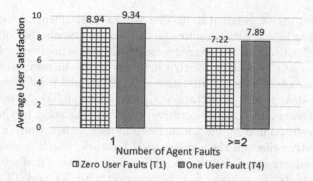

Fig. 4. Comparison of user satisfaction in T1 and T4, providing a proof of concept for the success of designs that incorporate user's own faults effect over her satisfaction with the collaborative agent.

7.2 Incorporating User Faults in Agent Design - Proof of Concept

As a proof of concept for our ability to utilize the above findings for improving agent design, we provide a detailed comparative analysis of the results obtained in treatments T1 and T4. As explained above, the two treatments are identical, except that they differ in the way the agent allocates captchas to the user and to itself. In T1, the agent always allocate the more (human) challenging captchas to itself whereas the user gets the easier captchas. This is an optimal allocation design whenever the goal is to minimize the expected duration of the task— as explained above, the features that make the captcha challenging to human users in our experiments have no effect over the time it takes the agent to solve it, hence the agent's probability of correctly solve a captcha is constant. For the human user, however, receiving the less challenging captchas will definitely (positively) affect the time it takes to solve the captcha and the probability of solving it successfully. Alas, as shown above, minimizing the expected task duration does not necessarily guarantee maximizing user satisfaction. Instead, user satisfaction is also influenced by the faults made by the agent and by the user. Consequently, the agent in T4 will strive to increase the chance the user will be responsible to a captcha-solving fault (preferably a single fault overall, to keep the increase in task duration minimal). As mentioned earlier (see also in Fig. 6), this is achieved by allocating a challenging captcha to the user after an agent mistake, in case that a user fault has not occurred so far in the task. Indeed, this is highly non-intuitive in terms of overall task duration considerations. Yet with the increased chance for a user fault, the user will likely fail, and this in turn is likely to substantially increase her satisfaction with the collaborative agent. All in all, users in T1 in our experiments made 0.19 faults on average, compared to 1.04 in T4. The average number of agent faults was 1.375 and 1.4 in T1 and T4, respectively, reflecting a non-significant change as the agent is not affected by the type of captcha assigned to it. The average task duration increased from 75 s in T1 to 89 s in T4. Still, despite the increase in the number of faults made and in the task duration, the average user satisfaction increased from 8.52 in T1 to 8.84 in T4, reflecting a relative increase of 4% in user satisfaction and relative decrease of 22% in user dissatisfaction. This latter difference in average satisfaction is statistically significant with t-test ($p < 0.05$). Meaning that the proposed (and somehow counter-intuitive) captcha allocation mechanism in the agent design resulted in a significant increase in user satisfaction. In fact, we can see that a negative effect in the average task duration, which is one of the forms of the agent's effectiveness, led to a positive effect in user satisfaction.

Figure 4 depicts the average user satisfaction with T1 and T4, for the case the agent made a single fault and two or more faults, demonstrating that the improvement achieved is consistent between different number of agent faults. Notice that for any number of agent faults we compare the average satisfaction of those that had no fault in T1 and those that had a single fault in T4 as a result of being assigned a challenging captcha.[5] Therefore the comparison made in the figure is over the population of subjects actually being affected by the new captcha assignment rule. In this sense, we ultimately validate that it is the agent's decision to push the user to become faulty in T4 that accounts for the improved satisfaction. This completely equalizes all parameters in the two treatments except for the intervention. Here we find that the average satisfaction of subjects increased due to the additional user fault (98% of those that received a challenging captcha indeed failed in solving it), regardless of the number of agent faults. With one agent fault, the average user satisfaction increased from 8.94 (among 83 subjects) to 9.34 (among 80 subjects), reflecting a relative increase of 4% in satisfaction and relative decrease of 38% in dissatisfaction. With two and more agent faults (32 subjects who made no faults, and 44 subjects who made one fault) the relative increase in user satisfaction and relative decrease in dissatisfaction are 9% and 24%, respectively. Overall, the change in the agent's captcha allocation rule increased user satisfaction by 4% and decreased dissatisfaction by 23% in the population to which the change actually applied (statistically significant, $p < 0.04$).

As a final means of validation that the change in user satisfaction is the result of the captcha allocation method, we compare the average user satisfaction between those that had a single fault in T1 (i.e., had a fault not due to our intervention) and those from T4 that had a single fault following our intervention. The average satisfaction of subjects from the two populations is 8.89 and 8.82 for T1 and T4, respectively. The difference is not statistically significant ($p = 0.38$) suggesting that the fact that the fault in T1 was natural and in T4 it resulted from a challenging captcha had no effect. The effect therefore can be fully attributed to the fact that the agent managed to push the user to fail.

---

[5] Meaning that subjects that already had a single fault from earlier rounds (hence did not receive a challenging captcha as no intervention was needed) are excluded from this analysis.

7.3 Participants' Qualitative (Textual) Responses

Alongside the quantitative data (satisfaction, competence and recommendation ratings), we have received much qualitative (textual) data from subjects, in the form of responses to open-ended questions. The analysis of this data reveals much insight related to participants' rating motives and perceptions of the agent and the entire experimental experience. Overall, we identify four high-level aspects influencing the user's satisfaction with the collaborative agents: agent competence per se, agent competence vs. participant competence, expectations from an automated agent and the complexity of the agent's tasks. Further details regarding each aspect as well as representative quotes from the participants can be found in Appendix F.

The first aspect (agent competence) is indeed a recurring theme in prior work [24, 25] and is also supported in our own results analysis above. The second aspect (agent competence vs. participant competence) is the one studied in the current paper and its high recognition rate in subjects' explanations strengthens the importance of incorporating this aspect in agent design. The last two aspects (expectations from an automated agent and the complexity of its task) are definitely a recurring theme in human-agent research, however sure require further study of the topic in the context of managing agent faults and the influence over user satisfaction with the collaboration.

8 Discussion, Conclusions and Future Work

The results analysis as reported in the previous section aligns with and strengthens prior work suggesting that agent's faults directly affect user satisfaction with the agent, in collaborative setting, beyond the indirect effect of such faults in the form of influencing task execution success measures [25]. Indeed, in all treatments used, an increase in the number of agent faults resulted in a decreased user satisfaction. However, and more importantly, the results analysis reveals that user's own faults also play a key role in establishing user's satisfaction with the agent. Specifically, we find that user's faults, and primarily the transition to a single fault (as opposed to performing faultlessly), substantially improve user satisfaction with the agent (as well as her estimate of its competence and willingness to recommend it to others, as reflected in the additional graphs that relate to these measures in Appendix E). To the best of our knowledge, we are the first to report and demonstrate this phenomenon.

The above findings carry much importance for collaborative agent designers, as they suggest that an effective agent design should not aim merely to optimize task performance measures but also take into consideration the direct effect of outcomes classified as faults. In particular, agents should not necessarily aim to keep the user from making faults (e.g., assign them with simple tasks) as a user fault can completely (positively) change her attitude and tolerance towards the agent's own faults. One such design which was tested in our experiments as a proof of concept suggests the deliberate assignment of a complex task to the user to implicitly encourage user fault.

The results indicate substantial improvement in user satisfaction with the proposed agent, compared to the agent which aims to minimize task duration. We emphasize that the improvement achieved is actually a lower bound for how much one can achieve from designs that take into consideration the effect of user faults over user satisfaction, as we took a very basic approach in our agent design. Further improvement can be potentially achieved by tuning the mechanism and including additional fault-related considerations such as fault timing.

One interesting question arising in light of the above results is whether there are psychological explanations supporting the phenomenon described. To the best of our knowledge, there is no prior work that can explicitly explain this phenomenon, however, there is other work suggesting some related effects that can implicitly support it. Among them are, for example, the role of subjective task complexity effects on performance [15], and the way people react to their own mistakes (though not in the context of human-agent interaction) [7]. Some explanation can also be found in the subjects' textual feedback, as some subjects seem to indicate that it is the breaking of the asymmetry between the flawless human player and the faulty agent that does the trick—when the user herself is faulty, becoming more tolerant to agent faults, in a way, reduces the self-blame for own fault. Still, this requires further research, possibly with mixed-methods approach (e.g., involving interviews and questionnaires).

The data collected in our experiments on cases where the user made more than a single fault seems to show that further increase in the number of user faults has little additional effect over user satisfaction if at all. Still, this data is limited in its extent, and further experiments are required in order to provide cohesive analysis of such cases.

Appendices

A   Experimental Framework Interface

Below is a screenshot of the experimental framework interface appeared to the participants (Fig. 5).

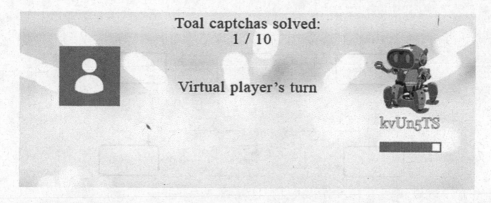

Fig. 5. A screenshot of the "Solve the Captcha" interface.

B Experimental Treatments Comparison

Table 1. Experimental treatments.

	T1	T2	T3	T4
Goal (# of captchas)	10	24	24	10
Score reduction due to agent fault	0	3	3	0
Turn switches upon	Solving captcha	Submitting answer	Solving captcha	Solving captcha
Number of agent faults	Chance (15% per turn)	2	1	Chance (15% per turn)
Captcha allocation method	Easier for user always	Regular queue	Regular queue	Usually easier for user

C State Machines

Figure 6 shows the state machines associated with each treatment flow. For simplicity, arrows pointing at the end state do not appear in all relevant states. However, the game does end when the pre-specified score is reached.

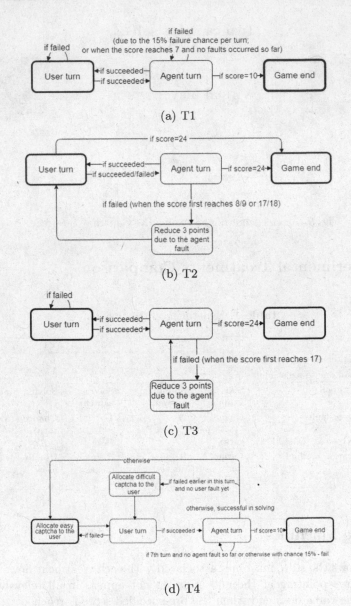

(a) T1

(b) T2

(c) T3

(d) T4

Fig. 6. State machines representing treatments flows.

D   Measures

- **Competence.** (Question: "To what extent did you find the virtual player to be a competent partner?") - this measure aims to capture the agent's degree of ability, from the participant's point-of-view.
- **Satisfaction.** (Question: "To what extent are you satisfied with the virtual player?") - this measure reflects the overall user experience and degree of happiness with the agent.
- **Recommendation.** (Question: "To what extent would you recommend the virtual player to a friend, as a partner to work with?") - this measure reflects the user's loyalty.

E   Complementary Graphs

While the graphs provided in the paper applied to the main measure of interest, user satisfaction, the equivalent graphs for the other two measures (user's estimate for the agent's competence and her willingness to recommend the agent to other users) reveal similar phenomena, qualitatively. We therefore provide these figures here.

E.1   Competence

See Figs. 7, 8, 9 and 10.

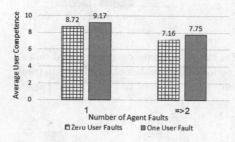
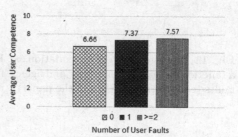

Fig. 7. Average user competence as a function of the number of user and agent faults (equivalent to Fig. 1a in the paper).

Fig. 8. Average user competence as a function of the number of user faults in treatment T2 (equivalent to Fig. 2 in the paper).

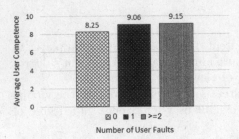

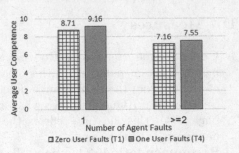

Fig. 9. Average user competence as a function of the number of user faults in treatment T3 (equivalent to Fig. 3 in the paper).

Fig. 10. Comparison of user competence in T1 and T4, providing a proof of concept for the success of designs that incorporate user's own faults effect over her competence from the collaborative agent (equivalent to Fig. 4 in the paper).

E.2   Recommendation

See Figs. 11, 12, 13 and 14.

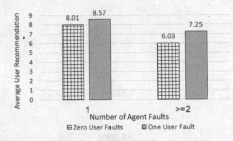

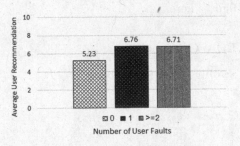

Fig. 11. Average user recommendation as a function of the number of user and agent faults (equivalent to Fig. 1a in the paper).

Fig. 12. Average user recommendation as a function of the number of user faults in treatment T2 (equivalent to Fig. 2 in the paper).

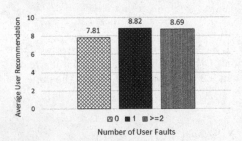

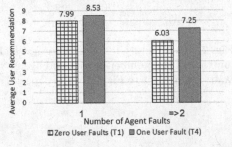

Fig. 13. Average user recommendation as a function of the number of user faults in treatment T3 (equivalent to Fig. 3 in the paper).

Fig. 14. Comparison of user recommendation in T1 and T4, providing a proof of concept for the success of designs that incorporate user's own faults effect over her recommendation from the collaborative agent (equivalent to Fig. 4 in the paper).

F   Participants' Qualitative (Textual) Responses

- Agent competence per se - several subjects justified their rating solely based on the agent performance, i.e., on how they perceived their agent's competency. Some of them expressed it explicitly (*"The virtual player made several mistakes"*, *"He only got one wrong and was quick"*), while others implicitly (*"This virtual player didn't do very well"*, *"My virtual partner was quick and accurate. A great team player!"*).

- Agent competence vs. participant competence - many subjects explained their rating by comparing the agent competence (mostly based on its number of faults) to their own. This pattern was observed in various cases, in which the number of the agent's faults was greater, equal, or smaller than that of the participant. For example, *"The player did some extra mistake compared to me."* (two agent faults vs. one participant fault), *"It was as competent as I was, only making one mistake."* (one agent fault vs. one participant fault), *"Both of us solved with just one error."* (one agent fault vs. one participant fault), *"played almost as good as me"* (one agent fault vs. zero participant faults).

- Expectations from an automated agent - participants also correlated their ratings with their initial assumptions about and expectations from the agent. In particular, most participants emphasizing this aspect mentioned they expected the agent to be flawless. For example, *"The virtual player only made one mistake, but I don't think a virtual player should have made any mistakes at all"*, *"Baffled how a machine could err on such a simple task"*, *"If it truly was virtual, as in AI, it shouldn't have missed any."*, *"I expect a computer to be competent"*. Several subjects compared the automated agent to a real person (*"Virtual players is like real player"*, *"It kind of felt like the virtual player was a real player even though I knew it wasn't."*), once again expressing their disappointment from agent faults.

- The complexity of the agent's tasks - some participants correlated their rating with the relative complexity of the agent's tasks compared to their own task. In particular, the fact that players (in T1 and T4) typically received the easier captcha seemed to had some influence on rating. For example, *"He did as well I did, and it looked like he had harder puzzles"*, *"He only made a single mistake and had what looked like more complex captchas"*, *"It only made one mistake and always seemed to have longer ones to solve than I did."*.

References

1. Asraf, R., Rozenshtein, C., Sarne, D.: On the effect of user faults on her perception of agent's faults in collaborative settings. In: Proceedings of the 9th International Conference on Human-Agent Interaction, HAI 2021, pp. 372–376. Association for Computing Machinery, New York (2021)
2. Azaria, A., Krishnamurthy, J., Mitchell, T.M.: Instructable intelligent personal agent. In: Thirtieth AAAI Conference on Artificial Intelligence (2016)

3. Azaria, A., Richardson, A., Kraus, S.: An agent for the prospect presentation problem. Technical report (2014)
4. Cassenti, D.N.: Recovery from automation error after robot neglect. In: Proceedings of the Human Factors and Ergonomics Society Annual Meeting, vol. 51, pp. 1096–1100. Sage Publications, Los Angeles (2007)
5. Correia, F., Guerra, C., Mascarenhas, S., Melo, F.S., Paiva, A.: Exploring the impact of fault justification in human-robot trust. In: Proceedings of the 17th International Conference on Autonomous Agents and Multiagent Systems, pp. 507–513 (2018)
6. Dabholkar, P.A., Spaid, B.I.: Service failure and recovery in using technology-based self-service: effects on user attributions and satisfaction. Serv. Ind. J. 32(9), 1415–1432 (2012)
7. Frost, R.O., Turcotte, T.A., Heimberg, R.G., Mattia, J.I., Holt, C.S., Hope, D.A.: Reactions to mistakes among subjects high and low in perfectionistic concern over mistakes. Cogn. Ther. Res. 19(2), 195–205 (1995). https://doi.org/10.1007/BF02229694
8. Giuliani, M., Mirnig, N., Stollnberger, G., Stadler, S., Buchner, R., Tscheligi, M.: Systematic analysis of video data from different human-robot interaction studies: a categorization of social signals during error situations. Front. Psychol. 6, 931 (2015)
9. Gompei, T., Umemuro, H.: A robot's slip of the tongue: effect of speech error on the familiarity of a humanoid robot. In: 2015 24th IEEE International Symposium on Robot and Human Interactive Communication (RO-MAN), pp. 331–336. IEEE (2015)
10. Hamacher, A., Bianchi-Berthouze, N., Pipe, A.G., Eder, K.: Believing in BERT: using expressive communication to enhance trust and counteract operational error in physical human-robot interaction. In: 2016 25th IEEE International Symposium on Robot and Human Interactive Communication (RO-MAN), pp. 493–500. IEEE (2016)
11. Honig, S., Oron-Gilad, T.: Understanding and resolving failures in human-robot interaction: literature review and model development. Front. Psychol. 9, 861 (2018)
12. Law, N., Yuen, A., Shum, M., Lee, Y.: Final report on phase (ii) study on evaluating the effectiveness of the "empowering learning and teaching with information technology" strategy (2004/2007). Education Bureau HKSAR, Hong Kong (2007)
13. Lee, M.K., Kiesler, S., Forlizzi, J., Srinivasa, S., Rybski, P.: Gracefully mitigating breakdowns in robotic services. In: 2010 5th ACM/IEEE International Conference on Human-Robot Interaction (HRI), pp. 203–210. IEEE (2010)
14. LeeTiernan, S., Cutrell, E., Czerwinski, M., Hoffman, H.G.: Effective notification systems depend on user trust. In: INTERACT, pp. 684–685 (2001)
15. Mangos, P.M., Steele-Johnson, D.: The role of subjective task complexity in goal orientation, self-efficacy, and performance relations. Hum. Perform. 14(2), 169–185 (2001)
16. Mendoza, J.P., Veloso, M., Simmons, R.: Plan execution monitoring through detection of unmet expectations about action outcomes. In: 2015 IEEE International Conference on Robotics and Automation (ICRA), pp. 3247–3252. IEEE (2015)
17. Mirnig, N., Stollnberger, G., Miksch, M., Stadler, S., Giuliani, M., Tscheligi, M.: To err is robot: how humans assess and act toward an erroneous social robot. Front. Robot. AI 4, 21 (2017)
18. Peltason, J., Wrede, B.: The curious robot as a case-study for comparing dialog systems. AI Mag. 32(4), 85–99 (2011)

19. Ragni, M., Rudenko, A., Kuhnert, B., Arras, K.O.: Errare humanum est: erroneous robots in human-robot interaction. In: 2016 25th IEEE International Symposium on Robot and Human Interactive Communication (RO-MAN), pp. 501–506. IEEE (2016)

20. Robinette, P., Howard, A.M., Wagner, A.R.: Timing is key for robot trust repair. In: ICSR 2015. LNCS (LNAI), vol. 9388, pp. 574–583. Springer, Cham (2015). https://doi.org/10.1007/978-3-319-25554-5_57

21. Rosenfeld, A., Zuckerman, I., Segal-Halevi, E., Drein, O., Kraus, S.: NegoChat-A: a chat-based negotiation agent with bounded rationality. Auton. Agent. Multi-Agent Syst. **30**(1), 60–81 (2016). https://doi.org/10.1007/s10458-015-9281-9

22. Ross, R., Collier, R., O'Hare, G.M.: Demonstrating social error recovery with agentfactory. In: Proceedings of the Third International Joint Conference on Autonomous Agents and Multiagent Systems, vol. 3, pp. 1424–1425 (2004)

23. Rossi, A., Dautenhahn, K., Koay, K.L., Walters, M.L.: Human perceptions of the severity of domestic robot errors. In: Kheddar, A., et al. (eds.) ICSR 2017. LNCS, vol. 10652, pp. 647–656. Springer, Cham (2017). https://doi.org/10.1007/978-3-319-70022-9_64

24. Salem, M., Lakatos, G., Amirabdollahian, F., Dautenhahn, K.: Would you trust a (faulty) robot? Effects of error, task type and personality on human-robot cooperation and trust. In: 2015 10th ACM/IEEE International Conference on Human-Robot Interaction (HRI), pp. 1–8. IEEE (2015)

25. Sarne, D., Rozenshtein, C.: Incorporating failure events in agents' decision making to improve user satisfaction. In: Bessiere, C. (ed.) Proceedings of the Twenty-Ninth International Joint Conference on Artificial Intelligence, IJCAI-2020, pp. 1549–1555. International Joint Conferences on Artificial Intelligence Organization, July 2020. Main track

26. Schütte, N., Mac Namee, B., Kelleher, J.: Robot perception errors and human resolution strategies in situated human-robot dialogue. Adv. Robot. **31**(5), 243–257 (2017)

27. Shiomi, M., Nakagawa, K., Hagita, N.: Design of a gaze behavior at a small mistake moment for a robot. Interact. Stud. **14**(3), 317–328 (2013)

28. Steinbauer, G.: A survey about faults of robots used in RoboCup. In: Chen, X., Stone, P., Sucar, L.E., van der Zant, T. (eds.) RoboCup 2012. LNCS (LNAI), vol. 7500, pp. 344–355. Springer, Heidelberg (2013). https://doi.org/10.1007/978-3-642-39250-4_31

29. Takayama, L., Groom, V., Nass, C.: I'm sorry, Dave: i'm afraid i won't do that: social aspects of human-agent conflict. In: Proceedings of the SIGCHI Conference on Human Factors in Computing Systems, pp. 2099–2108 (2009)

30. Wellman, M.P.: Putting the agent in agent-based modeling. Auton. Agent. Multi-Agent Syst. **30**(6), 1175–1189 (2016). https://doi.org/10.1007/s10458-016-9336-6

31. van der Woerdt, S., Haselager, P.: Lack of effort or lack of ability? Robot failures and human perception of agency and responsibility. In: Bosse, T., Bredeweg, B. (eds.) BNAIC 2016. CCIS, vol, 765, pp. 155–168. Springer, Cham (2017). https://doi.org/10.1007/978-3-319-67468-1_11

32. Zhou, S., Bickmore, T., Paasche-Orlow, M., Jack, B.: Agent-user concordance and satisfaction with a virtual hospital discharge nurse. In: Bickmore, T., Marsella, S., Sidner, C. (eds.) IVA 2014. LNCS (LNAI), vol. 8637, pp. 528–541. Springer, Cham (2014). https://doi.org/10.1007/978-3-319-09767-1_63

Behavioral Stable Marriage Problems

Andrea Martin[1]([⊠]), Kristen Brent Venable[1,2], and Nicholas Mattei[3]

[1] University of West Florida, Pensacola, FL 32514, USA
anm84@students.uwf.edu
[2] Institute for Human and Machine Cognition, Pensacola, FL 32502, USA
[3] Tulane University, New Orleans, LA 70118, USA

Abstract. The stable marriage problem (SMP) is a mathematical abstraction of two-sided matching markets with many practical applications including matching resident doctors to hospitals and students to schools. Several preference models have been considered in the context of SMPs including orders with ties, incomplete orders, and orders with uncertainty, but none have yet captured behavioral aspects of human decision making, e.g., contextual effects of choice. We introduce *Behavioral Stable Marriage Problems (BSMPs)*, bringing together the formalism of matching with cognitive models of decision making to account for multi-attribute, non-deterministic preferences and to study the impact of well known behavioral deviations from rationality on two core notions of SMPs: stability and fairness. We analyze the computational complexity of BSMPs and show that proposal-based approaches are affected by contextual effects. We then propose and evaluate novel ILP and local-search-based methods to efficiently find optimally stable and fair matchings for BSMPs.

Keywords: Stable marriage · Psychological decision-making

1 Introduction

The stable marriage problem (SMP) has a variety of applications in the context of two-sided markets, including matching doctors to hospitals and students to schools [24]. Typically, n men and n women express their preferences, via a strict total order, over the members of the other sex. Solving an SMP typically means finding a matching between men and women satisfying certain properties including *stability*, where no man and woman who are not married to each other would both prefer each other to their partners or to being single. Another desirable property is fairness, aiming at a balance between the satisfaction of the two groups [14]. A rich literature has been developed for SMPs [14], and many variants have been studied, including when there is uncertainty in the preferences [2] or where preferences are expressed according to multiple attributes [8].

We explore the connection between how people make choices, the process of matching, and the notions of stability and fairness. We assume that the preferences of each agent are encapsulated via a Multi-alternative Decision Field

© Springer Nature Switzerland AG 2022
J. Chen et al. (Eds.): DAI 2021, LNAI 13170, pp. 150–170, 2022.
https://doi.org/10.1007/978-3-030-94662-3_10

Theory (MDFT) model [23], that is, by a dynamic cognitive model of choice, capable of capturing behavioral aspects of human decision making. We choose this model as it has been shown to capture choice behavior accurately in human studies [6], it is designed to handle multiple options and attributes [23], and since it strikes a balance between the expressiveness of the underlying preference structure and its psychological underpinnings. One of the core characteristics of MDFT is that choices may change based on the particular subset presented at any given point. This raises questions for classical matching algorithms, such as Gale-Shapley [10], a proposal based method where an agent is selecting alternatives to propose to from an increasingly smaller subset.

From an AI point of view, we extend the state of the art on SMPs by introducing the first framework that incorporates simultaneously multi-attribute preferences with uncertainty and cognitive modeling of bounded-rationality. From a cognitive science perspective, our work provides a psychologically grounded computational model of how humans may respond to matching procedures.

Our work is related to that in Aziz et al. [1] where the authors consider SMPs with uncertain pair-wise preferences, equivalent to considering the choice probabilities induced on subsets of size two by MDFT. While the considered notions of stability are closely related, MDFTs also induce choice probability distributions over subsets of any size, which play an important role for proposal-based methods. The different models of uncertainty in preferences considered in [2] are less closely related as they do not consider choice probability distributions over all subsets of members of the opposite group, as in MDFTs. Preferences expressed via multiple attributes have also been considered in the literature and, more recently, in [22] and [8]. However, in both cases preferences are qualitative, rather than quantitative as in our case. Fairness in matchings has received new attention recently and new algorithms for different definitions of fairness [9], procedural approaches to enforce sex equal stable matchings [11,26,27] and new preference models including bounded lists [20] have been proposed. However, none of these works focus on behavioral models of choice, as in this paper.

Contribution. We define Behavioral Stable Matching Problems (BSMP), where agents express preferences via MDFTs and analyze the computational complexity of several problems related to stability and fairness. We study the impact of behavioral effects on proposal based matching algorithms. We propose novel algorithms for finding maximally stable and fair stable matchings in BSMPs, which we analyse experimentally in terms of the efficiency, stability, and fairness of the returned matchings.

2   Multialternative Decision Field Theory (MDFT)

MDFT [4] models preferential choice as an accumulative process in which the decision maker attends to a specific attribute at each time point in order to derive comparisons among options, and update his preferences accordingly. Ultimately the accumulation of those preferences forms the decision maker's choice.

In MDFT an agent is confronted with multiple options and equipped with an initial personal evaluation for them according to different criteria, called attributes. For example, a student who needs to choose a main course among those offered by the cafeteria will have in mind an initial evaluation of the options in terms of how tasty and healthy they look. More formally, MDFT, in its basic formulation [23], is composed of the following elements.

Personal Evaluation: Given a set of options $O = \{o_1, \ldots, o_k\}$ and set of attributes $A = \{a_1, \ldots, a_l\}$, the subjective value of option o_i on attribute a_j is denoted by m_{ij} and stored in matrix $\mathbf{M}$. In our example, let us assume that the cafeteria options are *Salad (S)*, *Burrito (B)* and *Vegetable pasta (V)*. Matrix $\mathbf{M}$, containing the student's preferences, could be defined as shown in Fig. 1 (left), where rows correspond to the options (S, B, V) and the columns to the attributes *Taste* and *Health*.

$$M = \begin{vmatrix} 1 & 5 \\ 5 & 1 \\ 2 & 3 \end{vmatrix} \quad C = \begin{vmatrix} 1 & -1/2 & -1/2 \\ -1/2 & 1 & -1/2 \\ -1/2 & -1/2 & 1 \end{vmatrix} \quad S = \begin{vmatrix} +0.9000 & 0.0000 & -0.0405 \\ 0.0000 & +0.9000 & -0.0047 \\ -0.0405 & -0.0047 & +0.9000 \end{vmatrix}$$

Fig. 1. Evaluation (M), Contrast (C) and Feedback (S) matrix.

Attention Weights: Attention weights express the attention allocated to each attribute at a particular time t during deliberation. We denote them by vector $\mathbf{W}(t)$, where $W_j(t)$ represents the attention to attribute a_j at time t. We adopt the common simplifying assumption that, at each point in time, the decision maker attends to only one attribute [23]. Thus, $W_j(t) \in \{0, 1\}$ and $\sum_j W_j(t) = 1$, $\forall t, j$. In our example, where we have two attributes, at any point in time t, we will have $\mathbf{W}(t) = [1, 0]$, or $\mathbf{W}(t) = [0, 1]$, representing that the student is attending to, respectively, *Taste* or *Health*. The attention weights change across time according to a stationary stochastic process with probability distribution $\mathbf{p}$, where p_j is the probability of attending to attribute a_j. In our example, defining $p_1 = 0.55$ and $p_2 = 0.45$ means that at each point in time, the student will be attending *Taste* with probability 0.55 and *Health* with probability 0.45; i.e., *Taste* matters slightly more than *Health* to this student.

Contrast Matrix: Contrast matrix $\mathbf{C}$ is used to compute the advantage (or disadvantage) of an option with respect to the other options. In the MDFT literature [5,6,23], $\mathbf{C}$ is defined by contrasting the initial evaluation of one alternative against the average of the evaluations of the others, as shown for the case with three options in Fig. 1 (center).

At any moment in time, each alternative in the choice set is associated with a **valence** value. The valence for option o_i at time t, denoted $v_i(t)$, represents its momentary advantage (or disadvantage) when compared with other options on some attribute under consideration. The valence vector for k options $o_1, \ldots, o_k$ at time t, denoted by column vector $\mathbf{V}(t) = [v_1(t), \ldots, v_k(t)]^T$, is formed by $\mathbf{V}(t) = \mathbf{C} \times \mathbf{M} \times \mathbf{W}(t)$. In our example, the valence vector at any time point in which $\mathbf{W}(t) = [1, 0]$, is $\mathbf{V}(t) = [1 - 7/2, 5 - 3/2, 2 - 6/2]^T$.

In MDFT, preferences for each option are accumulated across iterations of the deliberation process until a decision is made. This is done by using **Feedback Matrix S**, which defines how the accumulated preferences affect the preferences computed at the next iteration. This interaction depends on how similar the options are in terms of their initial evaluation expressed in **M**. Intuitively, the new preference of an option is affected positively and strongly by the preference it had accumulated so far, while it is inhibited by the preference of other options which are similar. This lateral inhibition decreases as the dissimilarity between options increases. Figure 1 (right) shows **S** computed for our running example following the MDFT standard method described in [16].

At any moment in time, the preference of each alternative is calculated by $\mathbf{P}(t+1) = \mathbf{S} \times \mathbf{P}(t) + \mathbf{V}(t+1)$, where $\mathbf{S} \times \mathbf{P}(t)$ is the contribution of the past preferences and $\mathbf{V}(t+1)$ is the valence computed at that iteration. Starting with $\mathbf{P}(0) = 0$, preferences are then accumulated for either a fixed number of iterations, and the option with the highest preference is selected, or until the preference of an option reaches a given threshold. In the first case, MDFT models decision making with a *specified* deliberation time, while, in the latter, it models cases where deliberation time is *unspecified* and choice is dictated by the accumulated preference magnitude.

Definition 1 (Multi-Alternative Decision Theory (MDFT) Model). *Given set of options $O = \{o_1, \ldots, o_k\}$ and set of attributes $A = \{a_1, \ldots, a_l\}$, an MDFT Model is defined by the n-tuple $Q = \langle \mathbf{M}, \mathbf{C}, \mathbf{p}, \mathbf{S} \rangle$, where: $\mathbf{M}$ is the $k \times l$ personal evaluation matrix; $\mathbf{C}$ is the $k \times k$ contrast matrix; $\mathbf{p}$ is a probability distribution over attention weights vectors; and $\mathbf{S}$ is the $k \times k$ feedback matrix.*

Moreover, we will denote with s-MDFT, resp. u-MDFT, models with specified, resp. unspecified, deliberation time. We will, however, omit such prefixes whenever the discussion applies to both types of models.

Different runs of the same MDFT model may return different choices due to the uncertainty on the attention weights distribution. The model can be run on a subset of options $Z \subseteq O$ of size $k' \leq k$, by eliminating from **M** all of the rows corresponding to options not in Z and resizing the contrast matrix and the feedback matrix to size k'. An MDFT induces a choice probability distribution over the options in a set. More formally:

Definition 2 (Choice probability distributions induced by an MDFT model). *Given an MDFT model $Q = \langle \mathbf{M}, \mathbf{C}, \mathbf{p}, \mathbf{S} \rangle$, defined over options set O and with attributes in A, we define the set of choice probability distributions $\{p_Z^Q | \forall Z, Z \subseteq O\}$, containing a probability distribution, denoted p_Z^Q, for each subset Z of O, where $p_Z^Q(z_i)$ is the probability that option $z_i \in Z$ is chosen when Q is run on subset of options Z.*

If we run the model a sufficient number of times on the same set, we obtain a proxy of its choice probability distribution. We note that these choice distributions may violate the regularity principle, which states that, when extra options are added to a set, the choice probability of each option can only decrease.

This allows MDFT to effectively replicate bounded-rational behaviors observed in humans [5] such as the *similarity effect*, by which adding a new similar candidate decreases the probability of an option to be chosen, and the *compromise effect* where including a diametrically opposed option may increase the choice probability of a compromising one [23].

There is an interesting relation between the type of MDFT models and the stochastic transitivity of the induced probabilistic preference relation.

Definition 3 (Stochastic Transitivity). *Given MDFT model Q, defined over option set O, and induced choice probability distributions p_Z^Q, consider every $A, B, C \in O$ such that $p_{\{A,B\}}^Q(A) \geq 0.5$ and $p_{\{B,C\}}^Q(B) \geq 0.5$. If $p_{\{A,C\}}^Q(A) \geq 0.5$, then Weak Stochastic Transitivity (WST) holds. If $p_{\{A,C\}}^Q(A) \geq min\{p_{\{A,B\}}^Q(A), p_{\{B,C\}}^Q(B)\}$, then Moderate Stochastic Transitivity (MST) holds. If $p_{\{A,C\}}^Q(A) \geq max\{p_{\{A,B\}}^Q(A), p_{\{B,C\}}^Q(B)\}$, then Strong Stochastic Transitivity (SST) holds.*

Clearly SST implies MST and MST implies WST. In [7] it is shown that pairwise choice probabilities induced by s-MDFT models satisfy MST (and thus also WST), while only WST holds for those induced by u-MDFT models [4]. SST is not satisfied by MDFT models in general and, indeed, a systematic violation of SST by humans as been demonstrated by several behavioral experiments [21].

3   Stable Marriage Problems (SMPs)

In a *stable marriage problem* (SMP), we are given a set of n men $M = \{m_1, \ldots, m_n\}$, and a set of n women $W = \{w_1, \ldots, w_n\}$, where each strictly orders all members of the opposite gender. We wish to find a one-to-one matching s, of size n such that every man m_i and woman w_j is matched to some partner, and no two people of opposite sex who would both rather be married to each other than to their current partners; also called a *blocking* pair. A matching with no blocking pairs always exists and is said to be *stable* [19].

The Gale-Shapley Algorithm. [10] is a well-known algorithm to solve an SMP. It involves a number of rounds where each un-engaged man "proposes" to his most-preferred woman to whom he has not yet proposed. Each woman must accept, if single, or choose between her current partner and the proposing man. GS returns a stable marriage in $O(n^2)$. Finding stable matching in variants of SMPs, such as with ties and incomplete lists, is, instead, NP-complete [19].

The pairing generated by GS with men proposing is male optimal, i.e., every man is paired with his highest ranked feasible partner, and female-pessimal [14]. Thus, it is desirable to require stable matchings to also be *fair*, for example, by minimizing the *sex equality cost (SEC)*: $SEC(s) = | \sum_{(m,w)\in s}(pr_m(w)) - \sum_{(m,w)\in s}(pr_w(m)) |$, where $pr_x(y)$ denotes the position of y in x's preference. For example, if we consider the SMP of size 3 with men preferences defined as $m_1 : w_1 > w_2 > w_3$, $m_2 : w_2 > w_1 > w_3$, and $m_3 : w_3 > w_2 > w_1$ and women preferences $w_1 : m_1 > m_2 > m_3$, $w_2 : m_3 > m_1 > m_2$ and $w_3 : m_2 > m_1 > m_3$, we have two stable matchings, $s_m = \{(m_1, w_1), (m_2, w_2), (m_3, w_3)\}$ and $s_w = \{(w_1, m_1), (w_2, m_3), (w_3, m_2)\}$, that are, respectively, male and female optimal and have a SEC of, respectively, 4 and 3.

Finding a stable matching with minimum SEC is strongly NP-hard and approximation techniques have been proposed for example in [17]. Local search approaches have been used extensively in SMPs to tackle variants for which there are no polynomial stability and/or fairness algorithms [11,12,19].

4 Behavioral Stable Marriage Problems (BSMPs)

Given a set of n men and n women where each women w_i (resp. man m_i) expresses her (resp. his) preferences over the men (resp. women) via an MDFT model $Q_{w_i} = \langle \mathbf{M_{w_i}}, \mathbf{C_{w_i}}, \mathbf{p_{w_i}}, \mathbf{S_{w_i}} \rangle$ (resp. $Q_{m_i} = \langle \mathbf{M_{m_i}}, \mathbf{C_{m_i}}, \mathbf{p_{m_i}}, \mathbf{S_{m_i}} \rangle$). Since, as described in Sect. 2, we adopt the standard definitions for contrast and feedback matrices $\mathbf{C}$ and $\mathbf{S}$, we will omit them for clarity, in what follows.

Definition 4 (Behavioral Profile). *A Behavioral Profile is a collection of n men and n women, where the preferences of each man and woman, x_i, on the members of the opposite group are given by an MDFT model $Q_{x_i} = \langle \mathbf{M_{x_i}}, \mathbf{p_{x_i}} \rangle$.*

While each individual can, in principle, use different attributes to express their preferences, similarly to the MDFT literature, we will assume two attributes for all MDFTs. Thus, for each group member x_i, his/her model expresses a (numerical) personal evaluation of each member of the opposite group with respect to two attributes in $\mathbf{M_{x_i}}$, and the importance of each attribute, $\mathbf{p_{x_i}}$ (see an example in Fig. 2). By running the MDFT models many times we can approximate the induced choice probabilities (Definition 2). For the profile in Fig. 2 we have $p_{\{w_1,w_2\}}^{Q_{m_1}}(w_1) = 0.485$, $p_{\{w_1,w_2\}}^{Q_{m_2}}(w_1) = 0.556$, $p_{\{m_1,m_2\}}^{Q_{w_1}}(m_1) = 0.495$, and $p_{\{m_1,m_2\}}^{Q_{w_2}}(m_1) = 0.562$.

$$M_{w_1} = \begin{bmatrix} A_1 & A_2 \\ 8 & 2 \\ 2 & 8 \end{bmatrix} M_{w_2} = \begin{bmatrix} A_1 & A_2 \\ 2 & 8 \\ 8 & 2 \end{bmatrix} M_{m_1} = \begin{bmatrix} A_1 & A_2 \\ 8 & 2 \\ 2 & 8 \end{bmatrix} M_{m_2} = \begin{bmatrix} A_1 & A_2 \\ 2 & 8 \\ 8 & 2 \end{bmatrix}$$

Fig. 2. A behavioral profile. Attention probability fixed to $p(A_1) = 0.55$.

As for SMPs, a *matching* is a one-to-one correspondence between men and women. However, the notion of blocking pair becomes probabilistic.

Definition 5 (β-blocking). *Let B be a behavioral profile, and s one of its matchings. Consider pair $(m, w) \notin s$ and let Q_m, Q_w, be the MDFT models of, respectively, m and w, and $s(m)$ and $s(w)$ be their respective partners in s. We say pair (m, w) is β-blocking if $\beta = p_{\{w,s(m)\}}^{Q_m}(w) \times p_{\{m,s(w)\}}^{Q_w}(m)$.*

In other words, we say that pair (m, w), unmatched in s, is β-blocking if β is the joint probability of m choosing w instead of $s(m)$ according to Q_m and of w choosing m instead of $s(w)$ according to Q_w. The higher the β, the higher the probability that m and w will break the current matching. For example, (m_1, w_2) is 0.29-blocking for matching $s = \{(m_1, w_1), (m_2, w_2)\}$ given the behavioral profile in Fig. 2.

Definition 6 (α-B-stable matching). *Let B be a behavioral profile, and s one of its matchings. We say that s is α-behaviorally-stable (α-B-stable), if $((1 - \beta_1) \times \ldots \times (1 - \beta_h)) \leq \alpha$, and α is the minimum value for which this holds, where β_i is the blocking probability of pair π_i, $i \in \{1, \ldots, h\}$, un-matched in s, and h is the number of blocking pairs, that is, $h = n \times (n - 1)$, if s has n pairs.*

Intuitively, a matching is α-B-stable if the probability that none of the unmatched pairs is blocking is smaller than or equal to α. We note that 1-B-stability corresponds to stability in the classical sense. Given the pair-wise probabilities described earlier, we see that matching $s = \{(m_1, w_1), (m_2, w_2)\}$ is 0.514-B-stable for the profile in Fig. 2.

Definition 7 (Behavioral Stable Marriage Problem (BSMP)). *Given behavioral profile B, the corresponding Behavioral Stable Marriage Problem (BSMP) is that of finding an α-B-Stable matching with maximum α.*

With abuse of notation, we will use BSMP and behavioral profile, as well as marriage and matching, interchangeably in what follows. Moreover, we will write s-BSMP, resp. u-BSMP, to denote a BSMP where all agents express their preferences via s-MDFTs, resp. u-MDFTs.

Given model Q_m of man m, we define the probability that m's choices will follow a particular linear order as follows.

Definition 8 (Induced probability on linear orders). *Consider MDFT model Q defined on option set O. Let us consider linear order $\omega = \omega_1 > \cdots > \omega_k$, $\omega_i \in O$, defined over O. Then, the probability of ω given Q is: $p^Q(\omega) = p_O^Q(\omega_1) \times p_{\{O - \{\omega_1\}\}}^Q(\omega_2) \times \cdots \times p_{\{\omega_{k-1}, \omega_k\}}^Q(\omega_{k-1})$.*

While one of several ways to obtain a linearization, the one in Definition 8 is particularly intuitive as the probability of a linear order is defined as the joint probability that the first element in the order will be chosen by the MDFT model among all of the options, the second element will be chosen among the remaining options, and so forth. We now define the expected position as follows.

Definition 9 (Expected position). *Consider BSMP B, man m and model Q^m. The expected position of w in m's preferences is defined as: $E[pr_m(w)] = \sum_{\omega \in L(W)} p^{Q^m}(\omega) \times pr_\omega(w)$, where $L(W)$ is the set of linear orders over the set of women W, and $pr_\omega(w)$ is the position of woman w in linear order ω.*

We can now define the sex equality cost for BSMPs.

Definition 10 (Sex equality cost (SEC)). *Given BSMP B and match-ing s, the sex equality cost of s is:* $SEC(s) = | \sum_{(m,w) \in s} E[(pr_m(w))] - \sum_{(m,w) \in s} E[(pr_w(m))] |$.

Clearly, the lower SEC the more fair the matching. Figure 3 provides two examples of BSMPs and SECs for matchings.

5   Complexity Results

In this section we study the complexity of several problems in the context of behavioral profiles. In particular, we reconsider some of the results presented in [1] in light of our setting. For all our results, we assume that we begin with the probabilities induced on all sets of size two by the agents' MDFTs. As noted in Sect. 2, s-MDFTs induce MST pairwise preferences, u-MDFTs induce WST pairwise preferences and, if we relax both the constraint on the specified deliber-ation time and neutral starting point, the induced probabilistic preferences may violate WST. While it is known that MDFTs are very successful in capturing choice distributions exhibited in humans, a theoretical analysis of their exact expressive power is still an open problem.

The problems we consider are the following: STABILITYPROBABILITY: Given a BSMP B and a matching s, find $\alpha \in [0, 1]$ such that s is α-B-stable; EXISTPOS-SIBLYSTABLEMATCHING: Does there exist an α-B-Stable matching with $\alpha > 0$?; MATCHINGWITHHIGHESTPROBABILITY: Compute an α-B-stable matching with maximum α; MAXIMALLYFAIRMATCHING: Find a matching s with minimum sex equality cost; MAXIMALLYFAIRSTABLEMATCHING: Find a matching with minimum SEC among those that are α-B-stable with maximum α.

We note that the complexity of obtaining induced probability distributions has been shown to be polynomial for s-MDFTs [7], where an analytical derivation from the parameters of the model is described. For u-MDFTs we leverage the fact that we can approximate such pairwise probabilities by running the model a sufficient number of times. When the size of the options set is fixed at two, the amount of time necessary to obtain these approximations for all of the agents in a u-BSMP grows linearly with the number of agents. Following [1], we define the certainly preferred relation where for agent w, $b \succ_w^{cert} c$ if and only if she chooses b over c with probability 1. All proofs are omitted due to lack of space.

Theorem 1. *For BSMPs,* STABILITYPROBABILITY *is polynomially solvable.*

This result derives directly from Theorem 1 in [1]. We note that the fact that pairwise probabilities in the context of MDFTs are defined in terms of choice distributions over subsets of size two, implies that they are independent. From this we derive the fact that the probabilities of each member of a blocking pair preferring the alternative options to their current match are also independent, this is also observed by [1].

Theorem 2. EXISTPOSSIBLYSTABLEMATCHING *is NP-complete even if one side of the market has linear preferences and the other side has weakly stochastic transitive (WST) pairwise probabilities.*

This result strengthens the statement of Theorem 2 in [1] by further restricting the preferences of one side of the market.

Lemma 1. *For s-BSMPs, an α-B-stable matching with $\alpha > 0$ always exists and can be found in polynomial time.*

This results is derived by linearizing the probabilistic preferences induced by the s-MDFTs in a specific way so to obtain and SMP the stable matchings of which are α-B-stable with $a > 0$ in the s-BSMP. An immediate consequence is:

Theorem 3. *For s-BSMPs,* EXISTPOSSIBLYSTABLEMATCHING *is polynomially solvable. (See proof in Appendix)*

We now consider the complexity of MATCHINGWITHHIGHESTPROBABILITY.

Theorem 4. MATCHINGWITHHIGHESTSTABILITYPROBABILITY
is NP-hard, even if the certainly preferred relation is transitive for one side of the market and the other side has WST preferences.

In [1] it is shown that this problem is NP-hard even if the certainly preferred relation is transitive for one side of the market and the other side has deterministic linear orders. Our result for WST preferences is orthogonal, as WST does not imply transitive certainly preferred relation and vice-versa.

The complexity of this problem when one side has MST preferences remains an open problem. We conjecture NP-hardness remains as MST preferences are a subset of those where the certainly stable relation is transitive. We note that from Theorem 4 we can also immediately derive that MAXIMALLYFAIRSTABLEMATCHING is also NP-hard.

We conclude elaborating on MAXIMALLYFAIRMATCHING. Let us denote with $F(n)$ the time required to run the MDFT model on a set of options of size n.[1] The complexity of computing a linearization as described in Definition 8 is $O(nF(n))$. If we repeat this process a sufficiently large number of times, K, we can approximate the expected positions in $O(Kn^2F(n))$. Finding a maximally sex-equal, i.e., fair, stable matching is a well known NP-hard problem [17,20]. The question of finding such a matching not subject to stability constraints remains an important open problem in the literature. As we will see in Sect. 6, we formulated an ILP to solve this problem to judge our algorithms effectiveness. Our results are summarized in Table 1.

Table 1. Complexity results. Problem names are abbreviated.

	STABPROB	EXISTPOSSSTABMATCH	MATCHHIGHPROB	MAXFAIRMATCH	MAXFAIRSTABMATCH
WST	P	NP-complete	NP-hard	?	NP-hard
MST	P	P	?	?	?

---

[1] As in the MDFT literature, we can assume constant number of attributes and assume $F(n)$ polynomial in n for both halting modes.

6   Algorithms for BSMPs

In this section we outline several algorithms that find matchings with different properties. In particular we introduce two variants of Gale Shapley (B-GS and EB-GS), two integer linear program (ILP) formulations and two local search approaches. The details of the first three algorithms can be found in the Appendix.

Gale Shapley for BSMPs: B-GS and EB-GS. The Gale Shapley procedure can be extended in a straightforward way to BSMPs by invoking the relevant MDFT models when a proposal or an acceptance has to be made. We call this variant of GS, Behavioral Gale Shapley, denoted with B-GS. B-GS still converges, since the sets of available candidates shrink by one every time a proposal is made, but it is no longer deterministic and may return different matchings as a consequence of the non-determinism of the underlying MDFT models. We also define another variant of GS, that we call Expected Behavioral Gale Shapley (EB-GS), which runs GS on the SMP obtained considering the linear orders corresponding to expected positions (see Definition 9).

Algorithm FB-ILP. We developed an integer linear program (ILP) to find the most fair solution according to the SEC with no guarantees on stability. For each combination of man $m_i \in M$ and woman $w_j \in W$, $|M| = |W| = n$, we introduce a binary variable $m_i w_j$ that takes value 1 if m_i is matched with w_j and 0 otherwise. The FB-ILP formulation also includes two $n \times n$ matrices (pos_M and pos_W) modeling expected positions of respectively women and men in each others preferences. The solution with the lowest SEC is then obtained by minimizing $SEC = |\sum_{i,j \in n} pos_M[i,j] \cdot m_i w_j - \sum_{i,j \in n} pos_W[j,i] \cdot m_i w_j|$, leveraging an indicator variables approach [3] to bypass the non-linearity.

Algorithm B-ILP. To find the optimal α-B-Stable solution with B-ILP, we begin with the same setup of FB-ILP. In addition, the B-ILP formulation uses an $n \times n$ matrix Pr_{m_i} where entry $Pr_{m_i}[j,k]$ gives the probability that man m_i prefers w_j to w_k. This matrix can be computed by running the BSMP of man m_i a sufficiently large number of times. Then, to address the fact that the product of the probabilities is a convex not linear function, and stability is a pairwise notion over a given matching, we introduce $\forall((i,j),(k,l)) \in \binom{n}{2}$ possible combinations of pairs of pairs, an indicator variable $m_i w_j + m_k w_l$ to indicate that both $m_i w_j$ is matched and $m_k w_l$ is also matched. This allows us to compute the blocking probability of m_i and w_l as well as of m_k and w_j. Hence for every pair of possible marriages $m_i w_j + m_k w_l$ we can compute the probability that these four individuals are not involved in blocking pairs by taking the likelihood that they swap partners, formally let $block[(ij),(kl)] = (1 - Pr_{m_i}[l,j] * Pr_{w_l}[i,k]) * (1 - Pr_{m_k}[j,l] * Pr_{w_j}[k,i])$. To handle the convex constraint we simply take the log of this quantity and maximize using an indicator variable which we implement using the Gurobi *And* constraint.

The B-LS Algorithm. B-LS, is a local search approach [15] that explores the space of matchings to find one with maximum α-B-stability starting from a randomly generated one. Each matching s is evaluated by its level α of behavioral stability. When we find a matching, we compute for each non-matched pair its β-blocking level. The neighborhood of a matching s consists of all the matchings that can be obtained from s by rotating a blocking pair (i.e., swapping partners) and is explored in decreasing order of β until a matching with a higher α-B-stability is found or the neighborhood is exhausted and search restarts from a randomly generated matching. The search ends after a max number of iterations, returning the matching with maximum α found so far.

Algorithm FB-LS. Algorithm FB-LS is another local search approach designed to take in input a value α and return a matching with the lowest SEC that is also α-B-stable. Intuitively, FB-LS runs B-LS on the space of matchings meeting a certain level of fairness. We first run B-LS to compute the maximum level of α-B-stability achievable, denoted α_{max}. We also compute the SEC for the matching returned by this run of B-LS, called $se_{\alpha_{max}}$. We then fix the lowest level of behavioral stability that we consider reasonable, denoted α_{min}, with $\alpha_{min} \leq \alpha_{max}$. Then FB-LS performs an incremental search where for each SEC value, se, it launches B-LS to find the matching with maximum α-B-stability value, say α_{se} and with SEC cost se. FB-LS starts with $se = se_{\alpha_{max}}$ and decreases se until it no longer finds a matching with stability $\alpha_{se} \geq \alpha_{min}$.

7 Experimental Results

We first exemplify how contextual effects impact the α-B-stability of a matching returned by a proposal-based approach. The key point is that MDFT captures and replicates preference reversals that humans exhibit when options are added or deleted to a choice set. Thus, what may have emerged like a good choice among several options at proposal time, may not be dominating when only choice sets of size two are considered for stability. As seen in Fig. 3(a), on an instance of the compromise effect both B-GS and EB-GS return a matching which is sub-optimal w.r.t. α-B-stability with high probability. An analogous situation can be observed for the instance of the similarity effect shown in Fig. 3(b). These examples show that, in general, there is no guarantee that a matching returned by B-GS or EB-GS will be optimal w.r.t. α-B stability. In the second column of the tables in Figs. 3 we show the SEC of the matchings. Not surprisingly, there is no guarantee on the fairness nor, most importantly, on the "unfairness" (as instead is the case for GS for SMPs) of the returned matching, the latter being an effect of the non-deterministic behavioral models.

To test our algorithms in terms of efficiency and quality of the solutions we first generate 100 random BSMPs for each size n between 10 and 16 where the **M** matrices are of size $n \times 2$ and contain random preferences between 0 and 9. Attention weights probabilities are fixed to $p([0,1]) = 0.45$ and $p([1,0]) = 0.55$.

Figure 4 shows α-B-Stability and SEC values of matchings returned by the algorithms averaged over the 100 instances. Each point on the lines represents the size of the problems from $n = 10$ to $n = 15$ moving from left to right. For $n = 16$ the ILP formulations timed-out at 6 h while B-LS converges at around

$M_{w_1} = \begin{bmatrix} A_1 & A_2 \\ 8 & 2 \\ 2 & 8 \\ 5 & 5 \end{bmatrix}$ $M_{w_2} = \begin{bmatrix} A_1 & A_2 \\ 2 & 8 \\ 2 & 2 \\ 5 & 5 \end{bmatrix}$ $M_{w_3} = \begin{bmatrix} A_1 & A_2 \\ 5 & 5 \\ 5 & 5 \\ 9 & 9 \end{bmatrix}$

Matching	α	SEC	%B-GS
$\{(m_1, w_1), (m_2, w_2), (m_3, w_3)\}$	0.47	0.21	0.62
$\{(m_1, w_2), (m_2, w_1), (m_3, w_3)\}$	0.54	0.01	0.33
$\{(m_1, w_3), (m_2, w_2), (m_3, w_1)\}$	0.03	0.26	0.01
$\{(m_1, w_3), (m_2, w_1), (m_3, w_2)\}$	0.02	1.5	0.04

$M_{m_1} = \begin{bmatrix} A_1 & A_2 \\ 8 & 2 \\ 2 & 8 \\ 5 & 5 \end{bmatrix}$ $M_{m_2} = \begin{bmatrix} A_1 & A_2 \\ 2 & 8 \\ 8 & 2 \\ 5 & 5 \end{bmatrix}$ $M_{m_3} = \begin{bmatrix} A_1 & A_2 \\ 5 & 5 \\ 5 & 5 \\ 9 & 9 \end{bmatrix}$

(a)

$M_{w_1} = \begin{bmatrix} A_1 & A_2 \\ 8 & 2 \\ 2 & 8 \\ 9 & 1 \end{bmatrix}$ $M_{w_2} = \begin{bmatrix} A_1 & A_2 \\ 2 & 8 \\ 8 & 2 \\ 9 & 1 \end{bmatrix}$ $M_{w_3} = \begin{bmatrix} A_1 & A_2 \\ 5 & 5 \\ 5 & 5 \\ 9 & 9 \end{bmatrix}$

Matching	α	SEC	%B-GS
$\{(m_1, w_1), (m_2, w_2), (m_3, w_3)\}$	0.47	0.02	0.58
$\{(m_1, w_2), (m_2, w_2), (m_3, w_3)\}$	0.61	0.07	0.41
$\{(m_1, w_3), (m_2, w_2), (m_3, w_1)\}$	0.01	0	0.01

$M_{m_1} = \begin{bmatrix} A_1 & A_2 \\ 8 & 2 \\ 2 & 8 \\ 9 & 1 \end{bmatrix}$ $M_{m_2} = \begin{bmatrix} A_1 & A_2 \\ 2 & 8 \\ 8 & 2 \\ 9 & 1 \end{bmatrix}$ $M_{m_3} = \begin{bmatrix} A_1 & A_2 \\ 5 & 5 \\ 5 & 5 \\ 9 & 9 \end{bmatrix}$

(b)

Fig. 3. Compromise (a) and Similarity effect (b), impact on GS. Profile (left) and results (right), for α-B stability value (α), Sex Equality Cost (SEC) and % of times returned by B-GS (%B-GS) out of 100 runs. EB-GS result in blue. (Color figure online)

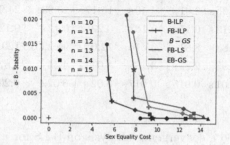

Fig. 4. Average α-B-Stability (y-axis) and SEC (x-axis) varying the number of agents. (Color figure online)

340 s (see Table 4). Not surprisingly, the quality of the solutions deteriorates as we move to larger problem sizes. The average results for B-ILP (dark blue-line) represent the optimal values for α-B-stability but exhibit average high SEC. In contrast, we can see how FB-LS (green line) allows to find matchings which have low SEC and are at most 30% less stable than optimal. As predicted, B-GS on average performs very poorly. At the bottom left corner we see the FB-ILP (red line) collapsed to a single point, as it always returns extremely unstable matchings of almost zero SEC. Our results showed very small variance in terms of α-B-stability, except for B-GS and EB-GS (see Table 3). Table 2 shows instead the SEC results with their standard deviations. All algorithms (except FB-ILP not shown since $\mu \cong 0$ and $\sigma^2 \cong 0$) have significant variance in terms of SEC, likely explained by the difference in preferences across instances. As expected, FB-LS exhibits the lowest SEC variance.

The B-GS time is the average over 100 runs on the same instance. While B-GS and EB-GS are significantly faster, for each n they returned a maximally behaviorally stable matching only around 30% of the time. B-ILP and B-LS have comparable running times up to $n = 16$, where B-ILP doesn't terminate. It should also be noted that B-ILP, when terminating, always returns a maximally B-stable matching while B-LS does so around 88% of the time.

Table 2. Sex equality cost

# Agents	B-ILP μ	σ^2	FB-LS μ	σ^2	B-GS $max(\alpha)$ μ	σ^2	EB-GS μ	σ^2
10	7.1	32.7	5.3	25.3	7.7	35.9	8.4	37.0
11	7.8	29.6	5.5	22.5	8.6	34.2	9.5	53.4
12	8.1	46.2	5.7	31.2	9.3	57.9	9.4	53.5
13	12.3	76.1	7.9	59.0	12.2	81.5	10.8	80.5
14	13.3	73.1	9.3	58.9	13.0	79.2	12.5	77.6
15	14.2	116.80	9.4	84.0	13.4	107.8	14.6	115.6

Table 3. α-B-Stability

# Agents	B-ILP μ	σ^2	FB-LS μ	σ^2	B-GS μ	σ^2	EB-GS μ	σ^2
10	0.0208	$5*10^{-4}$	0.0149	$3*10^{-4}$	0.0175	$5*10^{-4}$	$2*10^{-14}$	$6*10^{-26}$
11	0.0099	$2*10^{-4}$	0.0081	$2*10^{-4}$	0.0083	$2*10^{-4}$	$1*10^{-15}$	$1*10^{-28}$
12	0.0041	$3*10^{-5}$	0.0034	$3*10^{-5}$	0.0023	$2*10^{-5}$	$2*10^{-17}$	$5*10^{-32}$
13	0.0020	$5*10^{-6}$	0.0016	$4*10^{-6}$	0.0009	$2*10^{-6}$	$5*10^{-29}$	$2*10^{-55}$
14	0.0009	$3*10^{-6}$	0.0008	$2*10^{-6}$	0.0005	$2*10^{-6}$	$8*10^{-50}$	$4*10^{-97}$
15	0.0004	$3*10^{-7}$	0.0002	$2*10^{-7}$	0.0001	$5*10^{-8}$	$3*10^{-50}$	$5*10^{-98}$

Table 4. Average execution time for B-ILP, B-LS and B-GS varying n.

Algorithm	10	11	12	13	14	15	16
B-ILP	1.03 s	2.74 s	3.90 s	6.61 s	12.62 s	27.05 s	N/A
B-LS	0.66 s	2.01 s	4.40 s	15.17 s	20.93 s	24.94 s	342 s
FB-ILP	0.13 s	0.15 s	0.18 s	0.22 s	0.12 s	0.12 s	0.24 s
FB-LS	2.83 s	8.81 s	35.16 s	72.0 s	90.223 s	120.76 s	941 s
B-GS	1.93 s	2.81 s	3.18 s	4.04 s	4.55 s	5.87 s	7.2 s
EB-GS	0.01 s	0.015 s	0.017 s	0.02 s	0.022	0.26	0.028 s

We also performed a convergence analysis on B-LS for $n = 16$ which showed B-LS plateaus at 300 iterations, corresponding to approximately 340 s. We then tested B-LS on larger instances, generated under similar conditions, for $n \in \{20, 30, 40, 50\}$. Convergence was observed at, respectively, 500, 800, 1200 and 1900 iterations and average running times over 10 instance ranged from 896 s for instances of size 20 to 6433 s for size 50. We also note that, on average, the pre-processing times to compute the pairwise choice probabilities and the expected positions ranged between 16 s for size $n = 10$ to 1592.5 s for $n = 50$.

Our experimental results show that when the goal is to find a maximally stable matching, B-ILP is a viable and complete option for smaller problems. If fairness is also considered, then, FB-LS produces high quality solutions compromising between the two criteria while scaling reasonably well. This experimental study has also confirmed the negative impact of the underlying behavioral models on the quality of solutions returned by proposal based approaches.

8 Future Work

We plan to consider the impact of behavioral models in one-to-many and many-to-many matching problems and their integration with other algorithms such as the Boston Mechanism [18]. We also plan to study methods proposed to achieve fairness over time which ties particularly well with the concept of repeated choices underlying the MDFT models [25].

A   Proofs of Theorems

Theorem 1. *For BSMPs,* STABILITYPROBABILITY *is polynomially solvable.*

Proof. This result derives directly from Theorem 1 in [1]. Given the pairwise probabilities induced by the BSMPs, we can compute the probability that a unmatched pair is not blocking in constant time. We then take the product over such pairs which are quadratic in number.

Theorem 2. EXISTPOSSIBLYSTABLEMATCHING *is NP-complete even if one side of the market has linear preferences and the other side has weakly stochastic transitive (WST) pairwise probabilities.*

Proof. This results strengthens the statement of Theorem 2 in [1] by further restricting the preferences of one side of the market. There the authors reduce from EXISTCOMPLETESTABLEMATCHING in Stable Matching with Ties and Incompleteness (SMTIs) [19] to EXISTPOSSIBLYSTABLEMATCHING when men have linear preferences and by leveraging the ability to define a cycle of length three of certainly preferred relations in the women's preferences. WST pairwise probabilities do not allow for cycles of length three comprised of certainly preferred relations. However, they do allow for cycles of length four as the one shown in Fig. 5. This observation allows the proof to proceed in a very similar way as that of Theorem 2 in [1].

For the reader's convenience, we provide the complete proof below incorporating the extended cycle and associated modifications.

Given Theorem 1, we know that computing StabilityProbability for BSMPs is polynomially solvable. This implies that checking if a matching has a non-zero probability of being stable can be done in polynomial time, and thus the problem is in NP.

To prove NP-hardness, we follow the proof of Theorem 2 in [1] and we reduce from the problem of deciding whether an instance of SMTI admits a complete stable matching. This problem was shown to be NP-complete even if the ties appear only on the women's side, and each woman's preference list is either strictly ordered or consists entirely of a tie of size two [19].

Let $M = \{m_1, m_2, \ldots, m_n\}$ and $W = \{w_1, w_2, \ldots, w_n\}$ be the set of men and women in SMTI I. We create an instance of the pairwise probability model I' where women's preferences are WST as follows. We add 4 men and women: $m_{n+1}, m_{n+2}, m_{n+3}$ and m_{n+4} and $w_{n+1}, w_{n+2}, w_{n+3}$ and w_{n+4}. As in [1] we call acceptable partners in I proper partners in I'. For each man m_i, $i \in \{1, \ldots, n\}$, in the original instance I, we extend his strict preference ordering on his proper partners arbitrarily, by appending the four new women and his unacceptable partners in I in some arbitrary order. For every woman w_i, $i \in \{1, \ldots n\}$, in I, we create the pairwise preferences as follows. Firstly, w_i prefers every proper partner of hers to every new or unacceptable man. Secondly, w_i prefers each of the 4 new men to unacceptable men in I.

The pairwise preferences of w_i over her proper partners are defined in the same way as in [1]: w_i certainly prefers m_k to m_l if w_i strictly prefers m_k to m_l in I, and if w_i is indifferent between m_k and m_l in I then the corresponding pairwise probability is 0.5 in I'.

We then define the pairwise preferences of w_i over the 4 new men as in Fig. 5. More in detail, w_i certainly prefers m_{n+1} to m_{n+2}, m_{n+2} to m_{n+3}, and m_{n+3} to m_{n+4} and m_{n+4} to m_{n+1}, while she is indifferent between m_{n+1} and m_{n+3}, and m_{n+2} and m_{n+4}. We note that these preferences form a cycle of length 4 and respect WST.

The preferences of w_i over the unacceptable original candidates are arbitrary. Similarly to [1], we let each of the four new men have all the original women at the top of his preference list ordered according to their indices, followed with new women w_{n+1}, w_{n+2}, w_{n+3} and w_{n+4} (in this order). Moreover, the four new women have m_{n+1}, m_{n+2}, m_{n+3} and m_{n+4} at the top of their strict preference lists, followed by the original men in an arbitrary order. (Note that every complete linear order implies pairwise probability preferences and satisfies WST). At this point we can show that there exists a complete weakly stable matching in I if and only if there is a matching with positive stability probability in I' following the exact same reasoning as in [1]. To see the first direction, let μ be a complete weakly stable matching in I. It is easy to see that if we extend μ with pairs (m_{n+1}, w_{n+1}), (m_{n+2}, w_{n+2}), (m_{n+3}, w_{n+3}) and (m_{n+4}, w_{n+4}) then the resulting matching μ' has positive probability of being stable in I'. This is because there is no pair which would be certainly blocking for μ'. Conversely, suppose that μ' is a complete matching in I' with positive probability of being stable (i.e., it has no certainly blocking pair). It can be shown that every original woman has to be matched with a proper partner. Suppose for a contradiction that w_i is the woman with the smallest index who is not matched to a proper partner. If w_i is matched to an original man who was unacceptable to her in I then w_i would form a certainly blocking pair with any of the four new men. In fact, note that w_i certainly prefers either of the four new men to her partner. Moreover, as none of the new men are matched to a original woman with index smaller than i, hence they all certainly prefer w_i to their partners. Suppose now that w_i is matched to one of the four new men. Then w_i would form a certainly blocking pair with the subsequent new man according to her cyclical preference. (For instance, if w_i is matched to m_{n+1} in μ' then she forms a certainly blocking pair with m_{n+4}.) This is because the subsequent new man cannot have any better partner, since all the women with smaller indices than i are matched to a proper partner. So we arrive at the conclusion that every original woman is matched with a proper partner. Since μ' does not admit a certainly blocking pair and all original women are matched with proper partners, the restriction of μ' to the original agents is a stable and complete matching in I.

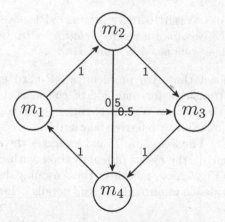

Fig. 5. Example of WST preferences with a cycle of length four. Directed edge represents dominance of source node on target node and the edges are annotated with the probabilities.

Theorem 3. *For s-BSMPs,* EXISTPOSSIBLYSTABLEMATCHING *is polynomially solvable.*

Proof. Consider BSMP B where all agents have s-MDFTs. For each man and woman, we extract a linear order from the pairwise probabilities induced by their s-MDFT thus obtaining an stable matching problem I. We then show that a matching is stable in I if and only if it is α-B-stable with $\alpha > 0$ in B. We illustrate the linearization for man m_i denoting with Q_i his s-MDFT model.

1. For every pair such that $P^{Q_i}_{\{w_k,w_j\}}(w_k) > 0.5$ we set $w_k >_{m_i} w_j$ in I. Note that, since the pairwise probabilities induced by an s-MDFt are MST, by doing this we cannot create any cycles in $>_{m_i}$ in I.
2. We perform the transitive closure adding all induced order relations.
3. At this point the only pairs that may still be not ordered in m_i's preferences in I must be such that $P^{Q_i}_{\{w_k,w_j\}}(w_k) = 0.5$. We order such remaining pairs (for example lexicographically) and we proceed in this order to pick one pair, order it in a random way, and then perform transitive closure.

It is easy to see that MST ensures that at the end of this process we obtain a linear order. Moreover each linearization requires polynomial time since in the worst case it performs a transitive closure $O(n^2)$ for each pair linearized in step 3, that is $O(n^2)$ times. Let μ be a complete stable matching in I. We know one exists [19]. Let's assume that μ is 0-B-stable in B. Then it must have a certainly blocking pair (m, w), where m prefers w to $\mu(m)$ and w prefers m to $\mu(w)$ with probability of 1 in B. If $P^m_{\{w,\mu(m)\}}(w) = 1$ in B then $w >_m \mu(m)$ in I. If $P^w_{\{m,\mu(w)\}}(m) = 1$ in B then $m >_w \mu(w)$ in I. That is, (m, w) is also a blocking pair in I, thus μ cannot be stable in I. This is a contradiction.

Theorem 5. MatchingWithHighestStabilityProbability
*is NP-hard, even if the certainly preferred relation is transitive for one side of
the market and the other side has WST preferences.*

Proof. In [1] it is shown that this problem is NP-hard even if the certainly
preferred relation is transitive for one side of the market and the other side
has linear orders. Our result for WST preferences is orthogonal, as WST does
not imply transitive certainly preferred relation and vice-versa. The proof is
adaptation of the one of Theorem 3 in [1] and leverages the same cycle described
in Fig. 5. Indeed, replacing the cycles of length three with cycles of length four,
as the one depicted in Fig. 5, does not affect the reasoning described in the proof.
For the reader's convenience we provide the full details below, clarifying why the
key steps still hold.

Following a similar reasoning as in [1], we derive the result by modifiyng
the proof of Theorem 2. Let SMTI I and pairwise probability model with WST
preferences I' be defined as in Theorem 2. We denote a new instance of a pairwise
probability model with WST preferences I'' as follows. Whenever some women
have cyclic certainly preferred relations in I', we modify the probabilities in these
pairwise comparisons by a small value ϵ. That is, whenever a woman w_i certainly
prefers man m_k to man m_l within a cycle in I', we modify the probability of the
relation to $1 - \epsilon$ in I''. For example, the probabilities of the perimeter edges in
Fig. 5 would be set to $1-\epsilon$. We note that, given how I' is defined, this modification
will not cause violations of WST. Thus, we have no certainly preferred relations
in any cycle in I''. However, as in [1] when considering the matching with the
highest stability probability in I'', we can still articulate our reasoning along
two cases with respect to the original NP-complete problem for I. Let's first
assume that we have a complete stable matching for I. In this case this matching,
extended with the four new pairs in I', will have a probability of being stable
at least $\frac{1}{2^n}$ in both I' and I''. This is because every woman who is indifferent
between some men has at most one tie of length two in her preference list in I by
definition, and so if this woman is matched to one of the men in her tie then only
the other man in this tie may block, which happens with 0.5 probability. Note
that this step is not affected by the fact that we are using cycles of length four
involving only the new men. On the other hand, if there exists no complete stable
matching for I then we know from the proof of Theorem 2 that there always
existed a certain blocking pair in I'. This certain blocking pair will now have a
probability of $1 - \epsilon$ to be blocking, implying that any matching in this case has
less than ϵ probability of being stable. Therefore, if we choose ϵ to be $0 < \epsilon < \frac{1}{2^n}$
we can use an algorithm which solves MatchingWithHighestStabilityProbability
to decide the existence of a complete stable matching for SMTI efficiently.

B   Algorithms

Algorithms B-GS and EB-GS. As we mentioned in the paper, the Gale Shapley procedure can be extended in a straightforward way to BSMPs by invoking

the relevant MDFT models when a proposal or an acceptance has to be made. When man m is proposing, model Q_m will be run to select the woman to propose to among the set of women to whom m has not proposed yet. In fact, an MDFT model can be run on any subset of options by simply removing irrelevant rows from the personal evaluation matrix and resizing the other matrices (contrast and feedback). Similarly, when woman w, currently matched with man $\sigma(w)$ receives a proposal from m, the choice will be picked by running Q_w on the set $\{m, \sigma(w)\}$. We call this variant of GS, Behavioral Gale Shapley, denoted with B-GS. While it is clear that B-GS still converges, since the sets of available candidates shrink by one every time a proposal is made, it is no longer deterministic and may return different matchings when run on the same BSMP. This is, of course, a consequence of the non-determinism of the underlying MDFT models.

We can also define another variant of GS that we call Expected Behavioral Gale Shapley (EB-GS). We first note that, given a man, we can extract a linear order from the expected positions of the women according to his MDFT model (breaking ties if needed). EB-GS corresponds to running GS on the profile of linear orders obtained in this fashion.

Algorithm FB-ILP. For each combination of $m_i \in M$ and $w_j \in W$, $|M| = |W| = n$, we introduce a binary variable $m_i w_j$ that takes value 1 if m_i is matched with w_j and 0 otherwise. We assume that for FB-ILP we have access to an $n \times n$ matrix $pos_M[i,j]$ where entry i,j gives us the expected position of w_j in the ranking of m_i, and the same matrix is available for the women, denoted pos_W.

Recall that finding the solution with lowest sex equality cost requires minimizing $SEC = |\sum_{i,j \in n} pos_M[i,j] \cdot m_i w_j - \sum_{i,j \in n} pos_W[j,i] \cdot m_i w_j|$. We cannot implement this absolute value directly as the optimization objective in Gurobi [13] as it is non-linear due to the presence of the absolute value. Since the SECs are always ≥ 0 we can overcome this using a standard trick in ILPs using indicator variables [3]. The SEC objective can be viewed as adding up the total man cost and the total woman cost, so we add indicator variables $tmc \geq 0$ and $twc \geq 0$ and minimize the difference between these two quantities. Hence, our full FB-ILP can be written as follows.

min	ind, s.t.,	
(1)	$\sum_{j \in n} m_i w_j = 1$	$\forall i \in n$
(2)	$\sum_{i \in n} m_i w_j = 1$	$\forall j \in n$
(3)	$\sum_{i,j \in n} m_i w_j = n$	
(4)	$\sum_{i,j \in n} pos_M[i,j] \cdot m_i w_j = tmc$	
(5)	$\sum_{i,j \in n} pr_W[j,i] \; m_i w_j - twc$	
(6)	$twc \geq 0$	
(7)	$twc \geq 0$	
(8)	$twc - tmc = ind$	

In the constraints above (1) and (2) ensures that every man m_i has exactly one match across all possible women and every woman w_j has one match across all possible men. The redundant constraint (3) ensures that we have exactly n matches, i.e., everyone is matched. Constraint (4) captures the total cost to the men by multiplying the expected position by the indicator variables for the matches. Likewise constraint (5) captures the total woman cost. Constraint (8) is necessary to ensure that Gurobi handles our absolute value constraint correctly. We know that both $tmc \geq 0$ and $twc \geq 0$ from constraints (6) and (7), hence when Gurobi uses the Simplex Algorithm to solve, it will set $tmc = ind$ and $twc = 0$ if $ind > 0$ and otherwise we will have $tmc = 0$ and $tmc = -ind$. In either case we have a bounded objective function and we can find a solution if one exists.

Algorithm B-ILP. To find the optimal α-B-Stable solution with B-ILP, we begin with the same setup. For each $m_i \in M$ and $w_j \in W$ we introduce a binary variable $m_i w_j$ defined as above. In addition, for B-ILP we assume that for each man and each woman we are given an $n \times n$ matrix Pr_{m_i} where entry $Pr_{m_i}[j, k]$ gives the probability that man m_i prefers w_j to w_k. This matrix can be computed by running the BSMP of man m_i a sufficiently large number of times.

There are two interrelated complications with formulating this probabilistic matching problem as an ILP: first we need the product of the probabilities which is a convex not linear function and, second, stability is a pairwise notion over a given matching. To deal with both of these issues we introduce $\forall((i,j),(k,l)) \in \binom{\binom{n}{2}}{2}$ possible combinations of pairs of pairs, an indicator variable $m_i w_j + m_k w_l$ to indicate that both $m_i w_j$ is matched and $m_k w_l$ is also matched. This allows us to compute the blocking probability of m_i and w_l as well as of m_k and w_j. Given the formulation in [2], we know that we want to maximize the probability that *no blocking pair exists*. Hence for every pair of possible marriages $m_i w_j + m_k w_l$ we can compute the probability that these four individuals are not involved in blocking pairs by taking the likelihood that they swap partners, formally let $block[(ij), (kl)] = (1 - Pr_{m_i}[l, j] * Pr_{w_l}[i, k]) * (1 - Pr_{m_k}[j, l] * Pr_{w_j}[k, i])$. To handle the convex constraint we simply take the log of this quantity and maximize using an indicator variable we which we implement using the Gurobi *And* constraint. We can write the full program as follows.

$$
\begin{array}{lll}
\max & \sum_{\forall((i,j),(k,l)) \in \binom{\binom{n}{2}}{2}} pair_{m_i w_j + m_k w_l} * log(block[(ij),(kl)]), s.t., & \\
(1) & \sum_{j \in n} m_i w_j = 1 & \forall i \in n \\
(2) & \sum_{i \in n} m_i w_j = 1 & \forall j \in n \\
(3) & \sum_{i,j \in n} m_i w_j = n & \\
(4) & AND(m_i w_j, m_k w_l) = pair_{m_i w_j + m_k w_l} & \forall((i,j),(k,l)) \in \binom{\binom{n}{2}}{2}
\end{array}
$$

In the constraints above (1) and (2) ensures that every man m_i has exactly one match across all possible women and every woman w_j has one match across all possible men. The redundant constraint (3) ensures that we have exactly n matches, i.e., everyone is matched. Constraint (4) uses the Gurobi [13] *AND* constraint to set the value of $pair\_m_iw_j + m_kw_l$ to be 1 if and only if both m_iw_j and m_kw_l are both 1. This allows us to capture all possible pairs of man/woman pairs and maximize the probability that no blocking pair occurs.

C   Convergence Analysis for B-LS

The convergence analysis performed for $n = 16$ is shown in Fig. 6. While we depict the results of for seven runs we performed a total of 50 runs. The results indicated that B-LS plateaus after 300 iterations, corresponding to approximately 340 s. B-LS does so around 88% of the time and returns a matching $1.006 * 10^{-6}$ far from optimal otherwise.

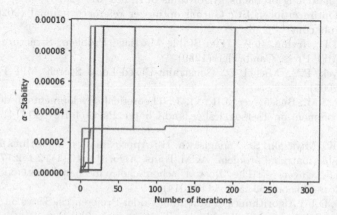

Fig. 6. Convergence of B-LS algorithm implementation with respect to α-B-Stability when $n = 16$.

References

1. Aziz, H., et al.: Stable matching with uncertain pairwise preferences. In: Proceedings of the 16th Conference on Autonomous Agents and MultiAgent Systems, AAMAS 2017, pp. 344–352. ACM (2017)
2. Aziz, H., Biró, P., Gaspers, S., de Haan, R., Mattei, N., Rastegari, B.: Stable matching with uncertain linear preferences. Algorithmica **82**(5), 1410–1433 (2020)
3. Bertsimas, D., Tsitsiklis, J.N.: Introduction to Linear Optimization, vol. 6. Athena Scientific, Belmont (1997)
4. Busemeyer, J.R., Diederich, A · Survey of decision field theory. Math. Soc. Sci. **43**(3), 345–370 (2002)
5. Busemeyer, J.R., Townsend, J.T.: Decision field theory: a dynamic-cognitive approach to decision making in an uncertain environment. Psychol. Rev. **100**(3), 432 (1993)
6. Busemeyer, J., Gluth, S., Rieskamp, J., Turner, B.: Cognitive and neural bases of multi-attribute, multi-alternative, value-based decisions. Trends Cogn. Sci. **23**(3), 251–263 (2019)

7. Busemeyer, J., Townsend, J.: Fundamental derivations from decision field theory. Math. Soc. Sci. **23**(3), 255–282 (1992)
8. Chen, J., Niedermeier, R., Skowron, P.: Stable marriage with multi-modal preferences. In: Proceedings of the 2018 ACM Conference on Economics and Computation (ACM:EC), pp. 269–286. ACM (2018)
9. Cooper, F., Manlove, D.: Algorithms for new types of fair stable matchings. arXiv preprint arXiv:2001.10875 (2020)
10. Gale, D., Shapley, L.S.: College admissions and the stability of marriage. Amer. Math. Monthly **69**, 9–14 (1962)
11. Gelain, M., Pini, M., Rossi, F., Venable, K., Walsh, T.: Procedural fairness in stable marriage problems. In: 10th International Conference on Autonomous Agents and Multiagent Systems (AAMAS 2011), pp. 1209–1210. IFAAMAS (2011)
12. Gelain, M., Pini, M.S., Rossi, F., Venable, K.B., Walsh, T.: Local search approaches in stable matching problems. Algorithms **6**(4), 591–617 (2013)
13. Gurobi Optimization, LLC: Gurobi optimizer reference manual (2020). http://www.gurobi.com
14. Gusfield, D., Irving, R.W.: The Stable Marriage Problem: Structure and Algorithms. MIT Press, Cambridge (1989)
15. Hentenryck, P.V., Michel, L.: Constraint-Based Local Search. MIT Press, Cambridge (2005)
16. Hotaling, J.M., Busemeyer, J.R., Li, J.: Theoretical developments in decision field theory: comment on Tsetsos, Usher, and Chater. Psychol. Rev. **117**(4), 1294–1298 (2010)
17. Iwama, K., Miyazaki, S., Yanagisawa, H.: Approximation algorithms for the sex-equal stable marriage problem. ACM Trans. Algorithms **7**(1), 2:1–2:17 (2010)
18. Kojima, F., Unver, M.: The "Boston" school-choice mechanism: an axiomatic approach. Econ. Theory **55**, 515–544 (2014)
19. Manlove, D.F.: Algorithmics of Matching Under Preferences, Series on Theoretical Computer Science, vol. 2. WorldScientific, Singapore (2013)
20. McDermid, E., Irving, R.W.: Sex-equal stable matchings: complexity and exact algorithms. Algorithmica **68**(3), 545–570 (2014)
21. Mellers, B., Biagini, K.: Similarity and choice. Psychol. Rev. **101**, 505–518 (1994)
22. Miyazaki, S., Okamoto, K.: Jointly stable matchings. J. Comb. Optim. **38**(2), 646–665 (2019). https://doi.org/10.1007/s10878-019-00402-4
23. Roe, R., Busemeyer, J., Townsend, J.: Multi-alternative decision field theory: a dynamic connectionist model of decision-making. Psychol. Rev. **108**, 370–392 (2001)
24. Roth, A.E.: Who Gets What - and Why: The New Economics of Matchmaking and Market Design. Houghton Mifflin Harcourt, New York (2015)
25. Sühr, T., Biega, A., Zehlike, M., Gummadi, K., Chakraborty, A.: Two-sided fairness for repeated matchings in two-sided markets: a case study of a ride-hailing platform. In: Proceedings of the 25th ACM SIGKDD International Conference on Knowledge Discovery & Data Mining, KDD 2019, pp. 3082–3092. ACM (2019)
26. Tziavelis, N., Giannakopoulos, I., Johansen, R.Q., Doka, K., Koziris, N., Karras, P.: Fair procedures for fair stable marriage outcomes. In: The Thirty-Fourth AAAI Conference on Artificial Intelligence, AAAI 2020, pp. 7269–7276. AAAI Press (2020)
27. Tziavelis, N., Giannakopoulos, I., Doka, K., Koziris, N., Karras, P.: Equitable stable matchings in quadratic time. In: Advances in Neural Information Processing Systems, pp. 457–467 (2019)

FUN-Agent: A HUMAINE Competitor

Robert Geraghty[✉], James Hale, and Sandip Sen

University of Tulsa, Tulsa, OK 74104, USA
{rrg053,jah6484,sandipv}@utulsa.edu

Abstract. Even prior to the recent surge of interest in industry and the general populace about the potential of human-AI collaboration [23], academic researchers have been pushing the frontier of new modalities of peer-level and ad-hoc human agent collaboration [12,24]. We are particularly interested in research on agents representing human users in negotiating deals with other human and autonomous agents [14,19,21]. Here we present the design for the conversational aspect of FUN-agent, our entry into the *HUMAINE* League of the 2020 Automated Negotiation Agent Competition (ANAC). We discuss how our seller agent utilizes conversational and negotiation strategies, mimicking those used in human negotiations, to maximize it's utility from negotiations with a human buyer in the presence of a competing seller. We leverage verbal influence tactics, offer pricing, and enhancing human convenience to entice the buyer, build trust, and thwarting the competition. We present analysis from in-house comparative evaluation of FUN-Agent against three well-known agent negotiators: Boulware, Conceder, and Linear Conceder.

Keywords: Negotiation · Human-agent interaction · Conversational agent

1 Introduction

Human-agent negotiation has emerged as a significant area of study in the domain of human-agent interaction and conversational AI. Contrary to inter-agent interactions, when designing agent interactions with human partners, agent designers must be keenly aware of how certain utterances or behaviors may impact a human competitor's perception of and alignment with the agent's goals. Some work has been done to evaluate the effectiveness of various emotional strategies. For example, research by van Kleef et al.-in negotiations between humans has suggested a hostile demeanor yields a higher concession from an opponent, whereas happiness causes one's exploitation [27]. It is also instructive to note that de Melo et al. confirmed that these results translated to a domain wherein a human competed against an agent [3].

The retail, e-commerce, legal, business, and industrial sectors all would benefit from an artificial proxy to secure the best possible deals in negotiation [20,26]. However, artificial agents and humans need not compete, they can work together to solve complex problems; with recent advents in this problem space, industry,

© Springer Nature Switzerland AG 2022
J. Chen et al. (Eds.): DAI 2021, LNAI 13170, pp. 171–184, 2022.
https://doi.org/10.1007/978-3-030-94662-3_11

researchers, and others have turned their attention to applications of human-agent collaboration [4,13,23,28,29].

In the current paper, we present the design for the conversational aspect of our entry into the *HUMAINE* League (https://cisl.rpi.edu/humaine2020) of the 2020 Automated Negotiation Agent Competition (ANAC) (http://web.tuat.ac.jp/~katfuji/ANAC2020/). We discuss how our agent utilizes conversational and negotiation strategies to maximize its utility as a simulated street vendor. We aim to achieve this goal by leveraging verbal influence tactics, offer pricing, and increasing human convenience to entice the buyer, build trust, and discourage exploitation. The tactics aim to mimic those used in human negotiations, including concepts such as

- Hard lining, Conceding, Expressing Frustration, Ultimatums [30]
- Positive Reinforcement, Splitting the difference [15].

Mimicking human behaviors has been shown to increase trust and likeability in agents, which conforms to our goal of being a friendly and helpful partner [23].

This is an extension of our work from the International Workshop of Multimodal Conversational AI (MuCAI) [10].

2 Related Work

Dybala et al. constructed a humorous, non-task oriented, conversational agent and listed several use cases for such a chatbot: a car navigation system keeping a driver engaged or supplementing an elderly person's societal necessities [7]. Actually, humor seems to be quite a focal point in the realm of human-agent interaction, as it generally seems to make the user feel better [8].

The field of human-agent interaction has applications in health-care [7,11,25]. Tapus et al. investigated the potential to use social robots as a means to comfort people with dementia and promote positive behavior [25]. While Gratch et al. investigated the benefits of using "virtual humans" to interview patients [11].

The ethics of conversational AI has also been explored; Ruane et al. emphasize transparency as an important step towards a user-centric experience [22]. This idea could be further expanded into the domain of explainable AI, in which an agent explains to the user its rationale for an offer.

Additionally, research has been conducted to have agents act as collaborators. Seeber et al. explore domains in which the complexity of a problem hampers the performance of a human team, and propose to improve performance by adding artificial teammates [23]. Our agent encapsulates these qualities.

Lastly, much literature exists regarding human-human, human-agent, and agent-agent interaction in various social contexts, including negotiation. Parks and Komorita emphasize the significance of reciprocity in a negotiation setting [18]. Axelrod and Hamilton mathematically investigate a tit-for-tat strategy in a 2-player prisoner's dilemma game, and its implications for the evolution of cooperative behavior [1]. We pull from this area of research in our design of FUN-agent; e.g. we adopt a reciprocal strategy (price-mirroring) in response to this research to elicit cooperation towards a fair deal.

3   2020 HUMAINE Competition

The 2020 HUMAINE (HUman Multi-Agent Immersive NEgotiation; https://cisl.rpi.edu/humaine2020) competition, collaboratively run by RPI and IBM's T.J. Watson labs, was a part of the ANAC collection, held in conjunction with the International Joint Conference on Artificial Intelligence. The HUMAINE competition scenario has a human buyer use natural language dialogues within a virtual immersive marketplace to procure items from competing seller agents [5, 6].

The HUMAINE competition consists of two selling agents, and one human buyer. Each agent's goal is to maximize it utility over a round of negotiation. During each round, the human has a set budget that they can spend on buying different ingredients used to either make cakes or pancakes. The seller agents compete with each other to obtain the human buyer's business, and try to maximize the profit from all sales. At the end of a round, each agent's score is based on their profit, and the human score is based on how they allocated their purchased ingredients into cakes and pancakes, where unspent budget and unused ingredients provide no utility.

To excel in the competition, the agent must understand the buyer preferences and earn the trust of the human player as well as infer the intentions of the opponent agent. We believe a key component of a competitive negotiation strategy will be to use inferred user preference to compose bundles to offer to the buyer that (a) increase local profits, (b) reduce the cognitive burden and time effort of the human, i.e., making things easier for the human to procure item bundles they need. This approach can facilitate win-win negotiation outcomes while also keeping the competitor agent at bay.

3.1   Language Processing

The agent makes use of IBM's Watson assistant for understanding human and competing agent speech [9]. This allows the agent to extract the intent and specifics of a message, including the goods in the bid, the quantity of each good in the bid, the addressee of the bid, and the proposed price for the bid. IBM's Watson allows for the consistent interpretation of user utterances with similar intent but with varying sentence structure and word choice. This information is then used by the strategy to arrive at a response message.

3.2   Formal Specification of the Repeated Task Negotiation Problem

We now present notations to describe the repeated task negotiation problem of the HUMAINE competition. We use the term *agent* and *AI* player interchangeably. The other competing player is referred to as the *Opponent*. Let $\mathcal{I}$ be the set of issues being negotiated. An offer O is a pair (O_p, O_I), containing a price, O_p, and a bundle O_I containing a set of pairs for each of the issues in the offer and the number of items of that issue being offered. For each issue j contained in the bundle being offered, O_j, is the number of units being offered. If the human accepts the bundle offered by our agent, the utility received is $U_{AI}(O) = O_p - C(O_I)$, where $C(O_I) = \sum_{j \in O_I} c_j O_j$ is the cost of the agent for providing offer O, and c_j is the per-unit cost of issue j to the AI player.

4   Strategy

4.1   Conversational Demeanor

A large part of our strategy relies on our agent's sales-oriented speech. The agent uses a set of modular sentences that can be included in the agent's offers and counteroffers to display a willingness to compromise and an enthusiasm about the negotiation process. Examples of this include bringing attention to low prices, stating our intent to cooperate, and expressing approval upon agreement.

The FUN-Agent utilizes problem statement knowledge to offer support for the buyer's goals, namely in buying cakes and pancakes. We use language and phrasing that is informative and persuasive, but not overly aggressive.

Finally, when the buyer is reluctant or unwilling to compromise, our agent will revert to a more assertive tone. For example, if the participant goes back on a previously stated by position by decreasing the amount they will pay for a bundle, or if they do not sufficiently increase that amount, our agent will "call out" the human for not negotiating in good faith citing counter-productive moves and encouraging more cooperation to reach a mutually acceptable deal. As mentioned previously, research substantiates this strategy; an angry negotiator often generates a larger concession [2, 27]. Our agent only uses hostility when the participant's counter-offers plateau, and that too citing lack of cooperation to achieve a deal as the underlying reason for the heightened rhetoric, so as not to portray unwarranted hostility to the user. We believe that the culmination of these design considerations will provide human buyers with more incentive to do business with our agent, even if the competition undercuts our pricing strategy.

4.2   Bundling

As humans find it cognitively challenging to compute the utilities of large quantities of discrete items, we opted to push for the participant to purchase functional bundles of ingredients, instead of buying each individually. We believe this approach is both easily understandable and reduces the cognitive load for the human, as well as having the benefit of being significantly more time efficient. In short, by channelizing the negotiation to pricing bundles of goods we hope to strike better deals in less time with more satisfied customers.

To maximize the convenience that our agent provides to the human buyer, our agent provides the buyer with guidance about how to phrase such a request. We believe maximizing the relative convenience of interacting with the FUN-agent, in comparison to the competing agent, will result in more buyers choosing to do business with our agent even at a more profitable deal for us. An associated benefit of the bundling strategy is that the other agent may be shut out of competing, as our agent will try to obtain the largest bundle deal possible given

Table 1. Examples of FUN-Agent conversational categories

Greeting: *"Hello! Welcome to Watson-mart!"*
Concern for User (prompt for bundles): *"I've encountered several people around the market-place looking to make cakes or pancakes"*
Positive Reinforcement and Encouragement: *"That works, you got a really good deal there! Thank you for your business!"*   *"I believe that fits your budget!"*   *"... a significant discount on what was already a good deal!"*
Expressing Displeasure and Frustration: *"You bargain really hard: you didn't change your offer. I hope you realize I already discounted my price for you. I can't sell for $20.00, that will put me in the red"*
Conceding: *"Let's not draw this out, the absolute best I can do $26.95"*
Sniping: *"I can do better, how does <offer> sound to you?"*

the user budget, thus draining the buyer's budget in one large bundle deal. If any budget is left over, our agent offers incremental deals, such as adding toppings. We believe that having negotiated a large bundle with us would entice the user to spend their remaining budget with us.

4.3 State Tracking

In order to make context aware offers, our agent stores every price offered by each participant in the negotiation, for each set of items. This information is then used for two purposes, to cooperate or punish the buyer based on price movements (price mirroring), and to undercut the competing agent (undercutting).

Our agent uses **price mirroring** to loosely reciprocate–up to a point–the buyer's change in bid price. If the human is recalcitrant and refuses to move their price, we do not reward that stubborn behavior with a large concession. When the human offers a lower price than a previous offer, our agent does not move its price at all, halting progress until the human concedes. In the case of a minimally raised price, our agent exactly mirrors the move in price and brings attention to it to encourage the user to mirror our own concessions. This builds a reciprocal dynamic between the buyer and our agent that humans have been shown to respond to in human-agent interactions [17].

In the event the competing seller agent has previously provided a price for the current set of goods our agent focuses on **undercutting**. This is done by using the competing agent's previous price as the upper bound for our counteroffer, instead of the human's price.

FUN Agent

Ok, in response to your counter offer of $20.00, I will provide a great deal of $36.40, a significant discount on what was already a good deal! So, I can give you 10 pancakes for $36.40.

User

Agent, I want 10 pancakes for 20$

FUN Agent

You bargain really hard: you didn't change your offer. I hope you realize I already discounted my price for you. I can't sell for $20.00, that will put me in the red. Ok, in response to your counter offer of $20.00, I will provide a great deal of $34.58, a significant discount on what was already a good deal! So, I can give you 10 pancakes for $34.58. Let's seal the deal!

User

Agent, I want ten pancakes for 25$

FUN Agent

Ok, in response to your counter offer of $25.00, I will provide a great deal of $31.20, a significant discount on what was already a good deal! So, I can give you 10 pancakes for $31.20.

Fig. 1. Sample conversation to demonstrate price mirroring

4.4 Conceding

To preempt prolonged negotiation, after five back-and-forth exchanges with the human buyer, the asking price is lowered to the maximum of (a) the agent's minimum acceptable price and (b) the buyer's highest offer. This reduces the likelihood of the human buyer getting frustrated and the competing seller sniping the deal. *"Let's not draw this out, the absolute best I can do is $26.95."*

4.5 Obfuscation as a Competitive Strategy

Though the Watson assistant works reasonably well in most cases, there are some language constructs and phrases which it finds challenging. We take advantage of this shortcoming by wording our responses in such a manner that the competing agent will be unlikely to decipher, because of our use of multiple prices and bundles. As evident from the previous quotes, the FUN-Agent's offers often contain multiple prices. Whereas the buyer should be able to understand the phrasing without issue, the competing agent's language processing component may be at a loss. For example, consider Watson's final message in Fig. 1: a competing agent may not understand whether the price we are offering is $25 or $31.20. The agent also may not know what a pancake or cake is, given that the problem domain only requires knowledge of the individual ingredients, i.e. flour, milk, etc. This approach proved effective in testing, with the competing agent interpreting similar messages as FUN-Agent offering zero ingredients for $25.

4.6   Initial Offer Generation

Algorithm 1 describes the initial offer generated by our agent in response to a user request R, an offer, with the associated price signaling what the user is willing to pay for the requested bundle. In the algorithm, $\gamma_{init} > 1$ represents FUN-Agent's initial profit aspiration. In practice, the user typically asks for a single item. Our agent's goal is to respond both with an attractive price for that item and to channel the discussion to a larger target bundle offer, T, which includes all items the agent estimates the user needs.

Algorithm 1. Initial Offer Generation

1: **procedure** INITIALOFFER(R)
2:     **if** R_I is not empty **then**
3:         compute cost, C_R, of R_I and response bid $O^R = (\gamma_{init} \times C_R, R_I)$
4:     Also propose alternate target bundle T offer $O^T = (\gamma_{init} \times C_T, T)$

4.7   Counteroffer Generation

Now we present the approach followed by FUN-Agent, referred to as the AI player, to develop counteroffers to the user. Let O^{AI} be the last offer made by the AI player. Algorithm 2 describes how the AI player generates a counteroffer to a subsequent offer O^h made by the human user. Here, we assume that $O_I^h \subseteq O_I^{AI}$ if O^h contains a subset (not necessarily proper) of the issues in O^{AI} and if for each issue j contained in both, $O_j^h \leq O_j^{AI}$, i.e., the human is asking equal or less number of items compared to what the AI player offered. The AI player then accepts an offer if the human is ready to pay the price quoted. If the human asks for a lower price, the AI player counter offers with the maximum of the following three values:

- a fixed discount fraction, α, of its previous quoted price,
- a fixed discount fraction, β, of the difference of the AI player's last quote and the price offered by the human, and

---

Algorithm 2. Counter-offer Generation

1: **procedure** COUNTEROFFER(O^{AI}, O^h)
2:     **if** $O_h \geq O_{AI}$ & $O_I^h \subseteq O_I^{AI}$ **then**
3:         *Accept O^h*
4:     **else**
5:         **if** $O_I^h = O_I^{AI}$ **then**
6:             $p_1 = \alpha O_{AI}, \ p_2 = O_{AI} - \beta(O_{AI} - O_h)$
7:             Offer $(max(p_1, p_2, U_{AI}(O^h)), \gamma_{final} O_I^h)$
8:         **else**
9:             INITIALOFFER(O^h)

– a fixed minimum profit point which is γ_{final} times the cost of the bundle for the AI player.

The price calculation at a round i in a negotiation is as follows:

$$p_i = max(\alpha p_{i-1}, p_{i-1} - \beta(p_{i-1}h_i), h_i, c) \tag{1}$$

where i is the round, p_i represents the agent's price in a round, h_i represents a human's price in a round, α is some value between zero and one, α is the max discount (.9), β is the typical discount (.5), and c is the minimum acceptable value for our agent.

If the user were to opt for another item bundle, we resort to our INITIALOFFER algorithm (see Algorithm 1) to generate a response. This pricing scheme allows for price degradation while avoiding accepting a price that would cause a negative profit. In cases where punitive price mirroring is used, the counteroffer price presented in this section is not relevant. The combination of these considerations provides an adaptive price degradation scheme that supports differing human negotiation dispositions while maintaining boundaries to prevent hard-lining and negative utilities for our agent's sales.

4.8 Guarding Against Competitor Sniping

Currently, we do not model the competing agent in detail but plan to implement a modeling system in the next stage of this project. We have, however, tested initial designs for preventing our agent's interactions being used by the opponent to model our agent's preferences and thereby gain any strategic advantage. We employ filler language and "non-essential dialogues" to hide the progression of our negotiation and any concessions granted by our agent. In addition, the design of this competition does not allow for round-over-round agent learning, so we did not focus on creating dynamic strategies to protect against competing agents learning our pricing strategies. Such a system would be useful in real world scenarios but it is outside of the scope of this specific work.

5   Experimentation and Design

We present the design of representative competing agents and results from the evaluation of our agent against each of the competitors in a set of in-house tests.

5.1   Competitor Agents

To test the relative performance of our agent, we created a few variants of time-dependent agents (referred to as "test agents") to be deployed as the competing seller. These agents were not provided many of the unique functionalities of our agent, including (a) Speech mimicking human conversation, (b) Bundling,

(c) Item specific price tracking, (d) Obfuscation, and (e) Price mirroring. The strategy of competitor agents we chose focused on pricing offers as follows

$$p_R = \max(C_R \cdot (1 + \frac{t_r}{t_t}^{\frac{1}{E}} \gamma_{AI}), \gamma_{final} C_R) \tag{2}$$

where C_R is the agent's cost of the item(s) requested by the buyer, R; t_r is the time remaining in the round; t_t is the total round duration; γ_{AI} is the initial markup for the competitor agent; γ_{final} is the minimum markup the agent will offer; E is a parameter that controls how fast the competitor agents drop price.

When $0 < E < 1$, the agent will lower its price early and rapidly, when $E > 1$ the agent will lower its price late in the round and slowly, and when $E = 1$ the agent will lower it's price linearly. The agent will never offer below its minimum acceptable price, defined by $\gamma_{final} C_R$. For our testing we used three agents [16]:

Conceder Agent: An eager conceder agent with $E = 0.5$ that drops its price relatively early in the round.

Boulware Agent: A stubborn boulware agent with $E = 3$ that holds a high price for much of the round and concedes at the end. This agent is an entrenched negotiator, budging little on price until late in the round.

Linear Agent: A linear agent with $E = 1$ that lowers its price consistently. This is an intermediate strategy between Conceder and Boulware.

We expect our agent to perform best against the Boulware agent, as our agent will compromise to a much greater degree for most of the round. We expect the Linear and Conceder agents will provide more of a challenge as their initial offerings will be lower and more enticing than the FUN-agent's offers early in the round. However, the FUN-agent will still concede more during extended negotiation over a given set of goods due to its split-the-difference strategy. Human buyers may not appreciate the time dependent nature of the competing agents, as the best deals will come with little time left when human buyers will be most pressured. Additionally, the price points offered by these competing agents are determined primarily by time remaining in the round and less on strategic or opportunistic pricing that can possibly cement more profitable deals.

5.2   Human Negotiators

These in-house tests were conducted with 7 subjects from primarily within our lab–excluding the authors of this paper. We offered a ten dollar incentive to the participant who won the most utility.

5.3   Negotiation Rounds

Each round of testing lasted 5 min. Each subject played a total of three rounds, one round each where the FUN-Agent was competing against one of the three different agents described above. The human subject had a budget of $50 USD

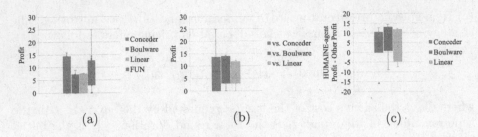

Fig. 2. (a) Profit distributions of the four agents for all tests (b) FUN-agent's profit distribution against the three types of agents. (c) Distributions of the difference between FUN-agent's profit and the competing agents' profits (i.e. $P_{FUN} - P_{other}$, so values above the x-axis imply better profit for FUN-agent).

in each round, and any unspent budget at the end of the round was considered worthless. Additionally, any ingredients purchased that were not used to make a cake or pancake were ignored in the end of round utility calculation. The break-even costs of items for sellers are drawn from the uniform distributions that follow: Eggs, [0.25, 0.50]; Flour, [0.50, 1.00]; Sugar, [0.50, 1.00]; Milk, [0.20, 0.40]; Chocolate, [0.20, 0.40]; Vanilla, [0.20, 0.40]; Blueberries, [0.25, 0.50]. More information on utility and round structure can be found at https://cisl.rpi.edu/humaine2020, which was excluded due to space considerations and marginal relevance to the discussion.

6   Results and Discussion

Figure 2 displays box-plots of agent profit distributions and show the bounds of the first and third quartiles, the means, and the minimum and maximum values. Figure 3a shows the distribution of the profit for all agents. Though the upper bound of the third quartile of the Conceder agent exceeds that of our FUN-agent, the FUN-agent's mean is the highest of all other agents. This may be due to a round where the Conceder agent won nearly all of the user's money. Due to the Conceder agent's superior pricing early in the round, which may convince the buyer to deal with it exclusively, FUN-agent occasionally performs worse. For similar reasons, the next best mean to the FUN-Agent is Linear agent.

Figure 3b illustrates the distribution of FUN-agent's profit against other agents. We see our agent performs best against the Boulware agent. The Boulware agent is the least generous time-dependent agent. Hence, users may be frustrated with its lack of concessions.

Figure 3c shows the distribution of the differences between the profit of FUN-agent and its competitors. The dashed line indicates the x-axis, which is the dividing line between FUN-agent outperforming its opponent and vice versa–any positive data point indicates superior performance by FUN-agent. The main item of note is that one can see the margin of victory–or defeat–of the FUN-agent. We see some of the distributions dip below the x-axis: indicating some

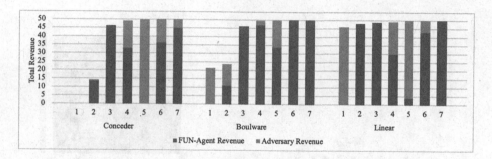

Fig. 3. Total revenue allocation to each agent over all seven trials.

games in which a competitor agent defeated FUN-agent. A time-dependent agent will lower its markup regardless of past negotiation. A time-dependent agent may offer a better price in negotiation over a novel bundle, which might be attractive to the user. However, if the FUN-Agent can entice them to make a deal immediately, this will not be of concern.

Figure 3 presents the total revenue to each agent over all trials. While the competing agent does trade exclusively with the buyer in a few cases, generally the FUN-Agent nabs a larger share of the pie or gets the entire deal.

Figure 4 (left) is a scatter plot, with linear regression, of an adversary's total revenue in a game versus FUN-agent's revenue. The region bounded by the x-axis and the line $y = x$, highlighted in blue, is the region in which FUN-agent's revenue exceeds that of its opponent. We note that the majority of points are placed in this region. Lastly, Fig. 4 (right) is a 2D histogram illustrating the distribution of the points in Fig. 4 (left). Though there are few exceptions, most points exist in the lower right corner, the region where FUN-agent outperforms its opponent.

We performed pairwise one-tail t-tests between the FUN-agent's and competing agents' profits and revenues. We found a significant difference between the FUN-agent's and Boulware agent's profits ($p = 0.011$) and revenues ($p = 0.007$). There was not a statistically significant difference between our agent and the Conceder agent in profit ($p = 0.21$) and revenue ($p = 0.10$). Similar comparisons were observed between our agent and the Linear agent in profit ($p = 0.10$) and revenue ($p = 0.10$). So, the FUN-agent does better, but not significantly better, when compared to the Conceder and Linear agents. Finally, we compared the average profits and revenues of the FUN-agent for each subject to the combined average of all other agents for each subject. There was a significant difference in both profits ($p = 0.033$) and revenues ($p = 0.013$). Hence, on average our agent still attracts more business compared to those competing agents, shown by the higher profit and revenue. The less than significant win margins against the Linear and Conceder agents could be due to the lower prices offered. Despite the lack of significance in some tests, user feedback suggests that the FUN-agent's negotiation style is more attractive to human buyers.

Many of the testers preferred the FUN-agent's style of negotiation, where it would lower prices based on user interaction, rather than the purely time

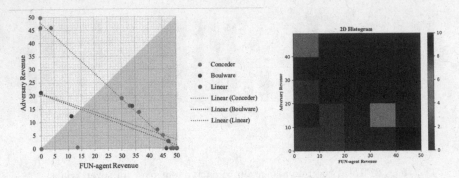

Fig. 4. In-house testing revenues: scatter (left) and density (right) plots. (Color figure online)

dependent discounting used by the test agents. The splitting-the-difference tactic was a major advantage of the FUN-agent as it could quickly bring its price down, which led to the subjects concluding that the FUN-Agent appeared more willing to find a compromise. The subjects cited the time constraint imposed on each round as a reason for using the cake and pancake bundles offered by our agent, as it was faster and easier to negotiate over these bundles. There were also some critiques of the FUN-agent: it was too verbose when compared to the simple and concise language used by the competing agents. Also, users at times found the initial counteroffer made by the Linear and Conceder agent to be lower and accept it outright, or only address the competing agent while ignoring our agent.

FUN-agent placed second out of five agents in the HUMAINE competition. Minor error fixes could have further improved its performance.

7   Conclusion

We outline the motivations and the strategy design, highlighting conversational aspects designed to engage with and convince human negotiators, of our FUN-Agent entrant to the 2020 HUMAINE competition. We leverage verbal influence tactics, offer pricing, and increasing human convenience to entice the buyer, build trust, and discourage exploitation. The tactics are inspired by those observed in human negotiations [15, 30] including hard-lining, strategic conceding, expressing frustration, ultimatums, positive reinforcement, splitting the difference, etc. Mimicking human behaviors has been shown to increase trust and likeability in agents, which conforms to our goal of being a friendly and helpful collaborator. Experimental results demonstrated the supremacy of our agent design compared with the competing agents. In the majority of negotiation games the FUN-agent outperforms its competitor agents both in revenue and typically also in profit.

We plan to further refine the language–basing it more on literature and testing–to make FUN-agent more human-like and to address verbosity concerns. For example, since others have found success in using humor [7], a joke generator could enhance our agent's interactions with a user.

References

1. Axelrod, R., Hamilton, W.D.: The evolution of cooperation. Science **211**(4489), 1390–1396 (1981)
2. de Melo, C., Carnevale, P., Gratch, J.: Social categorization and cooperation between humans and computers. In: The Annual Meeting of The Cognitive Science Society (CogSci 2014), pp. 2109–2114, July 2014
3. de Melo, C.M., Carnevale, P., Gratch, J.: The effect of expression of anger and happiness in computer agents on negotiations with humans. In: The 10th International Conference on Autonomous Agents and Multiagent Systems-Volume 3, pp. 937–944 (2011)
4. Dellermann, D., Calma, A., Lipusch, N., Weber, T., Weigel, S., Ebel, P.: The future of human-AI collaboration: a taxonomy of design knowledge for hybrid intelligence systems. In: Proceedings of the 52nd Hawaii International Conference on System Sciences (2019)
5. Divekar, R.R., Mou, X., Chen, L., De Bayser, M.G., Guerra, M.A., Su, H.: Embodied conversational AI agents in a multi-modal multi-agent competitive dialogue. In: IJCAI, pp. 6512–6514 (2019)
6. Divekar, R.R., et al.: HUMAINE: human multi-agent immersive negotiation competition. In: Extended Abstracts of the 2020 CHI Conference on Human Factors in Computing Systems, pp. 1–10 (2020)
7. Dybala, P., Ptaszynski, M., Maciejewski, J., Takahashi, M., Rzepka, R., Araki, K.: Multiagent system for joke generation: humor and emotions combined in human-agent conversation. J. Ambient Intell. Smart Environ. **2**(1), 31–48 (2010)
8. Dybala, P., Ptaszynski, M., Rzepka, R., Araki, K.: Humorized computational intelligence towards user-adapted systems with a sense of humor. In: Giacobini, M., et al. (eds.) EvoWorkshops 2009. LNCS, vol. 5484, pp. 452–461. Springer, Heidelberg (2009). https://doi.org/10.1007/978-3-642-01129-0_51
9. Ferrucci, D.A.: Introduction to "this is Watson". IBM J. Res. Dev. **56**(3.4), 1:1–1:15 (2012)
10. Geraghty, R., Hale, J., Sen, S., Kroecker, T.S.: FUN-Agent: a 2020 HUMAINE competition entrant. In: Proceedings of the 1st International Workshop on Multimodal Conversational AI, MuCAI 220, pp. 15–21. Association for Computing Machinery, New York, NY, USA (2020)
11. Gratch, J., Lucas, G.M., King, A.A., Morency, L.-P.: It's only a computer: the impact of human-agent interaction in clinical interviews. In: Proceedings of the 2014 International Conference on Autonomous Agents and Multi-agent Systems. International Foundation for Autonomous Agents and Multiagent Systems, Richland, SC, pp. 85–92 (2014)
12. Hafızoğlu, F.M., Sen, S.: The effects of past experience on trust in repeated human-agent teamwork. In: Proceedings of the 17th International Conference on Autonomous Agents and MultiAgent Systems. International Foundation for Autonomous Agents and Multiagent Systems, pp. 514–522 (2018)
13. Kamar, E.: Directions in hybrid intelligence: complementing AI systems with human intelligence. In: IJCAI, pp. 4070–4073 (2016)
14. Lin, R., Gal, Y., Kraus, S., Mazliah, Y.: Training with automated agents improves peoples behavior in negotiation and coordination tasks. Decis. Support Syst. (DSS) **60**, 1–9 (2014)
15. Maaravi, Y., Idan, O., Hochman, G.: And sympathy is what we need my friend-polite requests improve negotiation results. PloS One **14**(3), 1–22 (2019)

16. Mccalley, J., Zhang, Z., Vishwanathan, V., Honavar, V.: Multiagent negotiation models for power system applications (2020)
17. Nass, C., Moon, Y.: Machines and mindlessness: social responses to computers. J. Soc. Issues **56**(1), 81–103 (2000)
18. Parks, C.D., Komorita, S.S.: Reciprocity research and its implications for the negotiation process. Int. Negot. **3**(2), 151–169 (1998)
19. Peled, N., Gal, Y.K., Kraus, S.: An agent design for repeated negotiation and information revelation with people. In Proceedings of the Twenty-Seventh AAAI Conference on Artificial Intelligence. Association for the Advancement of Artificial Intelligence, , pp. 789–795, July 2013
20. Ransbotham, S., Kiron, D., Gerbert, P., Reeves, M.: Reshaping business with artificial intelligence: closing the gap between ambition and action. MIT Sloan Manage. Rev. **59**(1) (2017)
21. Rosenfeld, A., Zuckerman, I., Segal-Halevi, E., Drein, O., Kraus, S.: NegoChat-A: a chat-based negotiation agent with bounded rationality. Autonom. Agents Multi Agent Syst. **30**(1), 60–81 (2015)
22. Ruane, E., Birhane, A., Ventresque, A.: Conversational AI: social and ethical considerations. In: AICS, pp. 104–115 (2019)
23. Seeber, I., et al.: Machines as teammates: a research agenda on AI in team collaboration. Inf. Manage. **57**, 03174 (2020)
24. Stone, P., Kaminka, G.A., Kraus, S., Rosenschein, J.S., et al.: Ad hoc autonomous agent teams: collaboration without pre-coordination. In: AAAI (2010)
25. Tapus, A., Tapus, C., Mataric, M.J.: The use of socially assistive robots in the design of intelligent cognitive therapies for people with dementia. In: 2009 IEEE International Conference on Rehabilitation Robotics, pp. 924–929 (2009)
26. Tung, K.: AI, the internet of legal things, and lawyers. J. Manage. Analytics **6**(4), 390–403 (2019)
27. Van Kleef, G.A., De Dreu, C.K.W., Manstead, A.S.R.: The interpersonal effects of anger and happiness in negotiations. J. Pers. Soc. Psychol. **86**(1), 57 (2004)
28. Wang, D., et al.: Human-AI collaboration in data science: exploring data scientists' perceptions of automated AI. In: Proceedings of the ACM on Human-Computer Interaction 3(CSCW), pp. 1–24 (2019)
29. Zhang, J., Bareinboim, E.: Human-assisted agent for sequential decision-making. In Adaptive Learning Agents Workshop at AAMAS (2017)
30. Zohar, I.: "The art of negotiation" leadership skills required for negotiation in time of crisis. Procedia - Soc. Behav. Sci. **209**, 540–548 (2015)

Signal Instructed Coordination in Cooperative Multi-agent Reinforcement Learning

Liheng Chen[1], Hongyi Guo[1], Yali Du[2], Fei Fang[3], Haifeng Zhang[4],
Weinan Zhang[1]([✉]), and Yong Yu[1]

[1] Shanghai Jiao Tong University, Shanghai, China
wnzhang@sjtu.edu.cn
[2] United Kingdom University College London, London, UK
[3] Carnegie Mellon University, Pittsburgh, USA
[4] Institute of Automation, Chinese Academy of Sciences, Beijing, China

Abstract. In many real-world problems, a team of agents need to collaborate to maximize the common reward. Although existing works formulate this problem into a centralized learning with decentralized execution framework, their decentralized execution paradigm limits the agents' capability to coordinate. Inspired by the concept of correlated equilibrium, we propose to introduce a *coordination signal* to address this limitation, and theoretically show that following mild conditions, decentralized agents with the signal can coordinate their individual policies as manipulated by a centralized controller. To encourage agents to learn to exploit the coordination signal, we propose *Signal Instructed Coordination* (SIC), a novel coordination module that can be integrated with most existing MARL frameworks. Our experiments show that SIC consistently improves performance in both matrix games and popular testbeds with high-dimensional strategy space.

Keywords: Multi-agent learning · Reinforcement learning · Correlated equilibrium

1 Introduction

Multi-agent interactions are common in real-world scenarios such as traffic control [24] and smartgrid management [27]. A straightforward approach to solve cooperative multi-agent environments is the *fully centralized* paradigm, where a centralized controller is used to make decisions for all agents, and its policy is learned by applying successful single-agent RL algorithms. However, the fully centralized method suffers from exponential growth of the size of the joint action space with the number of agents. Therefore, decentralized execution approaches are proposed, including the *fully decentralized* paradigm and the *centralized training with decentralized execution* (CTDE) [22,25] paradigm. The fully decentralized method models each participant as an individual agent with its own policy and critic

J. Chen et al. (Eds.): DAI 2021, LNAI 13170, pp. 185–205, 2022.
https://doi.org/10.1007/978-3-030-94662-3_12

conditioned on local information. This setting fails to solve the non-stationary environment problem [16,23], and is empirically deprecated by [8,20]. In CTDE framework, agents can leverage global information including the joint observations and actions of all agents in the training stage, e.g., through training a centralized critic, but the policy of an agent can only be dependent on the individual information and thus they can behave in the decentralized way in the execution stage. This training paradigm bypasses the non-stationary problem, and can lead to some coordination among the cooperative agents empirically [22].

Despite the merits of CTDE, the feasible joint policy space with distributed execution is much smaller than the joint policy space with a centralized controller, limiting the agents' capability to coordinate. For example, in a two-agent traffic system with agents A and B, whose individual action space is {go, stop}, we cannot find a joint policy that satisfies P(A goes & B stops) = P(A stops & B goes) = 0.5 if both agents are making decisions independently. Previous works [26,28] adopts peer-to-peer communication mechanism to facilitate coordination, but they require specially designed communication channels to exchange information and the agents' capability to coordinate is limited by the accessibility and the bandwidth of the communication channel.

Inspired by the *correlated equilibrium* (CE) [1,17] concept in game theory, we introduce a *coordination signal* to allow for more correlation of individual policies and to further facilitate coordination among cooperative agents in decentralized execution paradigms. The coordination signal is conceptually similar to the signal sent by a correlation device to induce CE. It is sampled from a distribution at the beginning of each episode of the game and carries no state-dependent information. After observing the same signal, different agents learn to take corresponding individual actions to formulate an optimal joint action. Such coordination signal is of practical importance. For example, the previous traffic system example can introduce a traffic policeman that sends a public signal via his pose to each agent. The type of the pose may be dependent on the current time (state-free) as a traffic light is, but agents can still coordinate their actions without any explicit communication among them. In addition, we prove that for a group of fully cooperative agents, if the signal's distribution satisfies some mild conditions, the joint policy space is equal to the centralized joint policy space. Therefore, the coordination signal expands the joint policy space while still maintains the decentralized execution setting, and is helpful to find a better joint policy.

To incentivize agents to make full use of the coordination signal, we propose *Signal Instructed Coordination* (SIC), a novel plug-in module for learning coordinated policies. In SIC, a continuous vector is sampled from a pre-defined normal distribution as the coordination signal, and every agent observes the vector as an extra input to its policy network. We introduce an information-theoretic regularization, which maximizes the mutual information between the signal and the resulting joint policy. We implement a centralized neural network to optimize the variational lower bound [2,5,18] of the mutual information. The effects of optimizing this regularization are three-fold: it (i) encourages each agent to align its individual policy with the coordination signal, (ii) decreases the uncertainty of policies of other agents to alleviate the difficulty to coordinate, and (iii) leads to a more diverse joint policy. Besides, SIC can be easily incorporated with most models that follow the decentralized execution paradigm, such as MADDPG [22] and COMA [9].

To evaluate SIC, we first conduct insightful experiments on a multiplayer variant of matrix game *Rock-Paper-Scissors-Well* to demonstrate how SIC incentivize agents to coordinate in both one-step and multi-step scenarios. Then we conduct experiments on *Cooperative Navigation* and *Predator-Prey*, two classic games implemented in multi-agent particle worlds [22]. We empirically show that by adopting SIC, agents learn to coordinate by interpreting the signal differently and thus achieve better performance. Besides, the visualization of the distribution of collision positions in *Predator-Prey* provides evidence that SIC improves the diversity of policies. An additional parameter sensitivity analysis manifests that SIC introduces stable improvement.

2 Methods

2.1 Preliminaries

We consider a fully cooperative multi-agent game with N agents. The game can be described as a tuple as $\langle \mathcal{I}, \mathcal{S}, \mathcal{A}, T, r, \gamma, \rho_0 \rangle$. Let $\mathcal{I} = \{1, 2, \cdots, n\}$ denote the set of n agents. $\mathcal{A} = \langle \mathcal{A}_1, \cdots, \mathcal{A}_n \rangle$ is the joint action space of agents, and $\mathcal{S}$ is the global state space. At time step t, the group of agents takes the joint action $\mathbf{a}_t = \langle a_{1t}, a_{2t}, \cdots, a_{nt} \rangle$ with each $a_{it} \in \mathcal{A}_i$ indicating the action taken by the agent i. $T(s_{t+1}|s_t, \mathbf{a}_t) : \mathcal{S} \times \mathcal{A} \times \mathcal{S} \rightarrow [0, 1]$ is the state transition function. $r(s_t, \mathbf{a}_t) : \mathcal{S} \times \mathcal{A} \rightarrow \mathbb{R}$ indicates the reward function from the environment. $\gamma \in [0, 1)$ is a discount factor and $\rho_0 : \mathcal{S} \rightarrow [0, 1]$ is the distribution of the initial state s_0.

Let $\pi_i(a_{it}|s_t) : \mathcal{S} \times \mathcal{A}_i \rightarrow [0, 1]$ be a stochastic policy for agent i, and denote the joint policy of agents as $\pi = \langle \pi_1, \cdots, \pi_n \rangle \in \Pi$ where Π is the joint policy space. Let $J(\pi) = \mathbb{E}_\pi \left[\sum_{t=0}^{\infty} \gamma^t r_t \right]$ denots the expected discounted cumulative reward, where r_t is the reward received at time-step t following policy π. We aim to optimize the joint policy π to maximize $J(\pi)$.

2.2 Joint Policy Space with Coordination Signal

In the fully centralized paradigm, a centralized controller is used to manipulate a group of agents. We denote Π^C as the policy space of the centralized controller and $\pi^C : \mathcal{S} \times \mathcal{A} \rightarrow [0, 1]$ as a joint policy in Π^C. In the decentralized execution paradigm, the agents make decisions independently according to their individual policies $\pi_i^D : \mathcal{S} \times \mathcal{A}_i \rightarrow [0, 1]$. We define the policy space of agent i as Π_i^D and the joint policy space as $\Pi^D = \Pi_1^D \times \cdots \times \Pi_n^D$, i.e. the Cartesian product of the policy spaces of each agent. For a joint policy $\pi^D \in \Pi^D$, we have $\pi^D(\mathbf{a}|s) = \pi_1^D(a_1|s) \cdots \pi_n^D(a_n|s)$, $\forall s \in \mathcal{S}$ and $\forall \mathbf{a} = \langle a_1, \cdots, a_n \rangle \in \mathcal{A}$. We conclude the relation between Π^C and Π^D as the following proposition:

Proposition 1. Π^D *is a subset of* Π^C.

This proposition is obvious and we provide a proof in the appendix. This proposition reveals one critical issue: as the objective of optimization is $J(\pi)$ instead of $J(\pi_1 \times \cdots \times \pi_N)$, it is possible that current CTDE methods can never perform as good as centralized methods when the optimal policy is in the complement space, i.e., $\Pi^C \backslash \Pi^D$. An intuitive method to solve this problem is to assume agents act sequentially and those who act later condition their policies on previous ones, e.g., $\pi^C(a|s) = \pi_1(a_1|s)\pi_2(a_2|s,a_1)\cdots\pi_N(a_N|s,a_1,\cdots,a_{N-1})$. However, this method is not practical as it requires stable communication channel and large bandwidth to implement.

From a game-theoretic perspective, the decentralized agents try to reach a Nash equilibrium, with each individual policy as a best response to others' policies. Previous studies on computational game theory show that by following the signal provided by a *correlation device*, agents may reach a more general type of equilibrium, *correlated equilibrium* (CE) [17], which can potentially lead to better outcomes for all agents [1,7,13]. Inspired by CE, we propose a *signal instructed* framework. We introduce a coordination signal sent to every agent at the beginning of a game, which is conceptually close to the signal sent by the correlation device in CE. The usage of the signal changes Π^D to a different joint policy space, Π^S. Every agent observes the same signal $z \in \mathcal{Z}$ sampled from a conditional distribution P_z, where $\mathcal{Z}$ is the *signal space*, and learns an individual policy as $\pi_i^S(a_i|s,z)$. Therefore, the agents formulate a special joint policy, π^S, which suffices that $\forall s \in \mathcal{S}, \forall a = \langle a_1, \cdots, a_n \rangle \in \mathcal{A}$, and $\forall z \in \mathcal{Z}, \pi^S(a,z|s) = P_z(z|s) \cdot \pi_1^S(a_1|s,z)\cdots\pi_n^S(a_n|s,z)$. π^S is a conditional joint distribution of a and z, and differs from aforementioned types of joint policies. However, by regarding z as an extension of global state, we do not change the way we model the individual policy of each agent, which is still $\pi_i^S(a_i|s')$ with $s' = (s,z)$.

In signal instructed approach, all agents observe the same z, and we assume that every agent follows the instruction of z, i.e., takes only one specific corresponding action a_i^z. We denote the corresponding joint action as $a^z = \langle a_1^z, \cdots, a_n^z \rangle$. Intuitively, this assumption is like that agents make an "agreement" on which joint action to take in current state when observing z, which is common in real-world scenarios. For example, the cars in a traffic junction can tell whether they should accelerate or stop from the same observed pose of a policeman, and they can be regarded as reaching a CE. Following the assumption still results in a stochastic joint policy, with the stochasticity conditioned on z now. With this assumption, we derive Proposition 2:

Proposition 2. Π^S *is equal to* Π^C.

We provide a proof in the appendix. This proposition shows that the signal instructed method enlarged the joint policy space to the same size as Π^C, while still maintaining a mostly decentralized framework. Therefore, agents can still act with their decentralized individual policies, while exploiting a larger joint policy space. We illustrates the relationship among different joint policy spaces in Fig. 1.

A practical concern is that according to the assumption, agents need to assign every joint action to a specific z, which means that the size of $\mathcal{Z}$ can be very

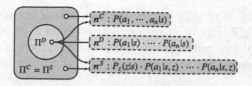

Fig. 1. The relationship among Π^C, Π^D and Π^S is $\Pi^C = \Pi^S \supseteq \Pi^D$. The white circle represents an element in the set.

large. Fortunately, the set of optimal joint actions is often small, and hence an optimal π^S needs only a small subspace of $\mathcal{Z}$ to instruct agents to take those optimal joint actions. Another concern is that agents might choose a non-optimal joint action when following the instruction of z, since the signal is drawn from a random distribution. Intuitively, a signal z serves as a "consensus", so that agents can infer others' actions and lean a good policy correspondingly. The signal itself does not explicitly tell each agent which action to take. Hence, when there is only one optimal joint action in current state, agents can learn to take the corresponding individual action whatever the signal is. In other words, the size of the signal space reduces to 1 in this case. In following sections, we show how this is achieved.

2.3 Signal Instructed Coordination

When a coordination signal is observed, how to incentivize agents to follow its instruction and coordinate is a critical issue. As a coordination signal is sampled from P_z, it is possible for agents to treat it as random noise and ignores it during the training process. Our idea is to facilitate the coordination signal to be entangled with agents' behaviors and thus encourage the coordination in execution. We name our method as *Signal Instructed Coordination (SIC)*. SIC introduces an information-theoretic regularization to ensure the signal makes an impact in agents' decision making. This regularization aims to maximize the mutual information between the signal z, and the joint policy π^S, given current state s, as

$$
\begin{aligned}
I(z; \pi^S) &= -H(\pi^S|z) + H(\pi^S) \\
&= -H(\pi^S_i|z) - H(\pi^S_{-i}|\pi^S_i, z) + H(\pi^S),
\end{aligned}
\tag{1}
$$

where π^S, π^S_i and π^S_{-i} are abbreviations for $\pi^S(\boldsymbol{a}, z|s)$, $\pi^S_i(a_i, z|s)$ and $\pi^S_{-i}(\boldsymbol{a}_{-i}, z|s)$, and π^S_{-i} and $\boldsymbol{a}_{-i}$ are the joint policy and the joint action of all agents except agent i respectively. The decomposition of π^S into π^S_i and π^S_{-i} in the second line holds in our decentralized approach.

Through decomposing the regularization term in Eq. (1), one can find that the effects for maximizing the mutual information between signal and policy are threefold. Minimizing the first term increases consistency between the coordination signal and the individual policy to suffice the assumption of Proposition 2. Minimizing the second term ensures low uncertainty of other agents' policies, which is beneficial to establish coordination among agents. Maximizing the third term encourages the joint policy to be diverse, which prohibits the opponents from inferring our policy in competition. These effects in combination improves the performance of the joint policy. However, directly optimizing Eq. (1) is intractable. Considering the symmetry property of mutual information, we aim to maximize

$$\begin{aligned} I(z; \pi^S(\boldsymbol{a}, z|s)) &= -H(z|\pi^S(\boldsymbol{a}, z|s)) + H(z) \\ &= \mathbb{E}_{z \sim P_z(\cdot|s), \boldsymbol{a} \sim \pi(\cdot|s,z)}[\mathbb{E}_{z' \sim P(\cdot|s,\boldsymbol{a})} \log P(z'|s, \boldsymbol{a})] + H(z) \quad (2) \\ &\geq \mathbb{E}_{z,\boldsymbol{a}}[\log U(z|s, \boldsymbol{a})] \quad (3) \end{aligned}$$

where $U(z|s, \boldsymbol{a})$ is an approximation of $P(z|s, \boldsymbol{a})$. A detailed derivation from Eq. (2) to Eq. (3) can be found in the appendix. The inequality sign holds due to $D_{KL}(\cdot) \geq 0$ and $H(\cdot) \geq 0$, and the equality sign in the last line holds as proved in Lemma 5.1 of [5]. Considering the visiting probability of s in sampled trajectories τ following π, we derive a mutual information loss (MI loss) from Eq. (3) as

$$L_I(\pi, U) = -\mathbb{E}_{s \sim \tau, z \sim P_z(\cdot|s), \boldsymbol{a} \sim \pi(\cdot|s,z)}[\log U(z|s, \boldsymbol{a})]. \quad (4)$$

Minimizing Eq. (4) facilitates agents to follow the instruction of the coordination signal.

2.4   Implementation Details

In the proposed signal instructed coordination method, agents optimize their joint policy to maximize expected returns and minimize the mutual information loss. To integrate SIC with existing models, the first problem is how to model P_z. Instead of using discrete signal space, we propose to adopt a continuous signal space, and approximate P_z in a Monte Carlo way. In detail, we sample a D_z-dimension continuous vector v from a normal distribution $\mathcal{N}(\boldsymbol{0}, I)$, and distribute it to all agents. v can be viewed as a proxy of the variable z. Agents learn to divide the $\mathbb{R}^{D_z}$ space into several subspaces, with each corresponding to one signal and hence one optimal joint action. The probability of sampling a specific z, $P_z(z|s)$, is approximated by the probability of sampling a vector v that belongs to the corresponding subspace. Therefore, we can replace z in Eq. (4) with v and change to compute $U(v|s, \boldsymbol{a})$.

To compute $U(v|s, \boldsymbol{a})$, we use a centralized multi-layer feed-forward network, named as U-Net, as a parameterized function, f_U. U-Net inputs s and $\boldsymbol{a}$, and outputs a continuous vector with the same dimension as v as the reconstructed value of the signal, $v' = f_U(s, \boldsymbol{a})$. Intuitively, we hope $v' = v$, which means that agents follow the instruction so well that we can infer what they see from their behaviors. Therefore, $U(v|s, \boldsymbol{a})$ is measured by the mean squared error between v' and v and minimized during training. One obvious advantage of SIC is that it can be easily integrated with most existing models with policy networks, as shown in Fig. 2. To pass gradients even when stochastic policy is used, we use a simplified approximation $U(z|s, \boldsymbol{h})$, where $\boldsymbol{h}$ is the concatenation of last layers of

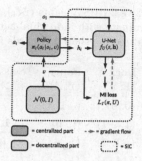

Fig. 2. Illustration of SIC. Black arrows indicate how variables are passed between components. Note that in U-Net, we use the concatenation of last hidden vectors in policy networks h, instead of a, to enable gradient flow when adopting stochastic policies. Besides, in practice we sample a continuous vector v from a fixed normal distribution $\mathcal{N}(0, I)$ as a proxy of z.

hidden vectors in policy networks. Parameters of the centralized U-net, ω, and parameters of decentralized policies, $\theta = \langle \theta_1, \cdots, \theta_n \rangle$, are jointly optimized as:

$$\max_{\omega, \theta} \mathbb{E}_{o \sim \tau, a \sim \pi_\theta, v \sim \mathcal{N}(0, I)}[Q_i(o, a) - \alpha L_I(\pi, U_\omega)], \tag{5}$$

where $Q_i(o, a|z)$ is the centralized critic function, and $\alpha > 0$ is the hyperparameter for information maximization term. By optimizing this objective, we aim to find a trade-off between *maximizing long-term returns* and *reaching consensus*. $Q_i(o, a|z)$ can also be substituted with the advantage function used by COMA. We do not share parameters among agents. Note that o_i may be a partial observation of agent i, and additional communication mechanism can be introduced to ensure theoretical correctness of Eq. (4). However, we empirically show that in some partially observable environments, e.g., particle worlds [22], where the agent can infer global state from its local observation, SIC can still work with o_i. When applied to Multi-agent Actor-Critic frameworks [22], the objective of updating critic remains unchanged.

3 Experiments

3.1 Rock-Paper-Scissors-Well (RPSW)

We use a 2 vs 2 variant of the matrix game, *Rock-Paper-Scissors-Well (RPSW)*. Each team consists of two independent agents, and the available actions of each agent are $Access(A)$ and $Yield(Y)$. The joint action space of each team consists of *Rock* ($\langle Y, Y \rangle$), *Paper* ($\langle Y, A \rangle$), *Scissors* ($\langle A, Y \rangle$), and *Well* ($\langle A, A \rangle$). The former three actions play as in the traditional *Rock-Paper-Scissors (RPS)* game, while *Well* wins only against *Paper* and is defeated by *Rock* and *Scissors*. We present the payoff matrix in Table 1. This game can reflect agents' capability to coordinate. Assume both teams are controlled by centralized controllers, the

Table 1. Payoff matrix of the 2 vs 2 RPSW game (M_4). Both row and column players consist of two agents, who coordinate with individual actions as Y or A to play a joint action, e.g., *Paper* ($\langle Y, A \rangle$), and receive shared rewards from (r_{row}, r_{col}).

	Rock $\langle Y, Y \rangle$	Paper $\langle Y, A \rangle$	Scissors $\langle A, Y \rangle$	Well $\langle A, A \rangle$
Rock $\langle Y, Y \rangle$	(0, 0)	(−1, 1)	(1, −1)	(1, −1)
Paper $\langle Y, A \rangle$	(1, −1)	(0, 0)	(−1, 1)	(−1, 1)
Scissors $\langle A, Y \rangle$	(−1, 1)	(1, −1)	(0, 0)	(1, −1)
Well $\langle A, A \rangle$	(−1, 1)	(1, −1)	(−1, 1)	(0, 0)

best π^C is $P(\langle Y, Y \rangle) = P(\langle Y, A \rangle) = P(\langle A, Y \rangle) = \frac{1}{3}$ and $P(\langle A, A \rangle) = 0$, since it is better to take $\langle A, Y \rangle$ instead of $\langle A, A \rangle$. To achieve this, agents within the same team need to coordinate to avoid the disadvantaged joint action $\langle A, A \rangle$, and choose others uniformly randomly. From a probabilistic perspective, the coordination requires high correlation between teammates, otherwise either $\langle A, A \rangle$ is inevitable to appear as long as $\pi_1(A) \cdot \pi_2(A) > 0$, or the joint action space degenerates to $\{\langle Y, Y \rangle, \langle Y, A \rangle\}$ or $\{\langle Y, Y \rangle, \langle A, Y \rangle\}$, and their joint policy may be easily exploited by the opponent.

We denote the matrix game in Table 1 as M_4, since the fourth joint action is undesired. By exchanging the fourth row with the i-th row, and the fourth column with the i-th column sequentially, we turn the i-th joint action to *Well* and denote the new matrix as M_i. In this way we obtain a set of matrices $\mathcal{M} = \{M_1, M_2, M_3, M_4\}$. We design a multi-step matrix game, where two teams play according to a random payoff matrix drawn from $\mathcal{M}$ in each step. Each agent can only observe the ID $i \in \{1, 2, 3, 4\}$ of the current matrix and the coordination signal. To simulate sparse rewards, we only give agents the sum of rewards on each step after an episode of game is finished, and train them with discounted returns. We use REINFORCE algorithm with fully independent agents as the baseline model, which we denote as **IND-RE**. We apply our SIC module to REINFORCE algorithm, and denote it as **SIC-RE**. The team-shared coordination signal $z \in \mathbb{R}^2$ is sampled from $\mathcal{N}(\mathbf{0}, I)$. Note that each agent takes a stochastic policy, and signals received by the two teams are different. We conduct experiments and observe that SIC-RE defeats IND-RE with averaged rewards close to 0.4, while both IND-RE vs IND-RE and SIC-RE vs SIC-RE settings gradually converge to a tie with averaged rewards equal to 0. We can see that although IND-RE also reaches an equilibrium with a game value as 0 in IND-RE vs IND-RE, its ability to coordinate is limited, as it can only formulate joint policy in Π^D. Therefore, in direct competition as SIC-RE vs IND-RE, IND-RE is outperformed and stuck in a disadvantaged equilibrium with a negative averaged reward.

To better study which kind of equilibrium agents have reached, we test how agents respond to 5000 randomly sampled signals before and after training on M_4 with SIC-MA vs SIC-MA, and plot results of row players in Fig. 3. Note that results of column players are similar to Fig. 3. Before training (a), the frequency

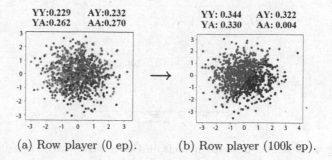

(a) Row player (0 ep).   (b) Row player (100k ep).

Fig. 3. Correlations between signal distribution and joint actions of row players. Each point represents a 2-dim signal, and red, green, blue, and cyan represent the corresponding joint action as $\langle Y, Y \rangle$, $\langle Y, A \rangle$, $\langle A, Y \rangle$, and $\langle A, A \rangle$ respectively. The frequency of different actions in these 5000 points is shown above each sub-figure.

of each joint action is roughly 0.25. each agent takes a random individual policy. In addition, the distribution of signals triggering different joint actions are quite spreading. After training (b), the signal space is roughly divided into three "zones", with each zone representing one joint action. The "area" of each zone, i.e., the probability of sampling one signal belonging to the zone, is roughly 1/3, which indicates that the result is close to the best performance a centralized controller can achieve.

3.2 Particle Worlds

In this section, we evaluate SIC on two particle world game, **Cooperative Navigation** and **Predator-Prey**, following the implementation in [22]. In cooperative navigation, $N = 3$ agents and $L = 3$ landmarks are randomly placed in a two-dimensional world. In each timestep, agents are rewarded with a common reward, which is the sum of the negative Euclidean distance between a landmark and the nearest agent. In addition, two agents are penalized simultaneously if they collide with each other. In Predator-Prey, M slow *predators* and M fast *preys* are randomly placed in a two-dimensional world with $L = 2$ large landmarks impeding the way, and predators need to collaborate to collide with another team of agents, preys. Note that when collisions happen, the predators are rewarded simultaneously, while the preys are penalized independently, which is different from the original setting in [22].

We use three popular models as our baselines: **MADDPG** [22], **COMA** [9] and **MAAC** [12]. All three models follow the CTDE framework, and MADDPG and COMA especially model only individual policies. Therefore, we integrate SIC with MADDPG and COMA and denote them as **SIC-MA** and **SIC-COMA**. We use hyper-parameters of neural networks in the original paper, and inherit them in SIC variants. For comparison, we also implement fully-decentralized and fully-centralized actor-critic methods based on DDPG, and denote them as **Dec-AC**

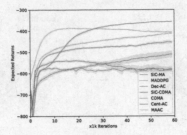

Fig. 4. Results of different models on Cooperative Navigation game.

Table 2. Results of different models on Predator-Prey game when playing against MADDPG preys. Results are reported in terms of predator scores with 95% confidence intervals in 30 repeated games.

Predator model	2 vs 2	4 vs 4
Dec-AC	6.70 ± 0.78	31.97 ± 2.08
COMA	1.50 ± 0.15	20.05 ± 2.46
MADDPG	12.11 ± 1.39	46.80 ± 3.29
MAAC	11.48 ± 2.01	50.33 ± 3.65
SIC-COMA	2.05 ± 0.12	26.33 ± 2.64
SIC-MA	**16.56** ± 1.50	**59.85** ± 2.89
Cent-AC	17.31 ± 1.14	40.33 ± 1.51

and **Cent-AC**. Note that to evaluate the performance of different models, we compare them with the same opponent prey model (MADDPG) and report predator scores on predator-prey games. We report results of Cooperative Navigation in Fig. 4 and Predator-Prey (2 vs 2 and 4 vs 4) in Table 2. We can see that

1. Although SIC requires additional learning budgets and slows down the learning speed in the initial stage, it stably improves performance when combined with a baseline model. In addition, SIC-MA significantly outperforms all other decentralized execution models including the complicated SOTA baseline (MAAC).
2. On 2 vs 2 Predator-Prey game, the performance of SIC-MA is close to that of the fully-centralized method. On 4 vs 4 Predator-Prey and Cooperative Navigation games, where the training of the centralized method suffers, SIC-MA learns better joint policies through decentralized paradigm. This shows that even in high-dimensional space, coordination signal can still facilitate coordination among decentralized agents.

To better understand how SIC improves performance, we conduct case study on the mixed cooperative-competitive environment, 2 vs 2 Predator-Prey game. We visualize the distribution of collision positions in Fig. 5. We repeat MADDPG vs MADDPG, and SIC-MA vs MADDPG for 10000 games with different

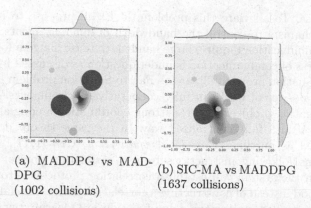

(a) MADDPG vs MAD-
DPG
(1002 collisions)

(b) SIC-MA vs MADDPG
(1637 collisions)

Fig. 5. The density and marginal distribution of positions of collision, (x, y), in 10000 repeated games initialized from the same environment. The orange, green and black circles representing predators, preys and landmarks at the start of each game. The result shows that compared to MADDPG, SIC-MA presents more diverse strategy and better performance.

seeds. We reset each game to the same state in each episode, and collect positions of collisions in the total 250000 steps. We visualize the positions in Fig. 5, where data points (x, y) with higher frequency are colored darker in the plane. In Fig. 5-a, most collisions happen around the lower prey, which reflects that both predators only collaborate to capture the lower prey. When SIC-MA plays predators as in Fig. 5-b, collisions appears in more diverse positions, and we observe two strategies: the two predators either chase the lower prey to the bottom part of the map, or move upward together to catch both preys. In this case, it is hard for preys to exploit the opponent strategies, but the two predators need high level of coordination to conduct such strategies, otherwise the lower prey may flee through the interval between them. Specifically, SIC-MA captures more preys (1637) than MADDPG does (1002). This evidences that SIC-MA learns better policies compared to MADDPG.

4   Related Works

Recent works [4] on MARL focus on complex scenarios with high dimensional state and action spaces like particle worlds [22] and StarCraft II [30]. Among different approaches to model the controlling of agents, *centralized training with decentralized execution* [22,25] outperforms others for circumventing the exponential growth of joint action space and the non-stationary environment problem [19]. Emergent communication [21] is proposed to enhance coordination and training stability, which allows agents to pass messages between agents and "share" their observations via communication vectors. [26,28] design special architectures to share information among all agents. The noisy channel problem arises when all other agents use the same communication channel to send information simultaneously, and the agent needs to distinguish useful information from useless or irrelevant noise. To alleviate this problem, [6,12,14] propose to introduce the attention mechanism to control the bandwidth of different agents dynamically. However, communication requires large bandwidth to exchange information, and the effectiveness of communication is under question as discussed by [21].

The coordination problem [3], or the Pareto-Selection problem [23], has been discussed by a series of works in fully cooperative environments. The solution to the coordination problem requires strong coordination among agents, i.e., all agents act as if in a fully centralized way. In the game theory domain, it can also be viewed as pursuing *Correlated equilibrium* (CE) [1,17], where agents make decisions following instructions from a correlation device. It is desired that agents in the system can establish correlation protocols through adaptive learning method instead of constructing a correlation device manually for specific tasks [11] proposes to replace the value function in Q-learning with a new one reflecting agents' rewards according to some CE. [31] maintains coordination sets and select coordinated actions within these sets.

A similar concept to our coordination signal is *common knowledge*, which refers to common information, e.g., representations of states, among partially observable agents. Common knowledge is used to enhance coordination [10,29] and combined with communication [15]. Among them, [10] proposes MACKRL which introduces a random seed as part of common knowledge to guide a hierarchical policy tree. To avoid exponential growth of model complexity, MACKRL restricts correlation to pre-defined patterns, e.g., a pairwise one, which is too rigid for complex tasks.

5   Conclusions

We propose a signal instructed paradigm to improve the popular decentralized execution framework, which theoretically manipulates decentralized agents as a centralized controller. Accordingly, we design a novel module named Signal Instructed Coordination (SIC) to enhance coordination of agents' policies by introducing a mutual information regularization. Our experiments show the performance improvement of popular centralized-training-decentralized-execution algorithms with the help of SIC.

Acknowledgements. Haifeng thanks the support from Strategic Priority Research Program of Chinese Academy of Sciences, Grant No. XDA27030401. Co-author Fang is supported in part by a research grant from Lockheed Martin and NSF grant IIS-2046640. The corresponding author Weinan Zhang is supported by "New Generation of AI 2030" Major Project (2018AAA0100900), Shanghai Municipal Science and Technology Major Project (2021SHZDZX0102) and National Natural Science Foundation of China (62076161, 61632017).

A   Algorithm

For completeness, we provide the SIC-MA algorithm as an example of application of SIC in Algorithm 1. The main body of SIC-MA is the similar to MADDPG, and the main change includes:

1. a common signal is sampled and every agent observes it before taking actions, and
2. a mutual information loss is computed to update parameters of U-Net and policy networks.

We can see that SIC is easy to implement with existing actor-critic-based algorithms.

B   Proof of Proposition 1

Proof. $\forall \pi^D \in \Pi^D$, we can construct $\pi^C \in \Pi^C$, which suffices that for $\forall s \in \mathcal{S}$, $\forall a = \langle a_1, \cdots, a_n \rangle \in \mathcal{A}$,

$$\pi^C(a|s) = \pi_1^D(a_1|s) \cdots \pi_n^D(a_n|s) = \pi^D(a|s).$$

Thus, we have every $\pi^D = \pi^C \in \Pi^C$ and $\Pi^D \subseteq \Pi^C$.

We use a counterexample to show that not every joint policy in Π^C is an element of Π^D. In a two-agent system where each agent has two actions x and y, $\forall s \in \mathcal{S}$, $\exists \pi^C \in \Pi^C$ that suffices $\pi^C(\langle a_1 = x, a_2 = x \rangle | s) = \pi^C(\langle a_1 = y, a_2 = y \rangle | s) = 0.5$, but $\pi^C \notin \Pi^D$, because there is no valid solution for $\pi_1(a_1 = x|s) \cdot \pi_2(a_2 = x|s) = (1 - \pi_1(a_1 = x|s)) \cdot (1 - \pi_2(a_2 = x|s)) = 0.5$. Since Π^C includes Π^D, the best joint policy in Π^C is superior or equal to the best joint policy in Π^D. However, due to computational complexity concerns, the decentralized execution paradigm is more practical in large-scale environments. Thus, we have motivation to propose a new framework which has centralized policy space Π^C and is executed in a decentralized way.

C   Proof of Proposition 2

Proof. We prove this proposition in two steps:

1. $\Pi^C \subseteq \Pi^S$: $\forall \pi^C \in \Pi^C$, we can construct $\pi^S \in \Pi^S$, which suffices that $\forall s \in \mathcal{S}$, $\forall a \in \mathcal{A}$, we assign a signal $z \in \mathcal{Z}$ to a with $P_z(z|s) = \pi^C(a|s)$, s.t.

$$\pi^S(z, a|s) = P_z(z|s)[\pi_1^S(a_1|s, z) \cdots \pi_n^S(a_n|s, z)]$$
$$= P_z(z|s)[1 \cdots 1] = \pi^C(a|s)$$

Thus, we have every $\pi^C = \pi^S \in \Pi^S$ and $\Pi^C \subseteq \Pi^S$.

2. $\Pi^S \subseteq \Pi^C$: $\forall \pi^S \in \Pi^S$, we can construct $\pi^C \in \Pi^C$, which suffices that $\forall s \in \mathcal{S}$, $\forall a \in \mathcal{A}$, $\forall z \in \mathcal{Z}$,

$$\pi(a|s) = \begin{cases} P_z(z|s) & a = a^z \\ 0 & \text{otherwise.} \end{cases}$$

Thus, we have every $\pi^S = \pi^C \in \Pi^C$ and $\Pi^S \subseteq \Pi^C$.

Given $\Pi^S \subseteq \Pi^C$ and $\Pi^S \supseteq \Pi^C$, we have $\Pi^S = \Pi^C$.

Algorithm 1. SIC-MADDPG Algorithm

---

for episode $= 1$ to M do

    Initialize a random process $\mathcal{N}$ for action exploration

    Receive initial state $\mathbf{x}$

    Generate random signal z according to pre-defined distribution.

    for $t = 1$ to max-episode-length do

        for each agent i, sample action a_i according to $\pi_i(a_i|o_i, z)$

        Execute actions $a = (a_1, \ldots, a_N)$ and observe reward r and new state $\mathbf{x}'$

        Store $(\mathbf{x}, a, r, \mathbf{x}', z^j)$ in replay buffer $\mathcal{D}$

        $\mathbf{x} \leftarrow \mathbf{x}'$

        for $k = 1$ to 2 do

            Sample a random mini-batch of S samples $(\mathbf{x}^j, a^j, r^j, \mathbf{x}'^j, z^j)$ from $\mathcal{D}$

            for agent i in team k do

                Calculate h_k^j by concatenating inputs to the last layer of policy networks of all cooperative agents in team k

                Calculate $U_k^j = U(z_k^j|\mathbf{x}_k^j, h_k^j)$

                Set $y^j = r_i^j + \gamma Q_i^{\pi'}(\mathbf{x}'^j, \pi_k)|_{a_k' = \pi_k'(\sigma_k^j)}$

                Update critic by minimizing the loss

$$\mathcal{L}(\theta_i) = \frac{1}{S} \sum_j (y^j - Q_i^{\pi}(\mathbf{x}^j, \pi_k^j))^2$$

                Update actor using the sampled policy gradient

$$\nabla_{\theta_i} J \approx \frac{1}{S} \sum_j \nabla_{\theta_i} \pi_i(o_i^j) \nabla_{a_i} [Q_i^{\pi}(\mathbf{x}^j, \pi^j) + \beta L_I(\pi, U)]|_{a_i = \pi_i(\sigma_i^j)}$$

                Update U-Net by minimizing

$$\mathcal{L}(w) = \frac{1}{S} \sum_j L_I(\pi, U)|_{a_i = \pi_i(\sigma_i^j)}$$

        end for

        end for

        Update target network parameters for each agent i:

$$\theta_i' \leftarrow \tau\theta_i + (1 - \tau)\theta_i'$$

    end for

end for

D   Derivation of the Lower Bound

The derivation of the lower bound mostly follows similar techniques used in variational inference domain [2,5,18]. Firstly, we have

$$I(z; \pi^S(a, z|s)) = -H(z|\pi^S(a, z|s)) + H(z)$$
$$= \mathbb{E}_{z \sim P_z(\cdot|s), a \sim \pi(\cdot|s,z)}[\mathbb{E}_{z' \sim P(\cdot|s,a)}$$
$$\log P(z'|s, a)] + H(z) \tag{6}$$

where $\pi(\cdot|s, z)$ is the Cartesian product of individual policies after observing a specific z, and $P(\cdot|s, a)$ is the posterior distribution estimating the probability of a specific signal z' after seeing the state s and the joint action a. Note that $P(\cdot|s, a)$ is not the same as $P_z(\cdot|s)$. Since we have no knowledge of the posterior distribution, we circumvent it by introducing a variational lower bound [2,5,18] which defines an auxiliary distribution $U(\cdot|s, a)$ as:

$$\text{Eq. } (6) = \mathbb{E}_{z \sim P_z(\cdot|s), a \sim \pi(\cdot|s,z)}[D_{KL}(P(\cdot|s, a)||U(\cdot|s, a))$$
$$+ \mathbb{E}_{z' \sim P(\cdot|s,a)} \log U(z'|s, a)] + H(z)$$
$$\geq \mathbb{E}_{z \sim P_z(\cdot|s), a \sim \pi(\cdot|s, Tz)}[\mathbb{E}_{z' \sim P(\cdot|s,a)} \log U(z'|s, a)],$$
$$= \mathbb{E}_{z \sim P_z(\cdot|s), a \sim \pi(\cdot|s,z)}[\log U(z|s, a)] \tag{7}$$

E   Experiment Details

E.1   Matrix Game Experiment

We conduct three multi-step matrix game experiments: SIC-RE vs SIC-RE, SIC-RE vs IND-RE, IND-RE vs IND-RE. For both SIC-RE and IND-RE models, we use the Adam optimizer with a learning rate of 0.0001. The policy network is parameterized by a one-layer ReLU MLP with 8 hidden units. We use a batch size of 100000. For SIC-RE models, we use, a two-layer ReLU MLP with 8 hidden units as U-Net, and set the coefficient of MI loss to be 0.01.

E.2   Particle World Experiment

We adopt 5 different models: MADDPG, SIC-MADDPG (SIC-MA), COMA, SIC-COMA, and MAAC. We provide details of our setups here:

1. **MADDPG**   We use the Adam optimizer with a learning rate of 0.001 and $\delta = 0.01$ (has the same meaning as in original MADDPG) for updating the target network. Both Actor and Critic are parameterized by a two-layer ReLU MLP with 64 units per layer. γ is set to be 0.95. We use a batch size of 1024 before making an update.

2. **SIC-MADDPG**   We use the Adam optimizer with a learning rate of 0.0005 and $\delta = 0.01$ for updating the target network. Gradient clipping is set to be 0.5. Both Actor and Critic are parameterized by a two-layer ReLU MLP with 64 units per layer. γ is set to be 0.95. We use a batch size of 1024 before making an update. The dimension of signals is 20 and the coefficient of information-theory regularization is 0.0001.

3. **COMA** We use the Adam optimizer with a learning rate of 0.00005 and $\delta = 0.01$ for updating the target network. Gradient clipping is set to be 0.1. Both Actor and Critic are parameterized by a two-layer ReLU MLP with 64 units per layer. γ is set to be 0.99 and λ is set to 0.8. We use a batch size of 1000 before making an update.

4. **SIC-COMA** We use the Adam optimizer with a learning rate of 0.00005 and $\delta = 0.01$ for updating the target network. Gradient clipping is set to be 0.1. Both Actor and Critic are parameterized by a two-layer ReLU MLP with 64 units per layer. γ is set to be 0.99 and λ is set to 0.8. We use a batch size of 1000 before making an update. The dimension of signals is 20.

5. **MAAC** We use the Adam optimizer and set the learning rate for policy and critic networks as 0.001 and 0.01 respectively. Both policy and critic networks adopt two-layer Leaky ReLU MLP with 128 units per layer. The number of attention head is set to 4. We use a batch size of 100, and set the reward rescaling factor to be 100 as in the original paper.

Except aforementioned models, we also implement two variants of MADDPG: a fully-decentralized actor-critic and a fully-centralized actor-critic. The former one can be viewed as MADDPG with a decentralized critic $Q_i(o_i, a_i)$, and the latter one as MADDPG with a centralized agent that takes joint actions directly $a = \pi(o)$.

We report results of games with 95% confidence intervals in 30 repeated games.

F   Visualization for Joint Policy of Multi-step Matrix Game

We plot the curves of joint policies of both row players and column players in multi-step matrix games in Fig. 6, 7, and 8.

G   Parameter Sensitivity

We conduct a parameter sensitivity analysis in 2 vs 2 Predator-Prey game on two crucial parameters of SIC, the coefficient of mutual information loss α and the dimension of signal D_z. We test SIC-MA vs MADDPG with different values of parameters in the 2 vs 2 Predator-Prey game, We find that when adopting signal without training U-Net ($\alpha = 0$), the performance of SIC-MA is close to MADDPG. Therefore, enforcing the mutual information constraint properly ($\alpha = 1e - 4$) is important in achieving better results. Besides, SIC-MA presents a stable improvement over MADDPG ($D_z = 0$) and, most importantly, approximation through neural networks can compress $\mathcal{Z}$ and ensure good performance.

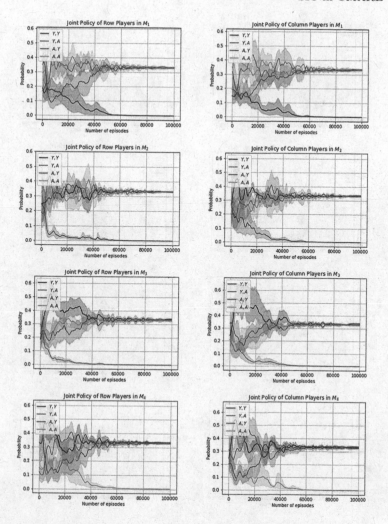

Fig. 6. Joint Policy of SIC-RE vs SIC-RE. During training, the i-th joint action in M_i is deprecated gradually, and all other joint actions are sampled uniformly randomly.

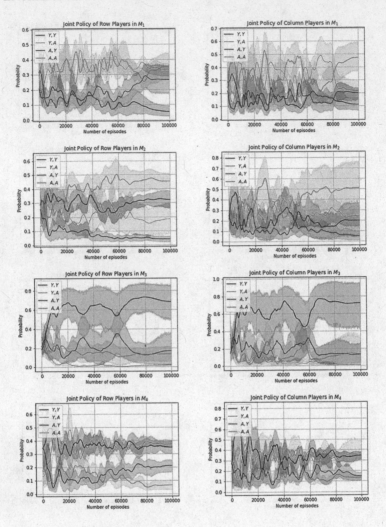

Fig. 7. Joint Policy of SIC-RE vs IND-RE. SIC-RE adjusts its joint policy to counter that of IND-RE, and achieves a positive game value.

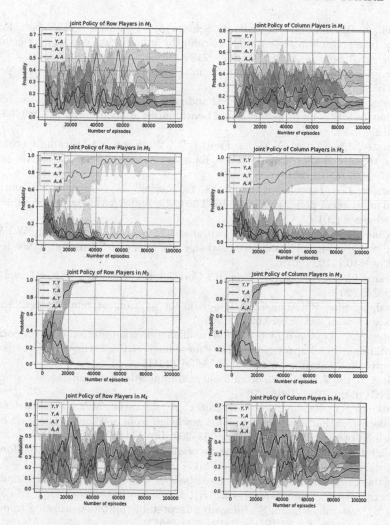

Fig. 8. Joint Policy of IND-RE vs IND-RE. IND-RE only finds worse joint policy in the team-policy space, and in some cases (M_2 and M_3), players play only one kind of joint action.

References

1. Aumann, R.J.: Subjectivity and correlation in randomized strategies. J. Math. Econ. **1**(1), 67–96 (1974)
2. Barber, D., Agakov, F.V.: The IM algorithm: a variational approach to information maximization. In: NIPS. p. None (2003)
3. Boutilier, C.: Sequential optimality and coordination in multiagent systems. In: IJCAI, vol. 99, pp. 478–485 (1999)
4. Busoniu, L., Babuska, R., De Schutter, B.: A comprehensive survey of multiagent reinforcement learning. IEEE SMC-Part C Appl. Rev. **38**(2), 2008 (2008)

5. Chen, X., Duan, Y., Houthooft, R., Schulman, J., Sutskever, I., Abbeel, P.: Infogan: Interpretable representation learning by information maximizing generative adversarial nets. In: NIPS, pp. 2172–2180 (2016)

6. Das, A., et al.: Tarmac: Targeted multi-agent communication. arXiv preprint arXiv:1810.11187 (2018)

7. Farina, G., Ling, C.K., Fang, F., Sandholm, T.: Correlation in extensive-form games: saddle-point formulation and benchmarks. arXiv preprint arXiv:1905.12564 (2019)

8. Foerster, J.N., Assael, Y.M., de Freitas, N., Whiteson, S.: Learning to communicate to solve riddles with deep distributed recurrent Q-networks. arXiv preprint arXiv:1602.02672 (2016)

9. Foerster, J.N., Farquhar, G., Afouras, T., Nardelli, N., Whiteson, S.: Counterfactual multi-agent policy gradients. In: Thirty-Second AAAI Conference on Artificial Intelligence (2018)

10. Foerster, J.N., de Witt, C.A.S., Farquhar, G., Torr, P.H., Boehmer, W., Whiteson, S.: Multi-agent common knowledge reinforcement learning. arXiv preprint arXiv:1810.11702 (2018)

11. Greenwald, A., Hall, K., Serrano, R.: Correlated Q-learning. In: ICML, vol. 3, pp. 242–249 (2003)

12. Iqbal, S., Sha, F.: Actor-attention-critic for multi-agent reinforcement learning. arXiv preprint arXiv:1810.02912 (2018)

13. Jiang, A.X., Leyton-Brown, K.: Polynomial-time computation of exact correlated equilibrium in compact games. In: Proceedings of the 12th ACM Conference on Electronic Commerce, pp. 119–126. ACM (2011)

14. Jiang, J., Lu, Z.: Learning attentional communication for multi-agent cooperation. In: NIPS, pp. 7254–7264 (2018)

15. Korkmaz, G., Kuhlman, C.J., Marathe, A., Marathe, M.V., Vega-Redondo, F.: Collective action through common knowledge using a Facebook model. In: Proceedings of the 2014 AAMAS, pp. 253–260. IFAAMAS (2014)

16. Lanctot, M., et al.: A unified game-theoretic approach to multiagent reinforcement learning. In: NIPS, pp. 4190–4203 (2017)

17. Leyton-Brown, K., Shoham, Y.: Essentials of game theory: a concise multidisciplinary introduction. Synth. Lect. Artif. Intell. Mach. Learn. **2**(1), 1–88 (2008)

18. Li, Y., Song, J., Ermon, S.: Infogail: interpretable imitation learning from visual demonstrations. In: NIPS, pp. 3812–3822 (2017)

19. Li, Y.: Deep reinforcement learning: an overview. arXiv preprint arXiv:1701.07274 (2017)

20. Li, Y.: Deep reinforcement learning. arXiv preprint arXiv:1810.06339 (2018)

21. Lowe, R., Foerster, J., Boureau, Y.L., Pineau, J., Dauphin, Y.: On the pitfalls of measuring emergent communication. In: Proceedings of the 18th AAMAS, pp. 693–701. IFAAMAS (2019)

22. Lowe, R., Wu, Y., Tamar, A., Harb, J., Abbeel, O.P., Mordatch, I.: Multi-agent actor-critic for mixed cooperative-competitive environments. In: NIPS, pp. 6379–6390 (2017)

23. Matignon, L., Laurent, G.J., Le Fort-Piat, N.: Independent reinforcement learners in cooperative Markov games: a survey regarding coordination problems. Knowl. Eng. Rev. **27**(1), 1–31 (2012)

24. Nunes, L., Oliveira, E.: Learning from multiple sources. In: Proceedings of the 3rd AAMAS. AAMAS 2004, pp. 1106–1113. IEEE Computer Society, Washington, DC, USA (2004). http://dl.acm.org/citation.cfm?id=1018411.1018879

25. Oliehoek, F.A., Spaan, M.T., Vlassis, N.: Optimal and approximate Q-value functions for decentralized POMDPs. J. Artif. Intell. Res. **32**, 289–353 (2008)
26. Peng, P., et al.: Multiagent bidirectionally-coordinated nets: emergence of human-level coordination in learning to play starcraft combat games. arXiv preprint arXiv:1703.10069 (2017)
27. Schneider, J.G., Wong, W.K., Moore, A.W., Riedmiller, M.A.: Distributed value functions. In: Proceedings of the 16th ICML. ICML 1999, pp. 371–378. Morgan Kaufmann Publishers Inc., San Francisco, CA, USA (1999). http://dl.acm.org/citation.cfm?id=645528.657645
28. Sukhbaatar, S., Fergus, R., et al.: Learning multiagent communication with back-propagation. In: NIPS, pp. 2244–2252 (2016)
29. Thomas, K.A., DeScioli, P., Haque, O.S., Pinker, S.: The psychology of coordination and common knowledge. J. Pers. Soc. Psychol. **107**(4), 657 (2014)
30. Vinyals, O., et al.: StarCraft II: a new challenge for reinforcement learning. arXiv preprint arXiv:1708.04782 (2017)
31. Zhang, C., Lesser, V.: Coordinating multi-agent reinforcement learning with limited communication. In: Proceedings of the 2013 AAMAS, pp. 1101–1108. IFAAMAS (2013)

A Description of the Jadescript Type System

Giuseppe Petrosino[1], Eleonora Iotti[2], Stefania Monica[1],
and Federico Bergenti[2(✉)]

[1] Dipartimento di Scienze e Metodi dell'Ingegneria,
Università degli Studi di Modena e Reggio Emilia, Modena, Italy
{giuseppe.petrosino,stefania.monica}@unimore.it
[2] Dipartimento di Scienze Matematiche, Fisiche e Informatiche,
Università degli Studi di Parma, Parma, Italy
{eleonora.iotti,federico.bergenti}@unipr.it

Abstract. Jadescript is an agent-oriented programming language that
benefits from JADE, the popular framework to develop multi-agent systems in Java. The major goal behind the introduction of Jadescript
is to simplify the implementation of real-world multi-agent systems by
devising a language with specific linguistic constructs for agent-oriented
abstractions. However, Jadescript needs to be coherent and consistent
across its several features to effectively achieve its goals. This paper discusses some of the most relevant decisions taken during the design of
Jadescript by means of an informal description of its type system. The
discussed type system includes ordinary types, as found in many other
programming languages, but it also includes a few types that are needed
to adequately support relevant agent-oriented abstractions.

Keywords: Agent-oriented software engineering · Agent-oriented
programming · Jadescript · JADE

1 Introduction and Motivation

AOP (*Agent-Oriented Programming*) (e.g., [38,39]) dates back to the initial
phases of the research on software agents and agent-based software systems.
One of the motivating ideas behind AOP is that agents should be treated
as autonomous and proactive extensions of the objects that characterize ordinary object-oriented programming. Hence, the languages to program agents are
expected to provide direct means to achieve the desired level of autonomy and
proactivity, as well as to enable the communication among agents. Today, software agents are much more than enhanced objects, and the research has evolved
in many different directions under the umbrella of *AOSE* (*Agent-Oriented Software Engineering*) [8]. Nonetheless, an *APL* (*Agent-oriented Programming Language*) is still expected to provide the needed primitives to treat agent-oriented
abstractions and to ensure their coherency and consistency (e.g., [20,21]).

J. Chen et al. (Eds.): DAI 2021, LNAI 13170, pp. 206–220, 2022.
https://doi.org/10.1007/978-3-030-94662-3_13

Jadescript [16,17,23,31–33] is an APL built on top of JADE (e.g., [1,7]), the popular framework to develop agents and *MASs (Multi-Agent Systems)* in Java. Since the initial phases of the project, the design of Jadescript was guided by decisions regarding which features were to be included in the language. The features included in the language, like declarative message handlers with pattern matching [32], were the features identified as essential or, at least, as very frequently needed. Other features, like the treatment of integer intervals, were identified as nonessential, and they were relegated to the features provided by general-purpose or application-specific libraries. As the number of features included in the language increased, the overall design of the language urged a review to check for coherency and consistency. Actually, the evolutionary process adopted to design Jadescript caused the language to offer several ways to perform similar operations on different data types. For this reason, and for similar weaknesses of the language, a complete review of the types that were included in the language was needed. The review of the types required an analysis of the language from the point of view of the type system, which, at the same time, provided a way to direct the design process toward a more coherent and consistent result. Such a review developed into a complete specification of the Jadescript type system. The specification is intended to evolve as the language evolves to be improved to respond to the needs of actual users. Nonetheless, the specification has to be coherent and consistent, and it has to be adequate for the peculiarities of the language, which are inherent to the abstractions related to agents and MASs. This paper provides a detailed, yet informal, overview of the current specification of the Jadescript type system.

In brief, the major characteristics of Jadescript can be summarized as follows. First, Jadescript is an APL, and for this reason the types that it provides should take into account abstractions that are strongly related to agents and MASs. For example, the activation of a behaviour (e.g., [2]) is a frequent operation that should be simplified by an effective linguistic construct. Second, Jadescript is a high-level language, and therefore it should provide users with effective means to deal with data at a high level of abstraction. For instance, the operations related to message passing should be provided by the language in terms of high-level constructs. Similarly, Jadescript should not emphasize low-level features like, for example, bitwise operations on integers, and it should relegate them to general-purpose or application-specific libraries. Third, Jadescript is a pseudocode-inspired language, which means that the adopted syntax mimics the syntax of the pseudocode found in textbooks, to increase readability and to promote naturalness in writing. Fourth, Jadescript, as its name suggests, is designed to target JADE and its MASs. This makes the language strongly related to *FIPA* (*Foundation for Intelligent Physical Agents*) (e.g., [34]). Therefore, the language is expected to treat abstractions like *ACL* (*Agent Communication Language*) messages as first class citizens of the language.

This paper discusses some of the most relevant decisions taken during the design of Jadescript, and it relates these decisions to the resulting type system. Actually, Sect. 2 discusses these decisions by means of a detailed, yet informal, description of the type system designed for Jadescript. Section 3 mentions relevant APLs and their respective approaches to data types. Finally, Sect. 4 briefly recapitulates the major contribution of the paper, and it outlines future developments of the language.

2   The Jadescript Type System

This section provides an informal description of the Jadescript type system, and it discusses the main ideas and decisions behind the design of the type system. Each type, or class of types, is described by enumerating its principal features and by listing its most relevant operations.

One of the most important peculiarities of the Jadescript type system is that there is no type that acts as the supertype of all other types. This means that the types of the language, together with their subtyping relationships, denote a non-lattice partially ordered set. Actually, it is not the case that, given an arbitrary set of types, the least upper bound of the types in the set is guaranteed to exist. Note that Jadescript is designed to support a limited form of type inference, and the compiler sometimes needs to compute the closest common upper bound of a set of types, if it exists. This is the case, for example, of the definition of a local variable as a list of items. In this case, the lack of the least upper bound of the partially-ordered set of the types of the items may result in a compilation error. Nonetheless, as discussed later in this section, some classes of types are structured in subtyping hierarchies for which it is guaranteed that the least upper bound of each class exists. However, trying to compute the least upper bound of two types belonging to two disjoint hierarchies results in a compilation error.

Jadescript satisfies the ordinary need of treating structured data by providing agent-oriented abstractions like concepts and predicates (e.g., [2]). This is the reason why the Jadescript type system treats the types for concepts, actions, predicates, ontologies, agents, behaviours, and messages as distinct classes of types, with distinct subtyping hierarchies, and, in some cases, distinct mechanisms to extend the hierarchies with user-defined types.

The remaining of this section describes the details of the various classes of types that Jadescript supports. First, basic and collection types, which are common to many other programming languages, are discussed. Then, the section continues with the types that support the agent-oriented abstractions that were included in the language, namely concepts, actions, predicates, propositions, ontologies, agents, behaviours, and messages.

2.1   Basic Types

Basic types constitute the fundamental building blocks of all other types. They are not parametric, and they are not composed of other types, because they represent the simplest types that can be manipulated. Jadescript supports eight basic types, and these types share some similarities. First, the values of basic types can be accessed using type-specific literals. Second, basic types are value types, which means that their values do not have identities. Therefore, basic types support dedicated operations for equality comparison, which are always based on values rather than on identities. Finally, it should be noted that all basic types have default values. Default values are used to automatically initialize declared properties that are not explicitly initialized using initialization expressions to ensure that properties are always bound to values.

Boolean. The type `boolean` is the simplest of basic types in Jadescript because it has only two values. These values are associated with the literals `true` and `false`. Jadescript provides, just like many other programming languages do, logical operations on values of this type, namely the prefix operator `not` and the infix operators `or` and `and`. Expressions of this type are the only expressions that can be used to test conditions in Jadescript. For example, expressions of this type can be used in the conditions of `if` statements, or in the looping conditions of `while` statements. The default value of the type `boolean` is `false`.

Integer and Real. The types `integer` and `real` are the only numeric types natively supported by Jadescript. Values of both types are represented by the target machine using 32 bits. The type `integer` denotes (signed) integers, and it supports wrap-around arithmetic and comparison operations. Note that integer division can cause runtime errors. Currently, runtime errors lead to the abortion of the execution of the running event handler, reporting the reason for the abortion in a message log. The type `real` denotes real numbers, which are represented using the single-precision floating-point format that is used by Java and by the Java virtual machine. The type `real` supports floating-point arithmetic and comparison operations. Note that `integer` and `real` are the only two types of the type system that provide implicit conversions for all operations. Specifically, procedures, functions, and constructors that expect values of type `real` can also be used with values of type `integer`, and viceversa. Moreover, when arithmetic and comparison expressions mix values of type `integer` and values of type `real`, all values of type `integer` are automatically promoted to values of type `real`. Such a conduct was chosen to allow simple and intuitive manipulations of numbers. The default value of the type `integer` is 0, and the default value of the type `real` is 0.0.

Text. Jadescript provides a type to treat sequences of characters, namely the type `text`. Note that characters does not have a dedicated type in the language because they are treated as texts of length one. The values of type `text` are immutable, and they can be expressed in Jadescript by means of literals delimited by single or double quotation marks. The values of all other types can be easily converted to values of type `text` by means of the infix operator `as`, which produces a textual representation of the provided value. The infix operator `as` can also be used to convert values of type `text` to values of other types, if the provided text follows the required format for the conversion. Conversions from values of type `text` cause runtime errors when texts are not formatted appropriately. Currently, runtime errors lead to the abortion of the execution of the running event handler, reporting the reason for the abortion in a log message. Multiple texts can be concatenated by means of the infix operator +. The default value of the type `text` is the empty text (`""`).

Performative. Performatives (e.g., [37]) represent the kinds of the communicative acts that agents perform when they send messages. A value of type `performative` is always included in a message, and it constrains the type of the content of the message (e.g., [2]). FIPA specifies twenty-two performatives, namely `accept_proposal`, `agree`, `cancel`, `cfp`, `confirm`, `disconfirm`, `failure`, `inform`, `inform_if`, `inform_ref`, `not_understood`, `propagate`, `propose`, `proxy`, `query_if`, `query_ref`, `refuse`, `reject_proposal`, `request`, `request_whenever`, `request_when`, and `subscribe`. Jadescript adds another performative, namely `unknown`, which is used as the default value of the type `performative` but causes a runtime error when used in a message.

Aid. The ability to manipulate references to agents is, expectedly, of primary importance in Jadescript. Agents are associated with unique *AIDs* (*Agent IDentifiers*), and Jadescript grants a type for AIDs. The type `aid` provides the values that are needed to refer to the agents in the MAS. Actually, Jadescript agents are distributed using JADE containers, which are federated into platforms determined by their respective main containers [2]. Therefore, an AID is always composed of two parts: an agent name, which is a name that uniquely identifies the agent in the platform, and a *HAP* (*Home Agent Platform*), which is an address that uniquely identifies the platform in which the agent is executing. Jadescript supports the type `aid` by providing two kinds of literals and simple ways to deconstruct a value of type `aid` into its parts. The default value of the type `aid` has an empty name and refers to the local platform.

Timestamp and Duration. The ability of agents to reason on temporal information is normally considered an important aspect of AOP. This is the reason why Jadescript provides two basic types, namely `timestamp` and `duration`, to manipulate time instants and time intervals, respectively. The precision of the values of both types is one millisecond. These types support a few arithmetic and comparison operations with intuitive semantics. The values of the type `timestamp` can be accessed using one of the seven forms of supported literals. The syntax of these literals is strongly inspired by the ISO 8601 standard. Actually, the change that Jadescript introduces with respect to the format described in the ISO 8601 specification is needed to allow expressing literals with no use of delimiters like quotation marks. In particular, the parts of a literal are separated by underscores instead of hyphens. Given a literal that represents a timestamp, the part that precedes T is a date expressed as an underscore-separated sequence year, month, day. Similarly, the part between T and Z is a time instant, which is expressed as a colon-separated sequence hour, minute, second, and millisecond. Finally, the part after Z is a time zone specification expressed as an offset with respect to UTC±00:00. The default value of the type `timestamp` is the Unix epoch. The literals of the type `duration` are simpler than the literals of the type `timestamp`. All parts of these literals are optional, and they are separated by underscores. Each part is composed of a number followed by the name of a unit of time. The supported units are days (`d`), hours (`h`), minutes (`m`), seconds (`s`), and milliseconds (`ms`). The default value of the type `duration` is 0 s. Table 1 lists a few examples of the literals of the types `timestamp` and `duration`.

Table 1. Examples of literals of the types `timestamp` and `duration` in Jadescript.

Literal	Description
now	The current date and time in the current time zone
today	The current date in the current time zone with time set at 00:00
2021_07_15T19:07:55:234Z+1:00	A complete timestamp (date, time, and time zone)
2021_07_15T19:07:55:234Z	A timestamp without a time zone specification
2021_07_15T	A timestamp for a date without a time zone specification
T19:07:55:234Z+1:00	A timestamp for a time instant with a time zone specification
T19:07:55:234Z	A timestamp for a time instant without a time zone specification
1d_12h_30m_0s_0ms	A complete duration (day, hour, minute, second, and millisecond)

2.2 Collection Types

Jadescript provides a dedicated support for collection types. The language provides three collection types, namely `list`, `set`, and `map`, that can be used to aggregate data in collections. The three collection types are parametric. In particular, the actual type of a list or a set is obtained by specifying the type of its elements, like, for example, `list of integer` and `set of text`. Similarly, the actual type of a map is obtained by specifying the types of its keys and values, like, for example, `map of text: aid`. Note that, in Jadescript, the parameters of collection types are invariant.

Collection types are designed so that their values are mutable, and Jadescript provides a set of operations to access and manipulate the contents of collections. The provided operations are expressed in terms of dedicated statements that mimic common English phrases. The choice of using English phrases for the operations that manipulate collections emphasizes the intended similarity between Jadescript and the pseudocode commonly used in textbooks. For example, the addition of an item e to a list l is achieved using the command add e to l.

The construction of collections is simplified by a set of literals provided natively by the language. The construction of a list is done by enumerating its elements as a comma-separated list of expressions enclosed in square brackets, like, for example, [1, 2, 3]. Sets are constructed with curly brackets containing a comma-separated list of expressions, like, for example, {1, 2, 3}. Finally, maps can be constructed by putting key-value pairs separated by commas in curly brackets, like, for example, {1 : ''one'', 2 : ''two'', 3: ''three''}.

Note that the type of a collection is automatically inferred by the compiler by computing the closest common upper bound of the types of the expressions enumerated in the literal used to initialize the collection, when a common upper bound exists. However, the compiler does not have sufficient information to infer the type of empty collections because there are no expressions in the literals to perform the inference. This is the reason why the construction of collections can be explicitly typed with the infix operator of, like, for example, [] of integer and {} of text: aid. The default values of the types list, set, and map are well defined only when the types are complete, either implicitly or explicitly, and they are the empty collections.

2.3   Concept, Action, Predicate, and Proposition Types

Agents need application-specific data to represent their knowledge to support reasoning and communication. Jadescript provides the following abstractions to represent knowledge:

- Concepts, which are values that represent objects in the world;
- Actions, which are values that represent the actions that agents can perform;
- Predicates, which are values used to represent facts about the world; and
- Propositions, which are atomic values to represent facts about the world.

Each one of these classes of values is associated with an appropriate set of types, which is structured as a hierarchy of types in which types are connected using single inheritance. As a matter of fact, the application-specific types for concepts, actions, predicates, and propositions are subtypes of the types Concept, Action, Predicate, and Proposition, respectively. The values of these types are accessed by means of type-specific constructors that are made available using the structures described in the declarations of the types. These values do not have identities, and the operations for equality comparison that these types provide are always based on values rather than on identities. Concepts, actions, and predicates, but not propositions, can have typed properties that can be accessed by means of the infix operator of. Note that, beside their use to represent the knowledge of agents, the values of concept, action, predicate, and proposition types can be used in messages.

2.4 Ontology Types

Concept, action, predicate, and proposition types are all collected in ontologies. Therefore, ontologies can be considered as namespaces that group declarations of concept, action, predicate, and proposition types. In addition, ontologies constitute a hierarchy of types in which types are connected using single inheritance. This hierarchy is rooted in the type Ontology. The type Ontology provides the basic building blocks to construct other ontology types, and it is strictly related to the basic ontology [2] that ships with JADE. Users can define an ontology type by means of an ontology declaration, which includes the name of the ontology, the name of its supertype, and a collection of concept, action, predicate, and proposition types that constitute the ontology. Note that each ontology type has only one value, which is used as default value and to make the ontology available to agents and behaviours.

2.5 Agent Types

In Jadescript, agent types constitute a hierarchy of types in which types are connected using single inheritance. This hierarchy is rooted in the type Agent, and it can be extended with the declarations of application-specific agent types. All Jadescript agents are programmed in terms of subtypes of the type Agent. Note that all Jadescript agents are also JADE agents.

The declaration of an agent type can associate the values of the declared type with a list of ontologies. The ontologies that are mentioned in the declaration of an agent type are immediately made available to all values of the declared type. Therefore, all agents of an agent type can access the ontologies mentioned in the declaration of their type.

In addition, the declaration of an agent type can be used to associate the values of the declared type with user-defined features that characterize the declared type. These features include user-defined properties that jointly represent the state of an agent. Every agent is associated with the values bound to the properties declared in its type, and the agent and its behaviours can read and write the values bound to these properties. Also, the user-defined features included in the declaration of an agent type comprise a set of procedures and functions that can be used in the scope of the procedural parts of the codes of agents and behaviours. Functions and procedures have a private reference to the agent, which is accessible via the keywords agent and this. Finally, the user-defined features that can be included in the declaration of an agent type can also include a set of event handlers for agent lifecycle events. Currently, only two handlers for these events are supported in Jadescript, namely on create, which is activated at the creation of an agent and can be used to initialize the state of the agent, and on destroy, which is activated just before the agent is destroyed and removed from its platform.

Note that the values of agent types cannot be used in messages. The reason for this limitation is that the access to a reference to an agent is sufficient to gain access to its properties, functions, procedures, and behaviours. Therefore, to preserve the autonomy and the independence that characterize agents, these references must be kept private to the agent and its behaviours.

2.6 Behaviour Types

Jadescript adopts the behavioral model of agents advocated by JADE (e.g., [2]), and therefore each agent in the MAS engages several behaviours that concurrently execute using a non-preemptive scheduler integrated in the agent. Like agent types, behaviour types constitute a hierarchy of types in which types are connected using single inheritance. As usual, the hierarchy can be extended with user-defined types. Note that the hierarchy forks at its root into two disjoint hierarchies. The most basic type of the hierarchy of behaviour types is `Behaviour`, and its direct subtypes are the types `CyclicBehaviour` and `OneShotBehaviour`. All user-defined behaviour types extend one of these two types, and therefore behaviours can be either cyclic or one shot.

The values of the type `Behaviour` are accessed by means of type-specific constructors that are made available using the structures described in the declarations of the types. Once created, a behaviour can be activated by means of the `activate` statement. This statement puts the behaviour in an internal collection of the agent, which is used by the behaviour scheduler of the agent to pick the next behaviour to execute following a round-robin policy. The `deactivate` statement removes a behaviour from the collection of active behaviours and sets its state to deactivated. A deactivated behaviour is no longer scheduled for execution by the agent, until another explicit activation occurs. Finally, the `destroy` statement sets the state of the behaviour to destroyed and releases its resources. A destroyed behaviour cannot be reactivated, and any attempt to reactivate a destroyed behaviour causes a runtime error.

Cyclic behaviours are designed for operations that are repeated over time, like message reception and event handling, and they are declared with `cyclic behaviour` declarations. On the contrary, one shot behaviours are designed for operations that are not meant to be performed repeatedly, and they are declared with `one shot behaviour` declarations. In both cases, the procedural task associated with the behaviour is defined by the **on execute** event handler. Other supported event handlers are:

- **on** `create`, which handles the creation of a behaviour and can be used to initialize the state of a behaviour;
- **on** `activate`, which handles the activation of a behaviour;
- **on** `deactivate`, which handles the deactivation of a behaviour;
- **on** `destroy`, which handles the destruction of a behaviour; and
- **on** `message`, which handles the reception of a message that matches the pattern declared together with the handler.

Note that the **on** `create` event handler can declare a list of named parameters that are used to generate the constructor of the behaviour type in which the handler is declared.

Besides event handlers, behaviour declarations, just like agent declarations, also contain property, function, and procedure declarations. In a behaviour declaration, the reference to the current behaviour is accessible by means of the keyword `this`. Instead, the keyword `agent` can be used to refer to the agent that created the behaviour. The `agent` reference can be of a specific agent type if the desired agent type is specified with `for agent` in the behaviour declaration. Note that the use of `for agent` constrains the behaviour type to be used only by the agents of the specified type or of one of its subtypes.

2.7 Message Types

The Jadèscript type system includes a set of types for messages. The message types are structured into a closed hierarchy of types. The root of this hierarchy is the type `Message`. Its direct, and only, subtypes are associated with the supported performatives. The compiler ensures that the content of a message respects the constraints imposed by FIPA standards. Therefore, for example, the type of the content of a `request` message can be only an action type. Similarly, the type of the content of an `inform` message can be only a predicate type or a proposition type. It is worth noting that `not_understood` messages are peculiar because their contents are messages. Another important constraint on the content of a message is related to the values bounds to the properties of the concepts, actions, and predicates included in the message. As a matter of fact, values of agent, behaviour, and ontology types cannot be sent as part of messages, and the compiler ensures that this constraint is respected.

The access to the values of message types is possible only via `send` statements, to send messages, or via the activation of message handlers, to receive messages. Normally, the `send` statement has the following form:

```
1  send message <performative> <content> to <receivers>
```

The ontology of the message does not need to be specified because it is inferred by the compiler from the content type. Sometimes, the compiler cannot infer the ontology to use, and in those rare cases, the following variation of the send statement can be used:

```
1  send message <performative> <content> to <receivers>
2    with ontology = <ontology>
```

Messages are accessible using message handlers, which normally have the following form:

```
1  on message <performative> <pattern>
2    when <expression> do
3    <block>
```

Note that message handlers can use an optional expression to accept messages. If the supplied expression evaluates to false, the handler is not activated and the message can be accepted by another handler.

3 Related Work

The type system discussed in this paper was designed specifically for Jadescript, and it is barely comparable with the type systems of other languages. However, it is worth discussing how the designers of other relevant APLs dealt with the problem of typing codes. Note that, in the tradition of logic programming languages, several APLs are not statically typed and their type systems are very restricted with respect to the Jadescript type system.

To some extent, the APL that most closely resembles Jadescript is SARL [36], and the reason for the similarity is that their compilers are both implemented using Xtext [22]. Xtext provides all the necessary tools to build effective *DSLs* (*Domain-Specific Languages*), such as a parser and a code generator. In addition, Xtext provides Xbase [18], which is an expression language that can be embedded into the DSLs developed using Xtext. The use of Xbase is a straightforward way to make DSLs fully interoperable with Java and its type system. Unfortunately, the Xbase grammar severely constrains the grammar of the DSLs that use it. Therefore, although the Xbase grammar was initially chosen for the development of a DSL for JADE (e.g., [3,11]), the path that started with the JADEL programming language [3,9–14] was abandoned in favor of a complete redesign of the grammar, which eventually resulted in the launch of the Jadescript project. On the contrary, the designers of SARL took the opposite decision, and they adopted the Xbase grammar for the core of their language.

It is worth noting that the underlying approaches to agents and MASs advocated by Jadescript and SARL are very different because Jadescript is based on JADE and SARL is based on Janus [36]. While Jadescript proposes agents, behaviours, and ontologies as its core agent-oriented abstractions, SARL uses events, skills, and capacities in addition to behaviors and agents. In particular, SARL supports messages only in terms of their representations as events. In addition, all the agent-oriented abstractions that characterize SARL can be used in a way that is largely inspired by object-oriented programming, which remarks the strong interconnection between SARL and Java. Instead, Jadescript keeps its coding style closer to pseudocode than to object-oriented code, even if it still grants full interoperability with Java.

Another well-known and appreciated APL is Jason [19], which is based on AgentSpeak [35]. An illustrative example used to briefly compare Jadescript and Jason is discussed in [23]. Although Jason allows distributing a MAS with the help of JADE, the approaches of Jason and Jadescript are very different. Jason, as an accurate implementation of AgentSpeak, relies on Horn clauses and uses a Prolog-like syntax. The language is declarative and depends mostly on unification. In fact, Jason supports the *BDI* (*Belief-Desire-Intention*) model [40], and it provides the possibility to define plans. Plans are inherent to the BDI model because they are driven by intentions, that bring the agent towards its desires, taking into account its beliefs. A plan defines a scope in which the variables, that are initially unbound, can be bound by means of unification. Jadescript does not use unification, but it provides a support for pattern matching to associate appropriate handlers with messages [32]. Note that Jadescript provides a statically typed form of pattern matching so that received messages are typed and they can be safely used in handlers.

4 Conclusions

The major goal that drove the introduction of Jadescript was to simplify the development of real-world agent-based software systems by means of a language with specific linguistic constructs for agent-oriented abstractions. The adoption of Jadescript is expected to simplify the development, and to increase the quality, of some of the systems that have already been implemented using JADE. Some of these systems include serious games [4,5,15], a network management system [6], and several tools based on accurate indoor positioning [24–30].

This paper provided an informal presentation of the type system designed for Jadescript. In particular, the paper discussed the decisions taken in the design of the type system, and it motivated the adopted design choices. The work documented in this paper is very important for the development of a language that focuses on the use of specific linguistic constructs to provide direct support for agent-oriented abstractions. However, this paper is not only focused on agent-oriented abstractions, but it also provides insights on Jadescript basic types and collection types, which are essential for a scripting language. Finally, the intended adherence of Jadescript to pseudocode is discussed throughout the paper because it is one of the driving forces behind the development of the language.

The presented type system is not definitive because it is intended to evolve as new agent-oriented abstractions will be included in Jadescript. For example, Jadescript does not currently provide a direct support for the abstractions needed to describe flexible deployment mechanisms for agents and MASs. Agents are supposed to be launched at the activation of the underlying agent platform, or they are supposed to be programmatically launched by other agents. Future developments planned for the language include the possibility of declaratively describing how agents and agent containers should be activated to support agent mobility, fault tolerance, and load balancing. Such a support for flexible deployment mechanisms calls for dedicated extensions of the language, and therefore, it is expected that new and dedicated data types will be introduced.

References

1. Bellifemine, F., Bergenti, F., Caire, G., Poggi, A.: Jade — a Java agent development framework. In: Bordini, R.H., Dastani, M., Dix, J., El Fallah Seghrouchni, A. (eds.) Multi-Agent Programming. MSASSO, vol. 15, pp 125–147. Springer, Boston, MA (2005). https://doi.org/10.1007/0-387-26350-0_5
2. Bellifemine, F., Caire, G.: Dominic Greenwood. Developing Multi-Agent Systems with JADE. Wiley Series in Agent Technology. John Wiley & Sons, New York (2007)
3. Bergenti, F.: An introduction to the JADEL programming language. In: Proceedings of the 26th IEEE International Conference on Tools with Artificial Intelligence (ICTAI 2014), pp. 974–978. IEEE (2014)

4. Bergenti, F., Caire, G., Gotta, D.: An overview of the AMUSE social gaming platform. In: Proceedings of the Workshop "From Objects to Agents" (WOA 2013), vol. 1099. RWTH, Aachen (2013)

5. Bergenti, F., Caire, G., Gotta, D.: Agent-based social gaming with AMUSE. In: Proceedings of the 5th International Conference on Ambient Systems, Networks and Technologies (ANT 2014) and 4th International Conference on Sustainable Energy Information Technology (SEIT 2014), Procedia Computer Science, pp. 914–919. Elsevier (2014)

6. Bergenti, F., Caire, G., Gotta, D.: Large-scale network and service management with WANTS. In: Industrial Agents: Emerging Applications of Software Agents in Industry, pp. 231–246. Elsevier (2015)

7. Bergenti, F., Caire, G., Monica, S., Poggi, A.: The first twenty years of agent-based software development with JADE. Auton. Agents Multi-agent Syst. **34**(2), 1–19 (2020). https://doi.org/10.1007/s10458-020-09460-z

8. Bergenti, F., Gleizes, M.-P., Zambonelli, F. (eds.) Methodologies and Software Engineering for Agent Systems: The Agent-Oriented Software Engineering Handbook. Springer, Boston (2004). https://doi.org/10.1007/b116049

9. Bergenti, F., Iotti, E., Monica, S., Poggi, A.: A case study of the JADEL programming language. In: Proceedings of the 17th Workshop "From Objects to Agents" (WOA 2016), vol. 1664 of CEUR Workshop Proceedings, pp. 85–90 (2016)

10. Bergenti, F., Iotti, E., Monica, S., Poggi, A.: Interaction protocols in the JADEL programming language. In: Proceedings of the 6th ACM SIGPLAN International Workshop on Programming Based on Actors, Agents, and Decentralized Control (AGERE 2016) at ACM SIGPLAN Conference Systems, Programming, Languages and Applications: Software for Humanity (SPLASH 2016), pp. 11–20. ACM (2016)

11. Bergenti, F., Iotti, E., Monica, S., Poggi, A.: Agent-oriented model-driven development for JADE with the JADEL programming language. Comput. Lang. Syst. Struct. **50**, 142–158 (2017)

12. Bergenti, F., Iotti, E., Monica, S., Poggi, A.: A comparison between asynchronous backtracking pseudocode and its JADEL implementation. In: Proceedings of the 9th International Conference on Agents and Artificial Intelligence (ICAART 2017), vol. 2, pp. 250–258. SciTePress (2017)

13. Bergenti, F., Iotti, E., Monica, S., Poggi, A.: Overview of a formal semantics for the JADEL programming language. In: Proceedings of the 18th Workshop "From Objects to Agents", vol. 1867 of CEUR Workshop Proceedings, pp. 55–60. RWTH Aachen (2017)

14. Bergenti, F., Iotti, E., Poggi, A.: Core features of an agent-oriented domain-specific language for JADE agents. In: Trends in Practical Applications of Scalable Multi-Agent Systems, the PAAMS Collection, pp. 213–224. Springer, Cham (2016). https://doi.org/10.1007/978-3-319-40159-1

15. Bergenti, F., Monica, S.: Location-aware social gaming with AMUSE. In: Demazeau, Y., Ito, T., Bajo, J., Escalona, M.J. (eds.) PAAMS 2016. LNCS (LNAI), vol. 9662, pp. 36–47. Springer, Cham (2016). https://doi.org/10.1007/978-3-319-39324-7_4

16. Bergenti, F., Monica, S., Petrosino. G.: A scripting language for practical agent-oriented programming. In: Proceedings of the 8th ACM SIGPLAN International Workshop on Programming Based on Actors, Agents, and Decentralized Control (AGERE 2018) at ACM SIGPLAN Conference Systems, Programming, Languages and Applications: Software for Humanity (SPLASH 2018), pp. 62–71. ACM (2018)

17. Bergenti, F., Petrosino, G.: Overview of a scripting language for JADE-based multi-agent systems. In: Proceedings of the 19th Workshop "From Objects to Agents" (WOA 2018), vol. 2215 of CEUR Workshop Proceedings, pp. 57–62. RWTH, Aachen (2018)

18. Bettini, L.: Implementing Domain-Specific Languages with Xtext and Xtend. Packt Publishing, Birmingham (2013)

19. Bordini, R.H., Hübner, J.F., Wooldridge, M.: Programming multi-agent systems in AgentSpeak using Jason. Wiley Series in Agent Technology. John Wiley & Sons, New York (2007)

20. Cardoso, R.C., Ferrando, A.: A review of agent-based programming for multi-agent systems. Computers 10(2), 16 (2021)

21. Dastani, M.: A survey of multi-agent programming languages and frameworks. In: Shehory, O., Sturm, A. (eds.) Agent-Oriented Software Engineering, pp. 213–233. Springer, Heidelberg (2014). https://doi.org/10.1007/978-3-642-54432-3_11

22. Eysholdt, M., Behrens, H.: Xtext-implement your language faster than the quick and dirty way. In: Proceedings of the ACM International Conference on Object Oriented Programming Systems Languages and Applications Companion (OOPSLA 2010). ACM (2010)

23. Iotti, E., Petrosino, G., Monica, S., Bergenti, F.: wo agent-oriented programming approaches checked against a coordination problem. In: Dong, Y., et al. (eds.) Distributed Computing and Artificial Intelligence, 17th International Conference. DCAI 2020. Advances in Intelligent Systems and Computing, vol. 1237, pp. 60–70. Springer, Cham (2020). https://doi.org/10.1007/978-3-030-53036-5_7

24. Monica, S., Bergenti, F.: Location-aware JADE agents in indoor scenarios. In: Proceedings of the 16th Workshop "From Objects to Agents", volume 1382 of CEUR Workshop Proceedings, pp. 103–108. RWTH, Aachen (2015)

25. Monica, S., Bergenti, F.: A comparison of accurate indoor localization of static targets via WiFi and UWB ranging. In: Trends in Practical Applications of Scalable Multi-Agent Systems, the PAAMS Collection, pp. 111–123. Springer, Cham (2016). https://doi.org/10.1007/978-3-319-40159-1

26. Monica, S., Bergenti, F.:Experimental evaluation of agent-based localization of smart appliances. In: EUMAS 2016, AT 2016: Multi-Agent Systems and Agreement Technologies, vol. 10207, LNCS, pp. 293–304. Springer, Cham (2017). /DOIurl10.1007/978-3-319-33509-4

27. Monica, S., Bergenti, F.: Indoor localization of JADE agents without a dedicated infrastructure. In: Berndt, J.O., Petta, P., Unland, R. (eds.) MATES 2017. LNCS (LNAI), vol. 10413, pp. 256–271. Springer, Cham (2017). https://doi.org/10.1007/978-3-319-64798-2_16

28. Monica, S., Bergenti, F.: An optimization-based algorithm for indoor localization of JADE agents. In Proceedings of the 18th Workshop "From Objects to Agents", vol. 1867 of CEUR Workshop Proceedings, pp. 65–70. RWTH, Aachen (2017)

29. Monica, S., Bergenti, F.: Optimization based robust localization of JADE agents in indoor environments. In: Proceedings of the 3rd Italian Workshop on Artificial Intelligence for Ambient Assisted Living (AI*AAL.IT 2017), vol. 2061 of CEUR Workshop Proceedings, pp. 58–73. RWTH Aachen (2017)

30. Monica, S., Bergenti, F.: An experimental evaluation of agent-based indoor localization. In: Proceedings of the Computing Conference 2017, pp. 638–646. IEEE (2018)

31. Petrosino, G., Bergenti, F.: An introduction to the major features of a scripting language for JADE agents. In: Ghidini, C., Magnini, B., Passerini, A., Traverso, P. (eds.) AI*IA 2018. LNCS (LNAI), vol. 11298, pp. 3–14. Springer, Cham (2018). https://doi.org/10.1007/978-3-030-03840-3_1
32. Petrosino, G., Bergenti, F.: Extending message handlers with pattern matching in the Jadescript programming language. In: Proceedings of the 20th Workshop "From Objects to Agents" (WOA 2019), vol. 2404 of CEUR Workshop Proceedings, pp. 113–118. RWTH Aachen (2019)
33. Petrosino, G., Iotti, E., Monica, S., Bergenti, F.: Prototypes of productivity tools for the Jadescript programming language. In: Proceedings of the 22nd Workshop "From Objects to Agents" (WOA 2021), vol. 2963 of CEUR Workshop Proceedings, pp. 14–28. RWTH Aachen (2021)
34. Poslad, S.: Specifying protocols for multi-agent systems interaction. ACM Trans. Autonom. Adap. Syst. **2**(4), 15:–15:24 (2007)
35. Rao, A.S.: AgentSpeak(L): BDI agents speak out in a logical computable language. In: Van de Velde, W., Perram, J.W. (eds.) MAAMAW 1996. LNCS, vol. 1038, pp. 42–55. Springer, Heidelberg (1996). https://doi.org/10.1007/BFb0031845
36. Rodriguez, S., Gaud, N., Galland, S.: SARL: a general-purpose agent-oriented programming language. In: Proceedings of the IEEE/WIC/ACM International Joint Conferences of Web Intelligence (WI 2014) and Intelligent Agent Technologies (IAT 2014), vol. 3, pp. 103–110. IEEE (2014)
37. Searle, J.: Speech Acts. Cambridge University Press, Cambridge (1969)
38. Shoham, Y.: Agent-oriented programming. Artif. Intell. **60**(1), 51–92 (1993)
39. Shoham, Y.: An overview of agent-oriented programming. In: Bradshaw, J. (ed.) Software Agents, vol. 4, pp. 271–290. MIT Press, Cambridge (1997)
40. Wooldridge, M.: Reasoning About Rational Agents. MIT Press, Cambridge (2000)

Combining M-MCTS and Deep Reinforcement Learning for General Game Playing

Sili Liang, Guifei Jiang$^{(\boxtimes)}$, and Yuzhi Zhang

College of Software, Nankai University, Tianjin 300350, China
2120210551@mail.nankai.edu.cn, {g.jiang,zyz}@nankai.edu.cn

Abstract. As one of the main research areas in AI, General Game Playing (GGP) is concerned with creating intelligent agents that can play more than one game based on game rules without human intervention. Most recent work has successfully applied deep reinforcement learning to GGP. This paper continues this line of work by integrating the Memory-Augmented Monte Carlo Tree Search algorithm (M-MCTS) with deep reinforcement learning for General Game Playing. We first extend M-MCTS from playing the single game Go to multiple concurrent games so as to cater to the domain of GGP. Then inspired by Goldwaser and Thielscher (2020), we combine the extension with deep reinforcement learning for building a general game player. Finally, we have tested this player on several games compared with the benchmark UCT player, and the experimental results have confirmed the feasibility of applying M-MCTS and deep reinforcement learning to GGP.

Keywords: General Game Playing · Memory-Augmented Monte Carlo Tree Search · Deep reinforcement learning · UCT algorithm · General game player

1 Introduction

In the past two decades, several remarkable results about computer games have been achieved in Artificial Intelligence (AI) [3,13,16,18,19]. These results indicate the incredible improvements of technologies and algorithms in AI. On the other hand, as Feng-hsiung Hsu pointed out, each of these intelligent systems is designed for a specific game, and building systems to play specific games has limited value in AI [13].

To address this issue, a project called General Game Playing (GGP) has been launched at Stanford University since 2005. This project is concerned with creating intelligent systems that understand the rules of arbitrary new games and learn to play these games without human intervention [9]. Unlike specialised systems, a general game player cannot use an algorithm with specific knowledge or heuristics that do not apply to other games [10]. Such a system rather requires

© Springer Nature Switzerland AG 2022
J. Chen et al. (Eds.): DAI 2021, LNAI 13170, pp. 221–234, 2022.
https://doi.org/10.1007/978-3-030-94662-3_14

a form of general intelligence that enables it to autonomously play a new game based on its rules. After more than ten years of development, GGP has become an important research area in AI [22].

Designing and implementing strategy generation algorithms is a core technique in GGP. Nowadays, the main algorithms used in successful general game players are the Monte Carlo Tree Search (MCTS) algorithm and its variants. The key idea of MCTS is to construct a search tree of states evaluated by fast Monte Carlo simulations. As a general-purpose algorithm, MCTS is widely used in GGP. However, the main disadvantage of MCTS is that it makes very limited use of game-related knowledge, which may be inferred from a game description and play a part for game playing [21].

On the other hand, with the great success of AlphaGo and its successors, the role of reinforcement learning in game playing has become prominent [18–20], and several reinforcement learning approaches have been proposed to improve MCTS. For instance, a deep neural network can be used to learn domain knowledge and evaluate a state so as to guide the MCTS search [18]. In particular, an algorithm called Memory-Augmented Monte Carlo Tree Search (M-MCTS) was proposed by incorporating MCTS with a memory structure to exploit generalization in online real-time search. It showed that M-MCTS is useful for improving the performance of MCTS in both theory and practice [24].

Recently there have been works applying reinforcement learning approaches to the domain of GGP [2,7,11,12,23]. A project called RL-GGP was proposed by [2] to integrate GGP with RL-glue so as to compare reinforcement learning algorithms in the context of games. It only provided a framework and did not implement a general game player. In [23], the canonical reinforcement learning method Q-learning was used in GGP. Yet it showed that this method converges much slower than MCTS. Most recently, deep reinforcement learning was applied to GGP by extending the AlphaZero algorithm, and it showed that this can provide competitive results in several games except for a cooperative game Babel [11]. Inspired by this work, GGPZero was developed by [12] to further explore the learning architecture for GGP, and systematic experiments were done to confirm its feasibility. Our work continues this line of work by combining M-MCTS and deep reinforcement learning for GGP and shows that it can outperform the benchmark player in a variety of games, which has further confirmed the feasibility of applying deep reinforcement learning to GGP.

The rest of this paper is structured as follows: Sect. 2 provides the background including GGP, M-MCTS and deep reinforcement learning involved in this paper. Section 3 extends M-MCTS to cater to the domain of GGP and further combines the extension with deep reinforcement learning for building a general game player. Section 4 implements and evaluates the general game player with the benchmark UCT player in terms of various games. Finally, we conclude this paper and point out some future work.

2   Background

In this section, let us provide some preliminaries. We begin with an overview of GGP, and then introduce the Memory-Augmented Monte Carlo Tree Search algorithm and deep reinforcement learning we use for building a general game player.

2.1   General Game Playing

There are many strategy generation algorithms used in previous successful general game players. At the early stage, GGP players mainly took evaluation-based search approach and were mostly based on MiniMax algorithm and its variants. The typical examples are the GGP competition champions Cluneplayer and Fluxplayer [4,17]. Since the 2007 champion Cadiaplayer first adopted the upper confidence bound on trees (UCT) algorithm [6], the simulation-based search approach has gradually become the mainstream algorithm for building a GGP player. All the following winners such as Ary, TurboTurtle used either UCT or its variants [15], except the 2016 winner WoodStock, which took a method based on constraint satisfaction programming [14].

UCT is a game tree search algorithm by combining the Monte Carlo Tree Search method with the Upper Confidence Bound (UCB) [1]. The key idea of UCT is to construct a search tree of states evaluated by fast Monte Carlo simulations, and use the UCB formula to determine whether to continue the search or consider the possibility of other sibling nodes. Each node of the search tree s stores three parameters: a node count $N(s)$, a count of the number of times action a has been taken from node s $N(s,a)$, and a value for how good the current state is, $\hat{V}(s)$. The choice of action at a state is then determined by the following UCB formula [1].

$$\mathbf{argmax}_{a \in A(s)} \left(\hat{V}(\delta(s,a)) + C\sqrt{\frac{ln(N(s))}{N(\delta(s,a))}} \right)$$

where $A(s)$ denotes the set of all legal actions at state s, $\delta(s,a)$ is the state transition function, i.e., the next state after taking action a at state s, and C is a constant used for balancing exploration and exploitation. The higher the constant C, the more UCT will explore unpromising nodes.

2.2   Memory-Augmented Monte Carlo Tree Search

With large state spaces and relatively limited search time, the inaccurate value estimation of a state can mislead building the search tree and degrade the performance of UCT algorithm. To address this issue, the Memory-Augmented Monte Carlo Tree Search algorithm (M-MCTS) was proposed by Xiao *et al.* in [24]. M-MCTS is inspired from the core idea of neural network algorithms: generalization, that is, similar states can share information. Specifically, it exploited generalization in online real-time search by incorporating MCTS with a memory structure, where each entry contains information of a particular state. This memory is used to generate an approximate value estimation by combining the estimations of similar states, which allows UCT algorithm to improve the accuracy of value estimation.

The approximate value estimation is performed as follows: given a memory $\mathcal{M}$ and a state s, they find the set of most similar states $M_s \subset \mathcal{M}$ according to a distance metric $d(.,s)$, and compute a memory-based value estimation $\hat{V}_{\mathcal{M}}(s) = \sum_{i=1}^{M} w_i(s)\hat{V}(i)$ such that $\sum_{i=1}^{M} w_i(s) = 1$, where $\hat{V}(i)$ denotes the value estimation of state i from simulations, and $w_i(s)$ is called the weight of state i with respect to s.

Each entry of $\mathcal{M}$ corresponds to one particular state $s \in S$. It contains the state's feature representation $\phi(s)$ as well as its simulation statistics $\hat{V}(s)$ and $N(s)$. To integrate memory with MCTS, in each node of an M-MCTS search tree, they store an extended set of statistics

$$\{N(s), \hat{V}(s), N_{\mathcal{M}}(s), \hat{V}_{\mathcal{M}}(s)\}$$

where $N_{\mathcal{M}}(s)$ is the number of evaluations of the approximated memory value $\hat{V}_{\mathcal{M}}(s)$. Then in the action selection policy of UCT, the value $\hat{V}(s)$ of state s in UCB formula is replaced by $(1 - \lambda_s)\hat{V}(s) + \lambda_s\hat{V}_{\mathcal{M}}(s)$, where λ_s is the learning rate, a constant parameter to guarantee no bias asymptotically. This process is shown in Fig. 1. It has been shown that the memory based value approximation is better than the vanilla Monte Carlo estimation with high probability under mild conditions, and M-MCTS outperforms the original MCTS with the same number of simulations in the game Go based on a handcrafted neural network [24].

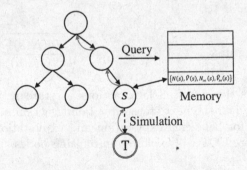

Fig. 1. An illustration of the architecture of M-MCTS, image based on [24]

2.3 Deep Reinforcement Learning

In AlphaGo, the tree search evaluated positions and selected moves using deep neural networks. These neural networks were trained by supervised learning from human expert moves, and by reinforcement learning from self-play [18].

This program was then upgraded to AlphaGo Zero which is trained solely by self-play reinforcement learning, starting from random play, without any supervision or use of human data [20]. Recently it has been generalized to AlphaZero that achieves superhuman performance in multiple games such as Go, Chess and Shogi by using a general-purpose reinforcement learning algorithm [19].

Specifically, AlphaZero uses a deep neural network which takes a board position as an input, and outputs a probability distribution over moves and an estimation of the expected outcome from the board position. AlphaZero learns these probability distributions and expected outcomes entirely from self-play. In each position, an MCTS search is executed, guided by the neural network. Each search consists of a series of simulated games of self-play that traverse a tree from root state until a leaf state is reached. It returns a probability distribution over moves. The triples of state, move distribution and winner are then sampled in order to train the neural network [19].

Most recently, Goldwaser and Thielscher have applied deep reinforcement learning to GGP by extending the AlphaZero algorithm [11]. To this end, they provided methods to deal with the limiting assumptions that AlphaZero makes about games that it plays [11].

1. To address the non zero-sum games, the expected outcome output of the neural network, i.e., 1 for a win, 0 for a draw, −1 for a loss, is replaced with an expected reward that is between 0 and 1. With this, each agent will simply try to maximize their reward with no need to consider the others. In this way, cooperative policies can be learned.
2. To deal with the asymmetric games, it uses a separate neural network for each player which is trained separately, and combines the early layers of all the players' neural networks since very similar features from the game state are extracted in these neural networks.
3. To cope with simultaneous games with multi-player, the transition from one state to another is no longer through a certain action of a certain player, but through the sequence of actions of all players in that turn. In order to generalize the turn-based games to the simultaneous, an action without any effect called "noop" is added. Thus, a player can only do "noop" when it is not her turn.
4. To remove the reliance on a board and a handcrafted neural network for each game, it uses the propositional network [5,10] as the input to the neural network. To this end, they take all nodes from the propositional network and convert them to vectors before feeding them into the network. The architecture of the neural network in [11] is shown in Fig. 2. That input is then passed through some automatically generated fully connected layers as follows: first all inputs pass through to a series of fully connected hidden layers, with each one half the size of the previous, finishing with a layer of size 50. Then each player's head comes off that layer as a common start, doubling the size of the layer with more fully connected layers until it reaches the number of legal actions. At this point, there is also a fully connected layer which goes to a single output for expected return prediction. All hidden layers use

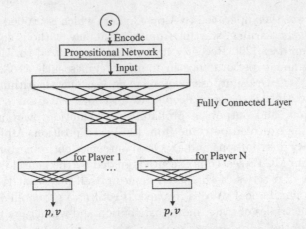

Fig. 2. The architecture of the neural network

ReLU activation, the expected reward uses sigmoid and policy output uses softmax. The output includes the probability distribution of this player p and the expectation of the final score of this player v in the state s.

To test the performance of their built learning agent, Goldwaser and Thielscher have conducted the comparison experiments with UCT as the benchmark due to its variants being state-of-the-art for many years. All hyperparameters were tuned on Connect-4 with a 6×7 board, and then evaluated on the typical games from past GGP competitions [8], namely Connect4, Babel, Breakthrough, Pacman3P. The experimental results show that with deep reinforcement learning the built agent performs better than the UCT benchmark agent in these games except the cooperative game Babel, as compared with their learning agent, the setup of the UCT agent tended to perform correlated actions which worked well as a joint strategy in Babel [11].

3   Method

In this section, we build a general game player by combining M-MCTS and deep reinforcement learning. First, we extend M-MCTS to suit the domain of GGP. Then we build a general game player by integrating the extension with deep reinforcement learning.

3.1   M-MCTS for GGP

As we mentioned before, M-MCTS was limited to game Go since a handcrafted neural network architecture for this game was used [24]. For board games such as Go, it is natural to extract features from the boards, while for GGP, the types of games are various and thus the method used in [24] is not applicable. In addition, it does not involve multi-player and simultaneous games. To remove these limitations, we extend M-MCTS in the following aspects so as to suit the domain of GGP.

1. To deal with the deep convolutional neural network specializing in Go,
   (a) we redefined the feature function based on the propositional network, since games in GGP are all defined by Game Description Language (GDL), and propositional networks serve as a natural graph representation of GDL [5,10]. A Propositional Network is a directed bipartite graph consisting of nodes representing propositions connected to either boolean gates, or transitions [5]. Based on this, similar to [11], we specify the feature of the state as the vector of the node corresponding to that state in the propositional network. In this way, each state can be specified by a different and uniquely corresponding vector.
   (b) we further modified the distance function over game states based on their vectors. As the dimension of the vector reflects the information about each state, the closer the bitwise comparison of the two vectors is, the more similar the information of their corresponding states should be. As shown in the **Algorithm**, given two vectors we did the bitwise comparison, and the proportion of vector differences is used as their similarity value after mapping to $[-1, 1]$ by a linear transformation.
2. To incorporate the multi-player and simultaneous games, we used the sum of the estimated score of action sequences including the other players' actions as the estimated score of the player's own action. Similarly, the "noop" action is also added to treat turn-based games as a special case of concurrent games.

Algorithm. Distance Function $d(\vec{s}, \vec{x})$

Input: Vectors $\vec{s}, \vec{x}$
Output: the similarity of $\vec{s}, \vec{x}$
$\quad n \leftarrow$ the dimensions of $\vec{s}$
$\quad t \leftarrow 0$
$\quad$ for $i = 1 \rightarrow n$ do
$\quad\quad$ if $s_i == x_i$ then
$\quad\quad\quad t \leftarrow t + 1$
$\quad\quad$ end if
$\quad$ end for
$\quad$ return $2 * (-t/n) + 1$

3.2 Combining M-MCTS with Deep Reinforcement Learning

Let us now combine the extended M-MCTS with deep reinforcement learning for building a general game player.

To this end, we keep the memory structure and the update strategy of M-MCTS, and replace the random simulations with the output of the neural network. The architecture is shown in Fig. 3. Although the architecture of the proposed player is similar to that of [11], there are following main differences between them: we used the extended M-MCTS algorithm instead of MCTS to generate the training set through self-play, since the former is proved to be better than the latter and thus the quality of the training set is supposed to be improved. As we will show in the experiments, this consequently speeds up the training efficiency. Moreover, to integrate with deep reinforcement learning, we also did the following modifications:

1. Besides the memory-based value approximation, the effects of the neural network were also considered in the selection function. Formally, the selection formula is adapted as follows:

$$\mathbf{argmax}_{a \in A(s)} \left(\lambda_s * \hat{V}_\mathcal{M}(\delta(s,a)) + (1 - \lambda_s) * \hat{V}(\delta(s,a)) + C * v * \frac{\sqrt{N(s)}}{N(\delta(s,a)) + 1} \right)$$

where v is referring to the output of the policy network, which means the expectation of the final score in state s.

2. To make the algorithm more robust and also encourage exploration [20], the Dirichlet noise was added to the root of the searching tree by replacing v with $((1 - w_{noise}) * v + w_{noise} * d)$, where d is a random variable from Dirichlet distribution which is the same as [11] and w_{noise} is the weight of the noise.

3. The estimation value of a state generated by random simulations is replaced by the output of the neural network, which is more accurate and more efficient to be calculated.

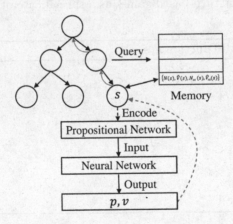

Fig. 3. The architecture of combining M-MCTS with deep reinforcement learning

4   Experimental Evaluation

In this section, we implement and evaluate the proposed GGP player with the benchmark UCT player which is commonly used in [6,8,11] in terms of a variety of games. We first introduce the evaluation methodology and the basic setting, and then provide the experimental results.

4.1 Evaluation Methodology

Similar to [11,12], we select the following four games in terms of complexity levels and game types to evaluate our built player extensively.

- Connect4 (6×7 board): two-player, zero-sum, player-symmetric, turn-based
- Breakthrough (6×6 board): two-player, zero-sum, player-symmetric, turn-based
- Babel: three-player, cooperative, player-symmetric, simultaneous
- Pacman 3P (8×8 board): three-player, cooperative/zero-sum, player-asymmetric, mixed turn-based/simultaneous

As we mentioned before, we also used the UCT player as the benchmark. Here we mainly focused on two aspects: the number of games trained on and the number of simulations per move during training. Both players had a limit of 1000 simulations per move during playing.

All hyperparameters were tuned on Connect4, and then evaluated on the other games. Here are the main parameters set in our experiments:

- $\mathcal{M}$, the size of memory structure, is set to 20.
- M, the top M similar states during the query, is set to 5.
- λ_s, the usage of memory, is set to 0.5.
- C, the parameter in UCB, is set to 2.
- w_{noise}, the weight of Dirichlet noise, is set to 0.25.

All the experiments were conducted on a server with Intel Xeon Gold 5218 running at 2.30 GHz.

4.2 Results and Analysis

Connect4. Our player scores 1 point for a win, 0.5 points for a draw, and no point for failure. As shown in Fig. 4, the performance of our player on Connect4 is better than that of UCT when selecting appropriate parameters and carrying out a certain amount of training. The winning rate can reach nearly 60% when the number of simulations per move during training is set to 100 and the number of games trained on is up to 4000. Given the same number of games trained on, the performance is much better when the number of simulations per move is 100 than 300.

Breakthrough. In this game, our player scores 1 point for a win and no point for failure. The results are shown in Fig. 5. The performance of our player on Breakthrough far exceeds that of the UCT. Especially, when the number of simulations per move during training is 300 and the number of games train on is up to 400, the winning rate is up to 95%.

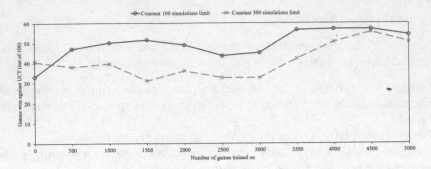

Fig. 4. Our player VS UCT in Connect4

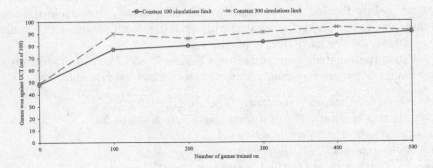

Fig. 5. Our player VS UCT in Breakthrough

Babel. There are three players in this game working together to build a tower. In a round, each builder had the same score, and we recorded the total score of each builder over 50 rounds. Each builder was given a score based on how well they reached the building level in one round. Each run consists of either three our players or three UCT players with no mixed teams. The results are shown in Fig. 6. Similar to the experimental results in [11], UCT players have performed much better on Babel. One of the most possible reasons is that the consistency policy which UCT players make has a great advantage on Babel. On the other hand, since we train the network individually for each role, it might not be easy to form a consistent policy for them. In addition, it can be found that our player performs better when the number of simulations per move during training is set to 300 rather than 100. It is likely that with some extra complexity in the neural network architecture and more training time it would be able to achieve a better performance.

Pacman 3P. In this game, one player controls Pacman to collect pellets and the other two players, each controlling a Ghost, work together to catch Pacman. In one round, for Ghosts, catching Pacman scores 1 point; for Pacman, there are 35 pellets, catching one pellet scores 1/35 points. For the evaluation, it was played both with Pacman being a UCT agent and both ghosts being our players

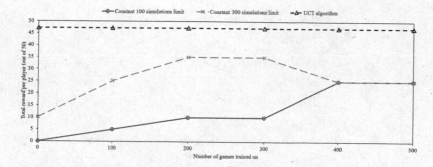

Fig. 6. Our player VS UCT in Babel

and the reverse. The results in terms of the number of simulations 100, 300 per move during training are shown in Fig. 7 and Fig. 8, respectively. As we can see, our player playing as Pacman and Ghosts can perform better than UCT when the number of simulations per move during training is set to 100. Yet when the number of simulations per move during training is increased from 100 to 300, the performance of our player shows a tendency of decline.

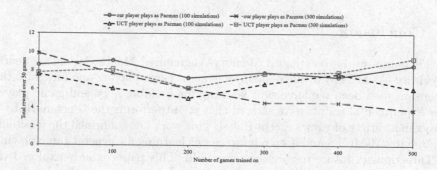

Fig. 7. Our player and UCT player play as Pacman in Pacman3P.

To conclude this section, on one hand, the experimental results show that our player with appropriate parameters and sufficient training can significantly outperform the benchmark UCT player on a number of games in both efficiency and effectiveness except the cooperative game Babel, which is similar to [11,12]. On the other hand, we also find that regarding the two parameters we concern it is not the case that the larger they are, the better our player performs. For instance, the performance of our player in game Pacman 3P shows a tendency of decline as the number of simulations per move during training increases. One possible reason is that at the very beginning the neural network could not make the right decision over actions due to the inaccurate probability distribution over actions and insufficient training data, which in turn might guide M-MCTS to

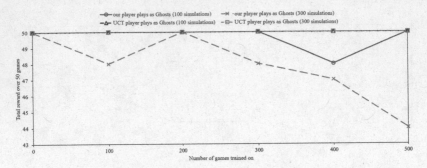

Fig. 8. Our player and UCT player play as Ghosts in Pacman3P.

generate the defective training set. This could make the neural network less effective than what we expect. Finally, it should be noted that without the details of their parameters, in this paper we did not conduct a direct comparison with Goldwaser and Thielscher's player. Yet as we can see from the above experimental results, to some extent, our algorithm has better results and fewer training rounds compared with theirs. A direct and detailed comparison is left for future work.

5   Conclusion

In this paper, we have extended Memory-Augmented Monte Carlo Tree Search algorithm (M-MCTS) to the domain of general game playing and integrate this extension with deep reinforcement learning for building a general game player. The experimental results have showed that it outperforms the benchmark UCT player in a variety of games except Babel. This work has confirmed the feasibility of applying M-MCTS and deep reinforcement learning for general game playing.

Directions of future research are manifold. This paper mainly explored the impacts of the number of games trained on and the number of simulations per move during training on the performance of our player. Besides the parameters we tuned on Connect4, we believe that the parameters related to M-MCTS algorithm we set will also affect the performance of our player and thus worth further investigating.

Regarding the feature function for M-MCTS, in this paper we coded a game state as a unique vector based on the propositional network. This method is natural and works well to extend M-MCTS from single game Go to various types of games. Yet it would be interesting to investigate other methods to improve the feature function so as to make the similarity of the states more accurate.

Last but not least, it is worth noting that our player is better than the UCT player in the simulation efficiency, especially at the beginning of a game situation. To further improve the performance of the player, we believe that larger and deeper networks for games are a potential approach, since it has been shown in [12] that deeper networks with more training tended to have better performance than shallower networks.

Acknowledgments. We are grateful to the reviewers of this paper for their constructive and insightful comments. The research reported in this paper was partially supported by the National Natural Science Foundation of China (No. 61806102), the Major Program of the National Social Science Foundation of China (No. 17ZDA026), and the National Key Project of Social Science of China (No. 21AZX013).

References

1. Auer, P.: Using confidence bounds for exploitation-exploration trade-offs. J. Mach. Learn. Res. **3**, 397–422 (2002)
2. Ayuso, J.L.B.: Integration of general game playing with RL-glue (2012). http://citeseerx.ist.psu.edu/viewdoc/download?doi=10.1.1.224.7707&rep=rep1&type=pdf
3. Brown, N., Sandholm, T.: Libratus: the superhuman AI for no-limit poker. In: Proceedings of the 26th International Joint Conference on Artificial Intelligence, pp. 5226–5228 (2017)
4. Clune, J.: Heuristic evaluation functions for general game playing. In: Proceedings of the 22nd AAAI Conference on Artificial Intelligence, pp. 1134–1139 (2007)
5. Cox, E., Schkufza, E., Madsen, R., Genesereth, M.: Factoring general games using propositional automata. In: Proceedings of the IJCAI Workshop on General Intelligence in Game-Playing Agents, pp. 13–20 (2009)
6. Finnsson, H., Björnsson, Y.: Simulation-based approach to general game playing. In: Proceedings of the 23rd AAAI Conference on Artificial Intelligence, pp. 259–264 (2008)
7. Finnsson, H., Björnsson, Y.: Learning simulation control in general game-playing agents. In: Proceedings of the 24th AAAI Conference on Artificial Intelligence (2010)
8. Genesereth, M., Björnsson, Y.: The international general game playing competition. AI Mag. **34**(2), 107–107 (2013)
9. Genesereth, M., Love, N., Pell, B.: General game playing: overview of the AAAI competition. AI Mag. **26**(2), 62–72 (2005)
10. Genesereth, M., Thielscher, M.: General game playing. Synthesis Lectures on Artificial Intelligence and Machine Learning **8**(2), 1–229 (2014). https://doi.org/10.1007/978-981-4560-52-8_34-1
11. Goldwaser, A., Thielscher, M.: Deep reinforcement learning for general game playing. In: Proceedings of the 34th AAAI Conference on Artificial Intelligence, pp. 1701–1708 (2020)
12. Gunawan, A., Ruan, J., Thielscher, M., Narayanan, A.: Exploring a learning architecture for general game playing. In: Gallagher; M., Moustafa, N., Lakshika, E. (eds.) AI 2020. LNCS (LNAI), vol. 12576, pp. 294–306. Springer, Cham (2020). https://doi.org/10.1007/978-3-030-64984-5_23
13. Hsu, F.H.: Behind Deep Blue: Building the Computer That Defeated the World Chess Champion. Princeton University Press, Princeton (2002)
14. Koriche, F., Lagrue, S., Piette, É., Tabary, S.: General game playing with stochastic CSP. Constraints **21**(1), 95–114 (2015). https://doi.org/10.1007/s10601-015-9199-5
15. Méhat, J., Cazenave, T.: A parallel general game player. KI-künstliche Intelligenz **25**(1), 43–47 (2011)
16. Moravčík, M., et al.: Deepstack: expert-level artificial intelligence in heads-up no-limit poker. Science **356**(6337), 508–513 (2017)

17. Schiffel, S., Thielscher, M.: Fluxplayer: a successful general game player. In: Proceedings of the 22nd AAAI Conference on Artificial Intelligence, pp. 1191–1196 (2007)
18. Silver, D., et al.: Mastering the game of Go with deep neural networks and tree search. Nature **529**(7587), 484–489 (2016)
19. Silver, D., et al.: A general reinforcement learning algorithm that masters chess, shogi, and Go through self-play. Science **362**(6419), 1140–1144 (2018)
20. Silver, D. et al.: Mastering the game of Go without human knowledge. Nature **550**(7676), 354–359 (2017)
21. Świechowski, M., Park, H., Mańdziuk, J., Kim, K.J.: Recent advances in general game playing. Sci. World J. **2015** (2015)
22. Thielscher, M.: General game playing in AI research and education. In: Bach, J., Edelkamp, S. (eds.) KI 2011. LNCS (LNAI), vol. 7006, pp. 26–37. Springer, Heidelberg (2011). https://doi.org/10.1007/978-3-642-24455-1_3
23. Wang, H., Emmerich, M., Plaat, A.: Monte carlo Q-learning for general game playing. arXiv preprint arXiv:1802.05944 (2018)
24. Xiao, C., Mei, J., Müller, M.: Memory-augmented Monte Carlo tree search. In: Proceedings of the 32nd AAAI Conference on Artificial Intelligence (2018)

A Two-Step Method for Dynamics of Abstract Argumentation

Xiaoxin Jing[1] and Xudong Luo[2($\boxtimes$)]

[1] College of Political Science and Law, Capital Normal University, Beijing, China
luoxd@mailbox.gxnu.edu.cn
[2] Guangxi Key Lab of Multi-Source Information Mining and Security,
School of Computer Science and Engineering, Guangxi Normal University,
Guilin, China

Abstract. This paper proposes a two-step method to handle the dynamics of argumentation. Firstly, we reach the conflict-free labellings of the updated argumentation framework by the intersection of expansions of the conflict-free labellings of argumentation frameworks before updating. Then we select the conflict-free labellings which have the least illegally labelled arguments when restricted to part of both argumentation frameworks in the update process. Finally, we prove the soundness and completeness of our method under the complete semantics. In other words, the complete labellings of the resulted argumentation framework after update is the same as that of our method.

1 Introduction

Argumentation has become a major research area in Artificial Intelligence over the last two decades [4]. Abstract argumentation is an elegant way to tackle reasoning problems in the presence of conflicting information. The seminal paper by Dung [12] defines an argumentation framework as a digraph whose nodes are abstract entities called arguments, and edges are attacks representing the conflicts between these arguments. The central concern of abstract argumentation is the evaluation of a set of arguments and their relations in order to be able to extract subsets of the arguments that can all be accepted together from some point of view. The criteria or methods used to settle the acceptance of arguments, on the other hand, are called "semantics".

Given that argumentation can be viewed as a process, there has been an increasing number of studies on different problems in the dynamics of argumentation frameworks [1,3,6,9–11,13]. The so-called enforcing problem [3,10,11] is concerned with whether and how an argumentation framework can be modified to make a certain set of arguments accepted. In [6,9], the change in argumentation is studied through belief revision. Coste-Marquis et al. [9] proposed rationality postulates which express conditions to be satisfied by the extensions of the revised argumentation framework.

© Springer Nature Switzerland AG 2022
J. Chen et al. (Eds.): DAI 2021, LNAI 13170, pp. 235–246, 2022.
https://doi.org/10.1007/978-3-030-94662-3_15

The addition (or removal) of arguments and attack relations of argumentation frameworks are extensively studied in [5,8,13], these kinds of problems are called "elementary change" in [11]. Cayrol, de Saint-Cyr, and Lagasquie-Schiex [8] focused on the addition of a new argument that interacts with previous arguments and studied the impact of such change on the outcome of the argumentation framework. Boella, Kaci, and Van der Torre [5] considered the dynamics of abstract argumentation framework for the evaluation of extension-based argumentation semantics. They do not consider individual approaches but define general principles that individual approaches may satisfy. Liao, Jin, and Koons [13] presents a general theory (called a division-based method) to cope with the dynamics of argumentation frameworks.

With the changing of arguments and/or attacks of an argumentation framework, the status of arguments in the argumentation framework may change, one of the challenging problems is how to efficiently compute the dynamics of argumentation frameworks. We may simply recompute the semantics of new argumentation framework afresh. However, this method is inefficient and difficult. In this paper, we focus on the problem of the impact on the outcome of an argumentation framework when a new argumentation framework comes into play. More specifically, we propose a method to compute the labelling-based semantics for the dynamics of argumentation frameworks. This is a two-step method, firstly, we reach the conflict-free labellings of the updated argumentation framework by the intersection of expansions of the conflict-free labellings for the two argumentation frameworks before updating, then we reach the complete labellings by selecting the conflict-free labellings which have the least illegally labelled arguments when restricted to part of each argumentation framework before the update.

The rest of the paper is organized as follows. Section 2 recaps Dung's abstract argumentation framework. Section 3 presents our method to update an argumentation framework when another argumentation framework comes into play. Section 4 proves the soundness and completeness result of our method. Section 5 discusses the related work to show how our work advances the state-of-art in the research area. Finally, Sect. 6 summarises our work with further work.

2   Preliminaries

We start with some preliminaries concerning Dung's argumentation theory [12]. In [12], argumentation frameworks are directed graphs, where nodes correspond to arguments and arcs to attacks between arguments. We limit ourselves to the abstract setting, meaning that we do not specify the content of arguments in a formal way. Meanwhile, throughout this paper, we assume that an argumentation framework is generated by an agent at a given time point, and therefore is finite. We assume an arbitrary finite non-empty domain $\mathcal{U}$ of arguments, then:

Definition 1. *An argumentation framework (AF for short) F is a pair (A, R) where $A \subseteq \mathcal{U}$ is a non-empty set of arguments and $R \subseteq A \times A$ is an attack relation. $a_i R a_j$ (which is also denoted by $(a_i, a_j) \in R$) means that a_i attacks a_j.*

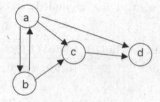

Fig. 1. An argumentation framework

We denote the collection of all AFs as $AF_{\mathcal{U}}$.

Definition 2. *Given an AF $F = (A, R)$, an argument a is unattacked if and if only $\nexists b \in A$, such that $(b, a) \in R$, and an argument a is attacked, if and only if there exists at least one argument b, such that $b \in A$ and $(b, a) \in R$.*

We denote the set of unattacked arguments of A in F as $Ua(A)$ and the set of attacked arguments of A in F as $At(A)$. It is obvious that $Ua(A) \cup At(A) = A$.

Example 1. The AF $F = (A, R)$ where $A = \{a, b, c, d\}$ and $R = \{(a, b), (b, a), (a, c), (b, c), (c, d), (a, d)\}$ defines the graph in Fig. 1. In addition, $At(A) = \{a, b, c, d\}$ and $Ua(A) = \emptyset$.

Given an argumentation framework, the acceptance status of arguments can be evaluated by applying several semantics. In other words, a semantic gives a formal definition of a method ruling the argument evaluation. A huge range of semantics have been defined so far [2]. They can be classified into two categories: extension-based semantics [12] and labelling-based semantics [7]. An extension-based semantics produces a set of extensions which is a set of sets of collectively acceptable arguments, and a labelling-based semantics produces a set of labellings which are functions assigning to each argument a label from a predefined set $\{in, out, undecided\}$, indicating that the argument is respectively accepted, rejected or neither. The two representations are essentially reformulations of the same idea as they can be mapped 1-to-1 such that extensions correspond to sets of in-labelled arguments [7]. For our purpose, we choose to adopt the labelling-based approach.

Definition 3. *A labelling of an AF $F = (A, R)$ is a function $L : A \to \{I, O, U\}$, where I, O and U are abbreviations of In, Out and Undecided, respectively. We denote by $I(L)$, $O(L)$ and $U(L)$ the set of all arguments $x \in A$ such that $L(x) = I$, $L(x) = O$ or $L(x) = U$, respectively, and by $\mathcal{M}_F$ the set of all labellings of F.*

In the following of this paper, we denote labellings by strings of the form $ABC \cdots$, where $A, B, C, \cdots$ are the labels of arguments $a, b, c, \cdots$, and the arguments in the argumentation framework are in their lexicographical order.

Example 2. Let us consider the AF F of Fig. 1. L is a labeling of F, where $L(a) = I$, $L(b) = O$, $L(c) = O$, and $L(d) = I$. We denote L as IOOI.

Table 1. Labellings on F w.r.t.to the grounded, stable, preferred and complete semantics

Semantics σ	In (L)	Undec (L)	out (L)
Grounded	$\emptyset$	$\{a,b,c,d\}$	$\emptyset$
Stable	$\{a\}$	$\emptyset$	$\{b,c,d\}$
	$\{b,d\}$	$\emptyset$	$\{a,c\}$
Preferred	$\{a\}$	$\emptyset$	$\{b,c,d\}$
	$\{b,d\}$	$\emptyset$	$\{a,c\}$
Complete	$\{a\}$	$\emptyset$	$\{b,c,d\}$
	$\{b,d\}$	$\emptyset$	$\{a,c\}$
	$\emptyset$	$\{a,b,c,d\}$	$\emptyset$

As we have mentioned, in this paper, we aim to address the problem of characterizing the effects of an agent learning about new argumentation frameworks. It does not make sense to study the update of argumentation frameworks directly on the attack graph, independently of any semantic. Stated otherwise, the update of a given argumentation framework under two different semantics may easily lead to two different results. In this paper, we focus on the complete semantic, as it lies at the heart of all traditional Dung's semantics.

Definition 4. *Let $F = (A, R)$ be an AF and $L \in \mathcal{M}_F$ be a labeling. A labeling L is said to be complete if and only if, for every argument $x \in A$,*

1. *$L(x) = I$ if and only if $\forall y \in A$, if $(y, x) \in R$ then $L(y) = O$; and*
2. *$L(x) = O$ if and only if $\exists y \in A$, such that $(y, x) \in R$ and $L(y) = I$.*

Given an argumentation framework F, we denote the set of all its complete labellings as $\mathcal{L}^c$. Thus, under the complete semantics, the outcome of an argumentation framework consists of labellings in which an argument is in if and only if all its attackers are out and is out if and only if it has an attacker that is labelled in. Many of the other semantics proposed in the literature, such as the grounded, preferred, and stable semantics [12] are based on selecting particular subsets of the set of complete labellings.

Definition 5. *Let L be a complete labeling of the AF. L is:*

1. *grounded if and only if there is no complete labeling L' of F such that $I(L') \subset I(L)$;*
2. *preferred if and only if there is no complete labeling L' of F such that $I(L) \subset I(L')$; and*
3. *stable if and only if $U(L) = \emptyset$.*

Example 3. Let us consider the AF F of Fig. 1. It has three complete labellings, namely IOOO, OIOI and UUUU. Its labellings, with respect to the complete, grounded, stable and preferred semantics, are given in Table 1.

In the following, we give a definition of conflict-free labellings, which is a reference to the definition in [6].

Definition 6. *Given an AF $F = (A, R)$, a labelling L of F is said to be a conflict-free labelling if and only if $\forall x \in A$, if $L(x) = I$, then $\forall y \in A$, such that $(x, y) \in R$ or $(y, x) \in R$, $L(y) = O$. We denote all the conflict-free labellings of F as $\mathcal{L}^{cf}$.*

Intuitively speaking, conflict-free labellings guarantee no in-labelled argument attacks an (other or the same) in-labelled argument.

Example 4. Let us consider the argumentation framework F of Fig. 1, examples of conflict-free labellings are IOOO, OIOI, OIOO, OOIO, OIOU, OOOI, and OOOO, and examples of labellings that are not are IIOO and UUIO.

3   The Update of Argumentation Frameworks

This section considers the dynamics of argumentation frameworks. Given AF F_1 and AF F_2, we denote the result of updating F_1 with F_2 as $F_1 * F_2$. Intuitively, the arguments in $F_1 * F_2$ includes all the arguments in F_1 and F_2, the attack relations in $F_1 * F_2$ is simply by combing the attack relations R_1 in F_1 and the attack relations R_2 in F_2. We will not discuss more in detail the formal syntax definition of the update here since the process is somehow trivial. Instead, we aim to characterize the outcome of an update. More specifically, we consider how the labellings are modified under the update process.

Noted that, in this paper, the revocation of attack relations are not allowed, *i.e.*, if $a, b \in A_1$, $(a, b) \in R_1$, then $a, b \in A_2$ entails $(a, b) \in R_2$. When updating F_1 with F_2, the set of arguments in F_1 may be different from that in F_2, which result in the domains of their labellings are different, so does the domain of the labellings for $F_1 * F_2$. Therefore, firstly we give the following definition to expand the labellings of argumentation frameworks before updating such that their domains are the same as the domain of the argumentation framework after updating.

Definition 7. *Given argumentation frameworks $F_1 = (A_1, R_1)$ and $F_2 = (A_2, R_2)$, let L_i be a labeling of F_1, then an expansion of L_i given F_2, which is denoted as $exp(L_i, F_2)$ is any labelling L_i' from $A_1 \cup A_2$ to (I, O, U), such that $L_{i \downarrow A_1}' = L_i$, where $L_{i \downarrow A_1}'$ is a function defined by $L_{i \downarrow A_1}'(x) = L_i(x)$ for all $x \in A_1$.*

In order to be general enough, this definition does not impose any other constraints on the resulted labellings: what is essential is to preserve the labels of the arguments from the initial labelling while extending the labellings to the scope of the union of all arguments.

In this paper, we study the outcome of the updated argumentation framework when facing a new argumentation framework. We mean an operation that incorporates the new information while bringing minimal changes to the labelling of the original AF. This minimal change here is characterized by notions of illegally labelled and rationality order.

Definition 8. *Let $F = (A, R)$ be an AF and $L \in \mathcal{M}_F$ a labeling of F. An argument $x \in A$ is said to be:*

1. *illegally in if $L(x) = I$ and $\exists y \in A$, $(y, x) \in R$ and $L(y) \neq O$;*
2. *illegally out if $L(x) = O$ and $\nexists y \in A$, $(y, x) \in R$ and $L(y) = I$; and*
3. *illegally undecided if $L(x) = U$ and $\exists y \in A$, $(y, x) \in R$ and $L(y) = I$ or $\forall y \in A$, if $(y, x) \in R$ then $L(y) = O$.*

We denote by $Z_F^I(L), Z_F^O(L)$ and $Z_F^U(L)$ the sets of arguments that are respectively, illegally in, out and undecided in L for AF F. Intuitively, an illegally labelled argument indicates a violation of the label imposed on the argument according to the complete semantics. In other words, a complete labelling has no arguments illegally labelled [14]. It can also be checked that, in a conflict-free labelling, arguments are never illegally in.

As for the rationality order, we consider a restricted faithful assignment, to each AF F of a total pre-order $\leq_{F \downarrow B}$ over conflict-free labellings and we use the cardinality of the sets of illegally labelled arguments in the restricted set as the criterion to define the rationality order. Booth et al. gave a faithful assignment definition in [6].

Definition 9. *A faithful assignment assigns to each argumentation framework F a total pre-order $\leq_F \subseteq \mathcal{L}^{cf} \times \mathcal{L}^{cf}$, such that for any $L \in \mathcal{L}^{cf}$ and $F \in AF_\mathcal{U}$, it holds that: $L \leq_F L'$ if and only if*

$$|Z_F^O(L) \cup Z_F^U(L)| \leq |Z_F^O(L') \cup Z_F^U(L')|.$$

In this paper, we define a restricted faithful assignment in which the total pre-order assigned to an argumentation framework is concerned with only part of the set of arguments in the argumentation framework.

Definition 10. *Given an AF $F = (A, R)$, a faithful assignment of F restricted to B $(B \subseteq A)$ assign to AF F a total pre-order $\leq_{F \downarrow B} \subset \mathcal{L}^{cf} \times \mathcal{L}^{cf}$ such that for any $L, L' \in \mathcal{L}^{cf}$ and $F \in AF_\mathcal{U}$, it holds that: $L \leq_{F \downarrow B} L'$ if and only if*

$$|Z_F^O(L_{\downarrow B}) \cup Z_F^U(L_{\downarrow B})| \leq |Z_F^O(L'_{\downarrow B}) \cup Z_F^U(L'_{\downarrow B})|.$$

Here we use the cardinality of the sets $Z_F^U(L_{\downarrow B})$ and $Z_F^O(L_{\downarrow B})$ as the criterion to define the rationality order $\leq_{F \downarrow B}$, making the assumption that the agent believes conflict-free labellings that require less impact to be turned into a complete labelling are more rational. Given a set $M \subseteq \mathcal{L}^{cf}$, we define $\min_{\leq_{F \downarrow B}}(M) = \{L \in M \mid \forall L' \in M, L \leq_{F \downarrow B} L'\}$, i.e., $\min_{\leq_{F \downarrow B}}(M)$ denotes the set of labellings in M which are least according to $\leq_{F \downarrow B}$.

In the following, we will give our method for updating argumentation frameworks. Formally, we have:

Definition 11. *Let $F_1 = (A_1, R_1)$ and $F_2 = (A_2, R_2)$ are argumentation frameworks, we intend to update F_1 by F_2. Let L and L' are labellings of $F_1 * F_2$, then*

$$\mathcal{L}_{(F_1, F_2)}^u = \{L \mid L, L' \in exp(L_{F_1}^c, F_2) \cap exp(L_{F_2}^c, F_1), L_{\downarrow F_1} \leq_{F_1 \downarrow A_1 \backslash (A_1 \cap A_2 \cap At(A_2))} L'_{\downarrow F_1}$$
$$and\ L_{\downarrow F_2} \leq_{F_2 \downarrow (A_2 \backslash (Ua(A_2) \cap At(A_1)))} L'_{\downarrow F_2}\}. \tag{1}$$

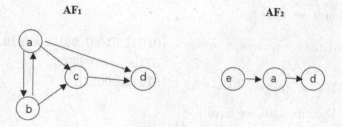

Fig. 2. Initial argumentation frameworks

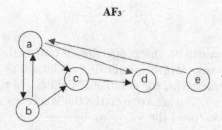

Fig. 3. Argumentation framework after updating

For a better understanding of the above definition, let us consider the following example.

Example 5. Let $F_1 = (A_1, R_1)$, $A_1 = \{a, b, c, d\}$ and $R_1 = \{(a, b), (b, a), (a, c),$ $(b, c), (c, d), (a, d)\}$, and $A_2 = \{a, d, e\}$ and $R_2 = \{(e, a), (a, d)\}$, we intend to update A_1 with A_2, and the result argumentation framework is $F_3 = \{A_3, R_3\}$ in which $A_3 = \{a, b, c, d, e\}$ and $R_1 = \{(a, b), (b, a), (a, c), (b, c), (c, d), (e, a)\}$. Figure 2 and Fig. 3 show the initial argumentation frameworks and the resulted argumentation framework after updating respectively.

By Definition 6, we have

$$L_{F_1}^{cf} = \{\mathtt{IOOO, OIOI, OIOO, OOIO, OOOI, OOOO}, \cdots, \mathtt{UUUU}\},$$
$$L_{F_2}^{cf} = \{\mathtt{OII, OOI, IOO, OIO, OOO}, \cdots, \mathtt{UUU}\},$$

where $\cdots$ represents all the remaining conflict-free labellings besides the one we have listed (for example, $\mathtt{OIOU, UOUU}$, and $\mathtt{UOUO}$ are in $L_{F_1}^{cf}$, and $\mathtt{OUI, OOU}$, and $\mathtt{OUU}$ are in $L_{F_2}^{cf}$).

By Definition 7, we have

$$exp(L_{F_1}^c, F_2) \cap exp(L_{F_2}^c, F_1)$$
$$= \{\mathtt{OIOII, OIOOI, OOIOI, OOOII, OOOOI, IOOIO, IOOOO, OIOIO, OIOOO,}$$
$$\mathtt{OOIOO, OOOIO, OOOOO}, \cdots, \mathtt{UUUUU}\}.$$

By Definition 9, we have

$$min \leq_{F_1 \downarrow A_1 \setminus (A_1 \cap A_2 \cap At(A_2))} (L_{\downarrow F_1}) = \{\texttt{OIOII}, \texttt{OIOOI}, \texttt{OOIOI}, \texttt{OIOIO}, \texttt{OIOOO},$$
$$\texttt{OOIOO}, \cdots, \texttt{UUUUU}\},$$
$$min \leq_{F_2 \downarrow (A_2 \setminus (Ua(A_2) \cap At(A_1)))} (L_{\downarrow F_2}) = \{\texttt{OIOII}, \texttt{OOOII}\},$$

Finally, by Definition 10, we have

$$\mathcal{L}^u(F_1, F_2) = \{\texttt{OIOII}\}.$$

4   Properties

In this section, we investigate the properties of our method for the update of argumentation frameworks. More specifically, we intend to prove that in updating argumentation frameworks, the complete labellings of the resulted argumentation framework after updating are exactly the labellings reached by using our method. To this end, we need the following lemma:

Lemma 1. $exp(L_{F_1}^c, F_2) \cap exp(L_{F_2}^c, F_1)$ *is the set of conflict free labellings for* $F_1 * F_2$.

Proof. Let $F_3 = F_1 * F_2$, by Definitions 6 and 7, we need to prove

(i) $\forall L_i$, if $L_i \in exp(L_{F_1}^c, F_2) \cap exp(L_{F_2}^c, F_1)$, then L_i is a conflict-free labelling for F_3; and

(ii) if L_i is a conflict-free labelling for F_3, then $L_i \in exp(L_{F_1}^c, F_2) \cap exp(L_{F_2}^c, F_1)$.

We first prove (i). Suppose not, *i.e.*, L_i is a labelling of F_3 but is not conflict-free. It means $\exists x \in A_1 \cup A_2$, $L_i(x) = I$ and $\exists y \in A_1 \cup A_2$, such that $(x, y) \in R_1$ or $(y, x) \in R_1$ or $(x, y) \in R_2$ or $(y, x) \in R_2$, and $L_i(y) \neq O$. If $(x, y) \in R_1$ or $(y, x) \in R_1$, then $L_{i \downarrow F_1} \notin L_{F_1}^{cf}$, which results $L_i \notin exp(L_{F_1}^c, F_2)$. If $(x, y) \in R_2$ or $(y, x) \in R_2$, then $L_{i \downarrow F_2} \notin L_{F_2}^{cf}$, which results $L_i \notin exp(L_{F_2}^c, F_2)$. Contradiction.

We then prove (ii). Suppose not, *i.e.*, L_i is a labeling of F_3 but $L_i \notin exp(L_{F_1}^c, F_2) \cap exp(L_{F_2}^c, F_1)$. If $L_i \notin exp(L_{F_1}^c, F_2)$, by Definitions 6 and 7, $L_{i \downarrow F_1} \notin \mathcal{L}^{cf}$. This means $\exists x, y \in A_1$, $L_i(x) = 1$, and $\exists y \in A_1$, $(x, y) \in R_1$ or $(y, x) \in R_1$ and $L_i(y) \neq O$. So L_i is not a conflict-free labelling for F_3. Contradiction. If $L_i \notin exp(L_{F_2}^c, F_1)$, the proof is similar. $\square$

This lemma reflects when updating an argumentation framework F_1 with argumentation framework F_2, the conflict-free labelling of the resulted argumentation framework is exactly the intersection of expansion of conflict-free labellings of F_1 and F_2 with respect to each other. Now we are ready to give the following theorem:

Theorem 1. *Given argumentation frameworks F_1 and F_2, when we update F_1 with F_2, $\forall L_i \in L^u(F_1, F_2)$, it holds that $L_i \in L^c(F_1, F_2)$.*

Proof. Since after update F_1 with F_2, the resulted AF is $F_3 = F_1 * F_2 = (A_1 \cup A_2, R_1 \cup R_2)$. By Definition 4, we need to prove $\forall L_i \in L^u(F_1, F_2)$ and $\forall x \in A_1 \cup A_2$,

(i) $L_i(x) = I$ if and only if $\forall y \in A_1 \cup A_2$, if $(y, x) \in R_1$ or $(y, x) \in R_2$, then $L_i(y) = O$,

(ii) $L_i(x) = O$ if and only if $\exists y \in A_1 \cup A_2$, such that $(y, x) \in R_1$ or $(y, x) \in R_2$, $L_i(y) = I$.

We first prove (i). ($\Rightarrow$) suppose not, i.e., $\exists y \in A$, $(y, x) \in R_1$ or $(y, x) \in R_2$, and $L_i(y) \neq O$, this means $L_i(y) = I$ or $L_i(y) = U$. Then we can find L_j by substituting x's label I by O (if $L_i(y) = I$) or U (if $L_i(y) = U$), then $L_{j\downarrow F_1} \leq_{F_1\downarrow A_1\backslash(A_1\cap A_2\cap At(A_2))} L_{i\downarrow F_1}$ and $L_{j\downarrow F_2} \leq_{F_2\downarrow(A_2\backslash(Ua(A_2)\cap At(A_1))} L_{i\downarrow F_2}$. Contradiction. ($\Leftarrow$) There are two cases:

(a) If $x \in A_1 \cup A_2 (At(A_1) \cup At(A_2))$, it means $\nexists y \in A_1 \cup A_2$ such that $(y, x) \in R_1$ or $(y, x) \in R_2$, by Definitions 6 and 7, $L_i(x) = I$ is trivial.
(b) If $x \notin A_1 \cup A_2 \backslash (At(A_1) \cup At(A_2))$, and $(y, x) \in R_1$ or $(y, x) \in R_2$, then $L_i(y) = O$, by Definitions 6 and 7, $L_i(x) = I$.

We then prove (ii). ($\Rightarrow$) Suppose not, if $\nexists y \in A$, such that $(y, x) \in R$ and $L_i(y) = I$. There are different cases:

(a) If $\nexists y \in A$ such that $(y, x) \in R_1$ or $(y, x) \in R_2$, this means $x \in A_1 \cup A_2 (At(A_1) \cup At(A_2))$, by Definition 2, $L_i(x) = I$.
(b) If $\forall y \in A$, $(y, x) \in R_1$ or $(y, x) \in R_2$ entails $L_i(y) \neq I$, i.e., $L_i(y) = O$ or $L_i(y) = U$. By Definition 13, we can find L_j by substituting x's label O with I (if $\forall y, L_i(y) = O$) or U (if $\exists y, L_i(y) = U$), we can see $L_j \in exp(L^p_{F',F}, F') \cap exp(L^p_{F,F'}, F')$ and $L_{j\downarrow F} \leq_{F\downarrow(A\backslash A\cap A'\cap At(A'))} L_{i\downarrow F}$ and $L_{j\downarrow F_2} \leq_{F_2\downarrow(A_2\backslash(Ua(A_2)\cap At(A_1))} L_{i\downarrow F_2}$.

($\Leftarrow$) By Definition 9, if $\exists y \in A_1 \cup A_2$, such that $(y, x) \in R_1$ or $(y, x) \in R_1$ and $L_i(y) = I$, it is obvious that $L_i(x) = O$. $\square$

The theorem above reflects that when in an updating of argumentation frameworks, the labellings which is reached by using our updating method are all complete labellings for the resulted argumentation framework. On the other hand, the following theorem shows in an updating of argumentation framework, all the complete labellings of the resulted argumentation framework can be reached by using our updating method.

Theorem 2. *Given argumentation frameworks F_1 and F_2, when we update F_1 with F_2, $\forall L_i \in L^c(F_1, F_2)$, it holds that $L_i \in L^u(F_1, F_2)$.*

Proof. By Definition 11, we need to prove that if $L_i \in L^c(F_1, F_2)$,

(i) $L_i \in exp(L^c_{F_1}, F_2) \cap exp(L^p_{F_2}, F_1)$;
(ii) $\forall j \in N$ and $j \neq i$, $L_{i\downarrow F_1} \leq_{F_1\downarrow A_1\backslash(A_1\cap A_2\cap At(A_2))} L_{j\downarrow F_1}$; and
(iii) $\forall j \in N$ and $j \neq i$, $L_{i\downarrow F_2} \leq_{F_2\downarrow(A_2\backslash(Ua(A_2)\cap At(A_1))} L_{j\downarrow F_2}$.

We first prove (i). Suppose not,

$$L_i \notin exp(L_{F_1}^c, F_2) \cap exp(L_{F_2}^c, F_1),$$

if $x \in A_1 \cup A_2 \setminus (At(A_1) \cup At(A_2))$, $L_i(x) \neq I$, since by Collary 1, $L_i \in exp(L_{F_1}^c, F_2) \cap exp(L_{F_2}^c, F_1)$ if L_i is conflict free, L_i is not conflict free. It means, there exists $x \in A_1 \cup A_2$, $L_i(x) = I$ and $\exists y \in A_1 \cup A_2$, $(y, x) \in R_1$ or $(y, x) \in R_2$, or $(x, y) \in R_1$ or $(x, y) \in R_1$, and $L_i(y) \neq O$. If $(y, x) \in R_1$ or $(y, x) \in R_2$ and $L_i(y) \neq O$, by Definition 4(1), contradiction. If $(x, y) \in R_1$ or $(x, y) \in R_1$, and $L_i(y) \neq O$, by Definition 4(2), contradiction.

We then prove (ii). Suppose not, $i.e.$, $\exists L_j \in exp(L_{F_1}^c, F_2) \cap exp(L_{F_2}^p, F_1)$ such that $L_{j \downarrow F_1} \leq_{F_1 \downarrow A_1 \setminus (A_1 \cap A_2 \cap At(A_2))} L_{i \downarrow F_1}$. For the sake of convenience, we let $B_1 = A_1 \setminus (A_1 \cap A_2 \cap At(A_2))$. By Definitions 9 and 10, we have

$$\mid Z_{F_1}^O(L_{j \downarrow B_1}) \cup Z_{F_1}^U(L_{j \downarrow B_1}) \mid < \mid Z_{F_1}^O(L_{i \downarrow B_1}) \cup Z_{F_1}^O(L_{i \downarrow B_1}).$$

It means there exists at least one argument $x \in A_1$, such that x is illegally labeled O or U in L_i while is not in L_j. By Definitions 6 and 2, $L_i \notin L^c$.

Finally, we can prove (iii) similarly as (ii).   □

5   Related Work

In recent years, dynamics of argumentation frameworks has been studied from different aspects by focusing on different problems. Our work is in line with the study of "elementary change" of argumentation frameworks which is given in [11]. Elementary change is concerned with the addition (or removal) of arguments and attack relations of argumentation frameworks [5, 8, 13].

Cayrol *et al.* [8] focus on the addition of a new argument that interacts with previous arguments and studied the impact of such change on the outcome of the argumentation framework, particularly on the set of its extension. Boella *et al.* [5] consider the dynamics of abstract argumentation framework for the evaluation of extension-based argumentation semantics. They both do not consider individual approaches to compute the status of arguments with the change of arguments as well as the attack relations but define general principles that individual approaches may satisfy. In contrast, this paper gives a specific method to compute the labelling of the updated argumentation frameworks more efficiently.

Liao, Jin, and Koons [13] focus on computing the semantics of updated argumentation frameworks, which is the closest concern with our aim. They presented a general theory to cope with the dynamics of argumentation frameworks. Our work differs from theirs in three aspects. (1) Their method of computing the semantics dynamically depends on the extension of the unaffected sub-framework and the extensions of a set of assigned CAFs, while our method depends on the attack relations of the initial AF and updating AF. (2) In the division-based method in [13], since the extension of the unaffected part, which is one part of the extension of updated AF, could be preserved, the complexity of computing the dynamics of argumentation might be decreased. We notice that if the unaffected part of AF is empty, this division-based method could not be applied to decrease the complexity of computing the dynamics of argumentation frameworks, while in our work, this problem could be avoided. (3) In their work, the acceptance status of arguments are evaluated by applying the extension-based semantics while our method is proposed based on the labelling-based semantics. Though the extension-based semantics and the labelling-based semantics can be mapped 1-to-1 such that extensions correspond to sets of in-labelled arguments, the labelling-based semantics are more intuitive by giving each argument of AF a label in $\{in, out, undecided\}$. In contrast, the extension-based semantics only present the in-labelled arguments in the extension of an AF. Therefore, we have to compute the status of the remaining arguments further when needed.

Our work is also related to the work of enforcing problem in [3,10,11]. The research on the problems focuses on how to change an argumentation framework to affect the acceptability of arguments, and how to modify an argumentation framework to guarantee that some arguments have a given acceptance status. Both our work and others' on the problems are concerned with the relations of changing argumentation framework with the status of arguments after changing. However, ours emphasises how to compute the status of arguments given the changing of argumentation frameworks, while they discuss how to change an argumentation framework given a set of arguments as an extension.

6   Conclusion and Further Work

We propose a two-step method to handle the dynamic of argumentation frameworks. Firstly, we reach the conflict-free labellings of the updated argumentation framework by the intersection of expansions of the conflict-free labellings for the argumentation frameworks before update; then we select the conflict-free labellings which have the least illegally labelled arguments when restricted to part of both argumentation frameworks in the update process. In addition, we proved the soundness and completeness of our method under the complete semantics. In essence, our method aims to compute the status of arguments separately in the two AFs before updating instead of the AF after updating. It is usually the case that the updated AF is complicated while the initial Afs are not. As a result, the computation of the status of arguments in AF is easier and much more efficient.

In the future, our work could be improved in several aspects. (1) In this paper, our method is given just under the complete semantic. Since the grounded, preferred, and stable semantics are just selecting a particular subset of the set of complete labelling, it seems our method probably could be applied under these semantics by adding some constraints. (2) In our work, the removal of the attacks are not allowed. To solve this problem, we may relax the requirement of the conflict-free labellings for the resulted argumentation framework by including the labellings in which the labels of the arguments concerning the removal attacks are conflicting. (3) it is interesting to study the computational complexity of our method. (4) The problem of iterated updates should be given into consideration in the dynamic of argumentation frameworks. (5) It is also interesting to design an algorithm to automatically generate the conflict-free labelling.

Acknowledgments. This work was supported by the National Natural Science Foundation of China (No. 61762016).

References

1. Alfano, G., Cohen, A., Gottifredi, S., Greco, S., Parisi, F., Simari, G.R.: Dynamics in abstract argumentation frameworks with recursive attack and support relations. In: Proceedings of the 24th European Conference on Artificial Intelligence, pp. 577–584 (2020)
2. Baroni, P., Caminada, M., Giacomin, M.: An introduction to argumentation semantics. Knowl. Eng. Rev. **26**(4), 365–410 (2011)
3. Baumann, R., Doutre, S., Mailly, J.G., Wallner, J.P.: Enforcement in formal argumentation. J. Appl. Logic **2**, 1623–1677 (2021)
4. Bench-Capon, T., Dunne, P.E.: Argumentation in artificial intelligence. Artif. Intell. **171**, 619–641 (2007)
5. Boella, G., Kaci, S., Van der Torre, L.: Dynamics in argumentation with single extensions: abstraction principles and the grounded extension. In: Proceedings of the 24th European Conference on Symbolic and Quantitative Approaches to Reasoning and Uncertainty, pp. 107–118 (2009)
6. Booth, R., Kaci, S., Rienstra, T., van der Torre, L.: A logical theory about dynamics in abstract argumentation. In: Liu, W., Subrahmanian, V.S., Wijsen, J. (eds.) SUM 2013. LNCS (LNAI), vol. 8078, pp. 148–161. Springer, Heidelberg (2013). https://doi.org/10.1007/978-3-642-40381-1_12
7. Caminada, M.: On the issue of reinstatement in argumentation. In: Fisher, M., van der Hoek, W., Konev, B., Lisitsa, A. (eds.) JELIA 2006. LNCS (LNAI), vol. 4160, pp. 111–123. Springer, Heidelberg (2006). https://doi.org/10.1007/11853886_11
8. Cayrol, C., de Saint-Cyr, F.D., Lagasquie-Schiex, M.: Change in abstract argumentation frameworks: adding an argument. J. Artif. Intell. Res. **38**, 49–84 (2010)
9. Coste-Marquis, S., Konieczny, S., Mailly, J.G., Marquis, P.: On the revision of argumentation systems: minimal change of arguments statuses. In: Proceedings of the 14th International Conference on the Principles of Knowledge Representation and Reasoning, pp. 52–61 (2014)
10. Coste-Marquis, S., Konieczny, S., Mailly, J.G., Marquis, P.: Extension enforcement in abstract argumentation as an optimization problem. In: Proceedings of the 24th International Joint Conference on Artificial Intelligence, pp. 2876–2882 (2015)
11. Doutre, S., Mailly, J.G.: Constraints and changes: a survey of abstract argumentation dynamics. Argument Comput. **9**(3), 223–248 (2018)
12. Dung, P.M.: On the acceptability of arguments and its fundamental role in nonmonotonic reasoning, logic programming and n-person games. Artif. Intell. **77**(2), 321–357 (1995)
13. Liao, B., Jin, L., Koons, R.C.: Dynamics of argumentation systems: a division-based method. Artif. Intell. **175**(11), 1790–1814 (2011)
14. Rahwan, I., Simari, G.R.: Argumentation in Artificial intelligence. Springer, Heidelberg (2009)

Author Index

Printed in the United States
by Baker & Taylor Publisher Services